Systems Biology of Tuberculosis

Johnjoe McFadden • Dany J.V. Beste
Andrzej M. Kierzek
Editors

Systems Biology of Tuberculosis

Springer

Editors
Johnjoe McFadden
Faculty of Health and Medical Sciences
University of Surrey
Guildford, Surrey, UK

Dany J.V. Beste
Faculty of Health and Medical Sciences
University of Surrey
Guildford, Surrey, UK

Andrzej M. Kierzek
Faculty of Health and Medical Sciences
University of Surrey
Guildford, Surrey, UK

ISBN 978-1-4614-4965-2 ISBN 978-1-4614-4966-9 (eBook)
DOI 10.1007/978-1-4614-4966-9
Springer New York Heidelberg Dordrecht London

Library of Congress Control Number: 2012951628

Printed on acid-free paper

Springer is part of Springer Science+Business Media (www.springer.com)

Introduction

Douglas Young

Reconstruction of evolutionary history by Bayesian analysis of extant genome sequences suggests that the *Mycobacterium tuberculosis* complex emerged as an infection of anatomically modern humans carrying the L3 mitochondrial haplotype around one hundred thousand years ago. Since then, *M. tuberculosis* has demonstrated a remarkable ability to persist amongst small, highly vulnerable populations of early humans migrating out of Africa and to thrive in response to changing demography and recent massive population expansion. The Global Plan to Stop TB proposes its elimination—defined as fewer than one case per million individuals—by 2050 [1]. Given our history of intimate companionship, how can we envisage a strategy to drive this microbe towards extinction?

While we can readily collect data on tuberculosis epidemiology, attempts to explain patterns of disease—and hence to design rational control strategies—reveal contributions from multiple variables. It is clear that disease results from some combination of host, microbial, and environmental factors, but it is hard to generate mechanistic models that tease these apart. This is an example of an "inverse problem", in which we have to try and infer the parts by examination of the whole. This is anathema to reductionist biology, and it is the territory to which systems biology aspires.

Sydney Brenner highlights the attempt to address ill-posed inverse questions as a lethal flaw in the systems approach, predicting a spiral into low-input, high-throughput, no-output biology [2]. Systems biologists counter by proposing an iterative process of modelling and forward testing of derived hypotheses. This involves an interesting mix of logical reasoning. Wikipedia pithily describes deduction (in which a conclusion is determined by a precondition) as the province of mathematicians, induction (in which a conclusion is a probable outcome of a

D. Young
MRC National Institute for Medical Research, Mill Hill,
London NW7 1AA, UK
e-mail: d.young@imperial.ac.uk

precondition) as the province of scientists, and abduction (in which a precondition is inferred from a conclusion) as the province of detectives. While Sherlock Holmes may dispute elements of this categorisation, the systems biology agenda strives towards some clear-sighted integration of all three methods.

A systems approach to tuberculosis can be envisaged as a series of nested problems (often inverse problems) ranging in scale from the level of populations, to individuals, to the cellular and molecular that can conveniently be branded as "systems epidemiology" [3], "systems vaccinology" [4], and so on. The present volume comprises a series of contributions from groups who are at the forefront of applying systems biology approaches to understand tuberculosis, working primarily at the cell biology end of this spectrum.

Beste and McFadden [5] and Jamshidi et al. [6] review advances in metabolic modelling. Metabolomics is particularly appropriate for network modelling. Connectivity between components in the network is direct—one metabolite changes into another—in contrast to the indirect spatial and temporal interactions used to build networks of genes and proteins. Jamshidi et al. describe fundamental processes of network modelling, while Beste and McFadden stress the importance of a robust experimental system to generate data for modelling, outlining the advantages of growth in a chemostat as a means to optimise relative homogeneity of the bacterial populations under study. Both groups stress the dependence of metabolic modelling on the availability of an accurately annotated genome. Contributions from a series of outstanding researchers established the foundations of mycobacterial biochemistry and metabolism during the first half of the twentieth century, but a relative neglect of tuberculosis research in the 1960s and 1970s left multiple gaps in our knowledge. In the genome era, there is a tendency to fill these gaps using sequence homologies with other organisms. Both papers stress the importance of caution in this, favouring an iterative process in which metabolic models are used both to formulate and test forward hypotheses as well as to track back and correct misannotations. Even core textbook metabolic pathways are found to differ between *M. tuberculosis* and "canonical" *E. coli*.

Lack of information about the physiological state of *M. tuberculosis* within infected humans presents a major challenge for modelling tuberculosis pathogenesis, and models derived from microbial culture systems have been usefully extended to metabolomics of intracellular *M. tuberculosis* in macrophages. With the exception of a limited range of microbe-specific metabolites, it is difficult to derive direct experimental data on mycobacterial metabolism in infected tissues; in fact it probably makes sense to view host and microbe as a single, integrated metabolic system. It is likely that the bacteria sample a highly diverse range of intracellular and extracellular microenvironments during infection, with availability of nutrients and oxygen varying widely over space and time [7]. A systems biology challenge will be to infer microbial physiological states on the basis of measurements of host metabolism.

Both of the metabolic modelling papers stress the importance of integrating metabolomics with transcriptional profiling and functional genomics. Waddell et al. [8] and Rao et al. [9] take up the story from a transcriptome perspective. Traditional transcriptional profiling is also contingent on accurate genome annotation to select genes that are interrogated in microarray platforms. More recent high-density tiling arrays and RNA sequencing approaches remove this limitation and are starting to uncover a considerable repertoire of non-coding RNA outside of annotated open reading frames [10]. This includes small intergenic RNAs, antisense transcripts, and *cis*-encoded untranslated regions that are likely to regulate the stability of mRNA transcripts and the efficiency of their interaction with ribosomes. Characterisation of this layer of post-transcriptional regulation will be an important element in integrating transcriptome and proteome data, and hence for inferring physiology from transcription profile. Both papers highlight a central role for the ability of transcriptional profiling to uncover crosstalk between host and microbe in macrophage infection models. As discussed for metabolomics, there is a trend in transcriptomic modelling to move from consideration of the isolated microbe to the infected cell or lesion as the system under study. Given the specificity and technical ease of nucleic acid amplification, transcriptome data represent the most accessible source of information about *M. tuberculosis* in lesions. Both papers stress the importance of linking transcriptional profiling to functional genomics, using transposon mutagenesis or formal gene deletion in the microbe side, and exploitation of RNAi screens for host manipulation.

Steyn et al. [11] contribute a review of mycobacterial proteomics, focusing in particular on different strategies for generating protein–protein interaction networks. They describe specific contributions to exploration of mechanisms that play a central role in host–pathogen interactions—two-component signalling, protein secretion, and DNA repair—and highlight the importance of studying post-translational modification events. Together with the chapters on metabolism and transcription, this builds an encouraging picture of a strong integrated 'omics platform that informs a systems biology understanding of *M. tuberculosis* and its interaction with host cells. Given our knowledge of the mycobacterial cell, it is obviously important that this platform includes the parallel advances occurring in the area of lipidomics.

Chandra [12] takes forward the application of a systems approach to identification of potential drug targets, integrating genetic essentiality with metabolic modelling to identify key choke-points. Target-driven approaches based on genetic essentiality have been profoundly disappointing in the field of antimicrobial drug discovery [13]. In part this reflects a series of technical limitations: compound libraries used in high-throughput screens may have inadequate representation of relevant chemical space, evidence from gene deletions does not stratify high- versus low-vulnerability targets, and potent enzyme inhibitors may not penetrate bacterial cell walls. There may also be a limitation in the general concept that simple inhibition of an essential enzyme is sufficient for bactericidal activity [14]. Events downstream of the initial drug–target interaction are probably crucial to the effectiveness of successful antibiotics, with accumulation of toxic effector molecules providing the actual trigger for cell death. Systems biology models that can predict lethal consequences of target of inhibition would provide an important advance. Given the large number of moderately potent hits arising from high-throughput screens against whole mycobacteria, a systems biology approach capable of identifying cell death parameters that are more experimentally tractable than measurement of colony forming units would also be of considerable use for prioritisation. Chandra outlines a concept of "polypharmacology", involving analysis of the interaction of a single drug with multiple targets and the effect of drug combinations. Drug combinations are central to tuberculosis treatment regimens, and there is a need for systems biology approaches to rationalise and predict positive and negative drug–drug interactions.

Ghosh et al. [15] describe an exciting international collaborative effort to exploit systems biology for TB drug discovery in the context of community engagement in a "big science" initiative. This is a joint programme between two Japanese systems biology institutes and the Indian Open Source Drug Discovery project. The emphasis is on novel communication systems, generating a virtual collaborative space that accommodates input from a wide community of researchers. This illustrates a key aspect of the systems biology agenda: stimulation of multidisciplinary interactions across a wide range of biology, engineering and mathematics. Other multidisciplinary consortia exploring the systems biology of tuberculosis have been established in the USA [16] and Europe [17]. There are formal similarities between social interaction networks and protein–protein interaction networks, and it is clear that social factors will be at least as important as molecular factors in the success of future strategies for tuberculosis control. Perhaps community-based approaches to enhance communication amongst systems biology researchers could be extended to enhance communication between scientists and the wider public?

Two papers address host–pathogen interactions from the perspective of the immune response. Pine et al. [18] present a comprehensive overview of the highly interconnected host immune network, picking out molecular and cellular biomarkers that could be used in combination to stratify the position of individuals with the tuberculosis infection spectrum. Fallahi-Sichani et al. [19] describe various techniques to model the development and heterogeneity of tuberculous granulomas, including a powerful and innovative agent-based modelling approach. Focusing on the role of TNFα concentration as an example, they illustrate a very important aspect of systems biology modelling as an aid to feature selection. This technique, commonly used in machine learning, distinguishes parameters whose variation has a more or less critical effect on overall behaviour of the system and therefore warrants higher or lower prioritisation for further experimental definition. This provides a framework for attractive synergy between modellers and experimentalists.

Rocco et al. [20] return to the issue of microbial modelling from a novel perspective, reviewing stochastic influences on gene expression and their relevance to population heterogeneity and persistence. Conventional 'omics approaches use high-throughput data generated from bulk populations that are assumed to display a homogeneous phenotype. This is certainly not the reality. Noise is an important element in bacterial physiology, and there is extensive evidence of heterogeneity in gene expression amongst bacterial cells grown in an identical strictly controlled environment. It can be anticipated that such effects are amplified multi-fold in the complex microenvironments encountered during infection. Rocco et al. highlight the potential links between population heterogeneity and the acutely practical issue of phenotypic tolerance to drugs. Addressing these issues requires alternative experimental and computational approaches, and it is crucial that strategies are developed to integrate single cell information with "mainstream" bulk population 'omics.

References

1. http://www.stoptb.org/global/plan/
2. Brenner S (2009) Interview with Sydney Brenner by Soraya de Chadarevian. Stud Hist Philos Biol Biomed Sci 40:65–71
3. Comas I, Gagneux S (2011) A role for systems epidemiology in tuberculosis research. Trends Microbiol 19:492–500
4. Pulendran B, Li S, Nakaya HI (2010) Systems vaccinology. Immunity 33:516–529
5. Beste DJV, McFadden J (2012) Metabolism of *Mycobacterium tuberculosis*. In: McFadden J (ed) Systems biology of mycobacteria. Springer
6. Jamshidi N, Bordbar A, Palsson B (2012) Modeling *Mycobacterium tuberculosis* H37Rv *in silico*. In: McFadden J (ed) Systems biology of tuberculosis. Springer
7. Barry CE 3rd, Boshoff HI, Dartois V, Dick T, Ehrt S et al (2009) The spectrum of latent tuberculosis: rethinking the biology and intervention strategies. Nat Rev Microbiol 7:845–855
8. Waddell SJ, von Kamp A, Klamt S, Neyrolles O (2012) Host-pathogen interactions. In: McFadden J (ed) Systems biology of tuberculosis. Springer
9. Rao KVS, Kumar D, Mande SC (2012) Probing gene regulatory networks to decipher host-pathogen interactions. In: McFadden J (ed) Systems biology of tuberculosis. Springer
10. Arnvig KB, Comas I, Thomson NR, Houghton J, Boshoff HI, et al (2011) Sequence-based analysis uncovers an abundance of non-coding RNA in the total transcriptome of *Mycobacterium tuberculosis*. PLoS Pathog 7:e1002342
11. Steyn AJC, Mai D, Saini V, Farhana A (2012) Protein-protein interaction in the –omics era: Understanding *Mycobacterium tuberculosis* function. In: McFadden J (ed) V. Springer
12. Chandra N (2012) Drug discovery. In: McFadden J (ed) Systems biology of tuberculosis. Springer
13. Payne DJ, Gwynn MN, Holmes DJ, Pompliano DL (2007) Drugs for bad bugs: confronting the challenges of antibacterial discovery. Nat Rev Drug Discov 6:29–40
14. Dick T, Young D (2011) How antibacterials really work: impact on drug discovery. Future Microbiol 6:603–604
15. Ghosh S, Matsuoka Y, Asai Y, Kitano H, Bhardwaj A, et al (2012) Software platform for metabolic network reconstruction of *Mycobacterium tuberculosis*. In: McFadden J (ed) Systems biology of tuberculosis. Springer
16. https://tbsystemsbiology.org/
17. http://www.systemtb.org/
18. Pine R, Bushkin Y, Gennaro ML (2012) Immunological biomarkers for tuberculosis: potential for a combinatorial approach. In: McFadden J (ed) Systems biology of tuberculosis. Springer
19. Fallahi-Sichani M, Marino S, Flynn JL, Linderman JJ, Kirschner DE (2012) A systems biology approach for understanding granuloma formation and function in tuberculosis. In: McFadden J (ed) Systems biology of tuberculosis. Springer
20. Rocco A, McFadden J, Kierzek AM (2012) Stochastic gene expression in bacterial pathogens: a model for persistence? In: McFadden J (ed) Systems biology of tuberculosis. Springer

Contents

Contributors

Yoshiyuki Asai Okinawa Institute of Science and Technology, Okinawa, Japan

Anshu Bhardwaj Council of Scientific and Industrial Research, Delhi, India

Dany J.V. Beste Faculty of Health and Medical Sciences, University of Surrey, Guildford, Surrey, UK

Kausik Bhattacharyya CSIR-Institute of Genomics and Integrative Biology, Delhi, India

Aarash Bordbar Department of Bioengineering, University of California, San Diego, La Jolla, CA, USA

Samir Brahmachari Council of Scientific and Industrial Research, Delhi, India

CSIR-Institute of Genomics and Integrative Biology, Delhi, India

Yuri Bushkin Public Health Research Institute, New Jersey Medical School, Newark, NJ, USA

Nagasuma Chandra Department of Biochemistry, Indian Institute of Science, Bangalore, India

Mohammad Fallahi-Sichani Department of Systems Biology, Harvard Medical School, Boston, MA, USA

A. Farhana Department of Microbiology, University of Alabama at Birmingham, Birmingham, AL, USA

JoAnne L. Flynn Department of Microbiology and Molecular Genetics, University of Pittsburgh School of Medicine, Pittsburgh, PA, USA

Maria Laura Gennaro Public Health Research Institute, New Jersey Medical School, Newark, NJ, USA

Samik Ghosh The Systems Biology Institute, Tokyo, Japan

Akanksha Jain Council of Scientific and Industrial Research, Delhi, India

Neema Jamshidi Department of Bioengineering, University of California, San Diego, La Jolla, CA, USA

Andrzej Kierzek Faculty of Health and Medical Sciences, University of Surrey, Guildford, Surrey, UK

Denise E. Kirschner Department of Microbiology and Immunology, University of Michigan Medical School, Ann Arbor, MI, USA

Hiroaki Kitano The Systems Biology Institute, Tokyo, Japan

Okinawa Institute of Science and Technology, Okinawa, Japan

Sony Computer Science Laboratories, Inc., Tokyo, Japan

Steffen Klamt Max Planck Institute for Dynamics of Complex Technical Systems Magdeburg, Germany

Dhiraj Kumar International Centre for Genetic Engineering and Biotechnology, New Delhi, India

Vikas Kumar CSIR-Institute of Genomics and Integrative Biology, Delhi, India

Jennifer J. Linderman Department of Chemical Engineering, University of Michigan, Ann Arbor, MI, USA

D. Mai Department of Microbiology, University of Alabama at Birmingham, Birmingham, AL, USA

Shekhar C. Mande National Centre for Cell Science, Pune, India

Simeone Marino Department of Microbiology and Immunology, University of Michigan Medical School, Ann Arbor, MI, USA

Yukiko Matsuoka The Systems Biology Institute, Tokyo, Japan

ERATO Kawaoka Infection-Induced Host Response Project, Japan Science and Technology Agency, Tokyo, Japan

Johnjoe McFadden Faculty of Health and Medical Sciences, University of Surrey, Guildford, Surrey, UK

Anupam Kumar Mondal CSIR-Institute of Genomics and Integrative Biology, Delhi, India

Olivier Neyrolles Centre National de la Recherche Scientifique, Institut de Pharmacologie et de Biologie Structurale, Toulouse, France

Institut de Pharmacologie et de Biologie Structurale, Université de Toulouse, Université Paul Sabatier, Toulouse, France

Bernhard Palsson Department of Bioengineering, University of California, San Diego, La Jolla, CA, USA

Anurag Passi Council of Scientific and Industrial Research, Delhi, India

Richard Pine Public Health Research Institute, New Jersey Medical School, Newark, NJ, USA

Priyanka Priyadarshini Council of Scientific and Industrial Research, Delhi, India

Kanury V.S. Rao International Centre for Genetic Engineering and Biotechnology, New Delhi, India

Andrea Rocco Faculty of Health and Medical Sciences, University of Surrey, Guildford, Surrey, UK

V. Saini Department of Microbiology, University of Alabama at Birmingham, Birmingham, AL, USA

Vinod Scaria CSIR-Institute of Genomics and Integrative Biology, Delhi, India

Anup Shah CSIR-Institute of Genomics and Integrative Biology, Delhi, India

Pratibha Sharma CSIR-Institute of Genomics and Integrative Biology, Delhi, India

Kumari Sonal CSIR-Institute of Genomics and Integrative Biology, Delhi, India

Adrie J.C. Steyn Department of Microbiology and Centers for AIDS Research, University of Alabama at Birmingham, Birmingham, AL, USA

KwaZulu-Natal Research Institute for Tuberculosis and HIV, Congella, Durban, South Africa

Rohit Vashisht Council of Scientific and Industrial Research, Delhi, India

Priti Vishnoi CSIR-Institute of Genomics and Integrative Biology, Delhi, India

Axel von Kamp Max Planck Institute for Dynamics of Complex Technical Systems, Magdeburg, Germany

Simon J. Waddell Brighton and Sussex Medical School, University of Sussex, Brighton, UK

Douglas Young MRC National Institute for Medical Research, Mill Hill, London, UK

Chapter 1
Modeling *Mycobacterium tuberculosis* H37Rv In Silico

Neema Jamshidi, Aarash Bordbar, and Bernhard Palsson

Abstract Network reconstructions and constraint-based modeling have been shown to be effective methods for understanding complex processes, such as metabolism. These reconstructions are in fact biologically structured knowledge-bases that can be queried through computations, and thus have become valuable tools for Systems Biology. Strengths of this approach include flexibility in incorporating "incomplete" data measurements, the ability to incorporate different types of data (high-throughput as well as physiological), simultaneously, as well as the ability to make predictions with minimal reliance on parameter and curve fitting. Thus, this approach aims to move away from fitting data to describe experimental results using the current understanding of metabolism in order to interpret the data, make predictions, and to identify the gaps and bridges in knowledge.

The critical components for creating genome-scale reconstructions of metabolism include a sequenced and annotated genome, reaction stoichiometry for the annotated enzymes, and a bibliome for the organism (combined primary and secondary literature sources). Network reconstructions of the devastating pathogen *Mycobacterium tuberculosis* have been developed and have enabled the ability to query functional capabilities using constraint-based modeling approaches. Since these networks are then structured in terms of "gene–protein–reaction" associations, these knowledge-bases can serve as biologically structured databases onto which various high-throughput data types can be directly mapped on.

This chapter will focus on the model reconstruction process, methods that have been employed for analysis, and predictive applications of modeling the pathogen H37Rv strain of tuberculosis. Employing the existing analysis methods and available datasets there have already been a large number of applications for modeling constraint-based modeling of H37Rv. The reconstruction process is a time and resource intensive procedure and to date high quality reconstructions have not been

N. Jamshidi (✉) • A. Bordbar • B. Palsson
Department of Bioengineering, University of California, San Diego,
9500 Gilman Drive, La Jolla, CA 92093-0412, USA
e-mail: neema@ucsd.edu

J. McFadden et al. (eds.), *Systems Biology of Tuberculosis*,
DOI 10.1007/978-1-4614-4966-9_1,

possible without manual curation. The benefit of having a detailed and quality controlled reconstruction procedure is to help determine a high quality model that will provide more meaningful predictions from simulations. Applications of *M. tuberculosis* models have included the prediction of growth rates, assessment of different growth media, prediction of gene knockouts, identification of new drug targets, identification of alternative drug targets for existing drugs, and modeling the interaction macrophages during different infectious states.Historically, technological advancements have driven biological discovery and have thus been a limiting factor in the development of methods to modify and alter biology, e.g., antibiotics. However, in the past decade with various high-throughput technologies (e.g., transcriptomics, proteomics, metabolomics, etc.) are being employed more frequently, thus there is a growing burden and need for means to integrate, interpret, and ideally make predictions for these datasets. Given the successes to date, with further development of new methods in conjunction with deeper experimental probing of tuberculosis in vitro and in vivo, constraint-based modeling will likely become even more important in the finding new targets and treatments for tuberculosis.

1 Introduction

The principle behind constraint-based modeling is to use physico-chemical or biological constraints to provide insights that are biologically insightful. The rich, detailed history of biochemistry during the past 60 years has resulted in the ability to describe metabolic networks in terms of elementally balanced biochemical reactions. Furthermore, thermodynamic and kinetic characterization of many of the enzymes in an effort to characterize the physical properties of these biological catalysts has enabled additional levels of characterization.

Flux Balance Analysis (FBA) has become the bread and butter of constraint-based modeling [1, 2]. Employing this approach involves the quasi-steady state assumption and knowledge of the identity and composition of the interacting components, as well as the set of the biochemical reactions that occur in the system of interest. If quantitative thermodynamic data is not available, qualitative thermodynamic data can also be incorporated, simply by specification of reversibility of a reaction. While this is a very simple encapsulation of the complex non-equilibrium thermodynamics within a cell, it can have significant implications on constraining a network and reducing (or expanding) the number of possible steady state solutions. At the most elementary level, FBA can be applied to a single biosynthetic pathway [3]. While this might be interesting in some cases, the benefits of this approach are really appreciated when one makes the jump to the organelle-, cell-, and genome-scale models [4, 5].

The data (in-)completeness problem is likely to always be present at all spatial and temporal hierarchies in biology. High-throughput technologies have been progressing to help close the data incompleteness gap. Genomics was the first “omics” field in biology and has been followed by numerous other high-throughput measurement technologies, including proteomics, and metabolomics and a seemingly innumerable array of other "omics" sub-fields. We focus the discussion on the technologically driven by high-throughput measurements.

Aside from technical challenges associated with the analysis of voluminous datasets, there is a more pressing challenge regarding the context and the manner in which the data are analyzed. There is a demonstrated need to move away from black-box modeling approaches to understand these data, towards mechanistic or partially mechanistic (gray box) models. The development of multiple "omics" data fields has further compounded this problem, further highlighting the need for analysis of large datasets that often include orthogonal types of data to be analyzed in a biologically relevant and biologically meaningful context.

Constraint-based modeling has provided one approach to organizing and analyzing this data from a biological viewpoint, while paying heed to physico-chemical constraints. The application of these methods to the deadly pathogen *Mycobacterium tuberculosis* [6] has resulted in advancements in the understanding of its metabolic capabilities and opened potential avenues for new or alternative treatments.

2 The Reconstruction Process

The quality and scope of metabolic network reconstructions have continued to evolve during the last 15 years, with current descriptions involving a detailed, iterative, quality-controlled process [7]. Progress has been made in the automation of reconstruction [8]. Nevertheless, a key hallmark of quality reconstructions however has been the need for manual curation, on some level. General steps employed in the process of network reconstruction are outlined in Fig. 1.1.

The starting point for genome-scale reconstructions is a sequenced and annotated genome. This serves as the scaffold onto which the biochemical transformations carried out by enzymes in the organism are mapped. Manual curation follows, which includes gathering evidence and critically evaluating the primary (and review) literature for information about the genes, proteins, and metabolites. There have been an increasing number of organism specific databases that have been developed during the past decade that were very helpful for fleshing out the general network architecture of *M. tuberculosis* [9–11].

Following manual curation, there is conversion of the set of biochemical reactions into a stoichiometric matrix. The stoichiometric matrix is unique compared to many other types of matrices in biological systems, as it has integer entries, thus there is no noise associated with the values [2, 12]. The conversion to a mathematical format also involves the application of (qualitative) thermodynamic constraints. Quantitative constraints have been explored [13, 14] and have been used for expanding the scope to dynamic models [15–17]. However, for the purpose of developing a basic model with which to carry out constraint-based modeling, only directionality needs to be specified (i.e., reversible or irreversible). The debugging and functionality testing stage is another step that is a critical step in the process, as it ensures network functionality. It is unfortunately also a time-consuming process.

For micro-organisms the primary functionality test involves biomass production. The biomass “reaction” is actually a pseudo-reaction that is included to account for the growth as well as non-growth associated demands [18]. Since this reaction requires the production and utilization of a large number of metabolites, the ability to produce biomass implicitly accounts for the biosynthesis of a large number of compounds. Construction of iNJ661 involved testing individual components (i.e., non-essential amino acids, mycolic acids, etc.) prior to testing the complete biomass function. Since the biomass function involves so many different compounds, in practice, many of the compounds (such as non-essential amino acids, vitamin products, etc.) are tested individually (see “Reaction and Pathway Function Testing” in Fig. 1.1). There are other considerations and issues to evaluate during the quality control process, such as revising constraints in order to eliminate “free energy” producing loops [19]. This issue and others are discussed in more detail by Thiele and Palsson in [7].

It should be noted that the iterative loop in Fig. 1.1 involves further manual curation and more detailed investigation into a particular functionality, in order to understand why the test failed. For iNJ661 this included multiple rounds of revisiting and reevaluating the primary literature as well as detailed evaluation of the relevant pathways. In some situations there may be no direct evidence to support incorporation of a particular reaction (such as a transport reaction), which is an intermediate in a pathway whose endpoint is known to occur in the organism. In order to produce a functional model, the transport reaction may need to be added. This is one example of why “confidence level” scores are an important quality control measure in network reconstructions, because they denote the type of evidence that was used to justify incorporation of the reaction. These can then be used to determine future experiments and to also re-evaluate model content when additional datasets are generated for the organism.

3 Network Characterization

There are a myriad of ways to test or assess functionality of a model [2, 20–22]. Once a functional model has been constructed, one of the first steps of analysis is to understand how a particular objective, such as growth, varies on varying substrate utilization. Phase-plane diagrams can address such questions by plotting two different network fluxes (uptake or exchange reactions) while optimizing for an objective. This can be used to assess the trade-off associated with the use of one substrate versus another. The fatty acid constitution of Mycobacteria and other acid fast organisms is complex and unique compared to most other prokaryotes. These fatty acids also constitute a significant portion of the biomass. Glycerol is a required substrate for this and the trade-off between glucose and glycerol is demonstrated in Fig. 1.2, while optimizing for biomass.

The ability to carry out genome-wide screening of gene essentiality in microbes, enables testing of in silico predictions using network models. Results can be categorized

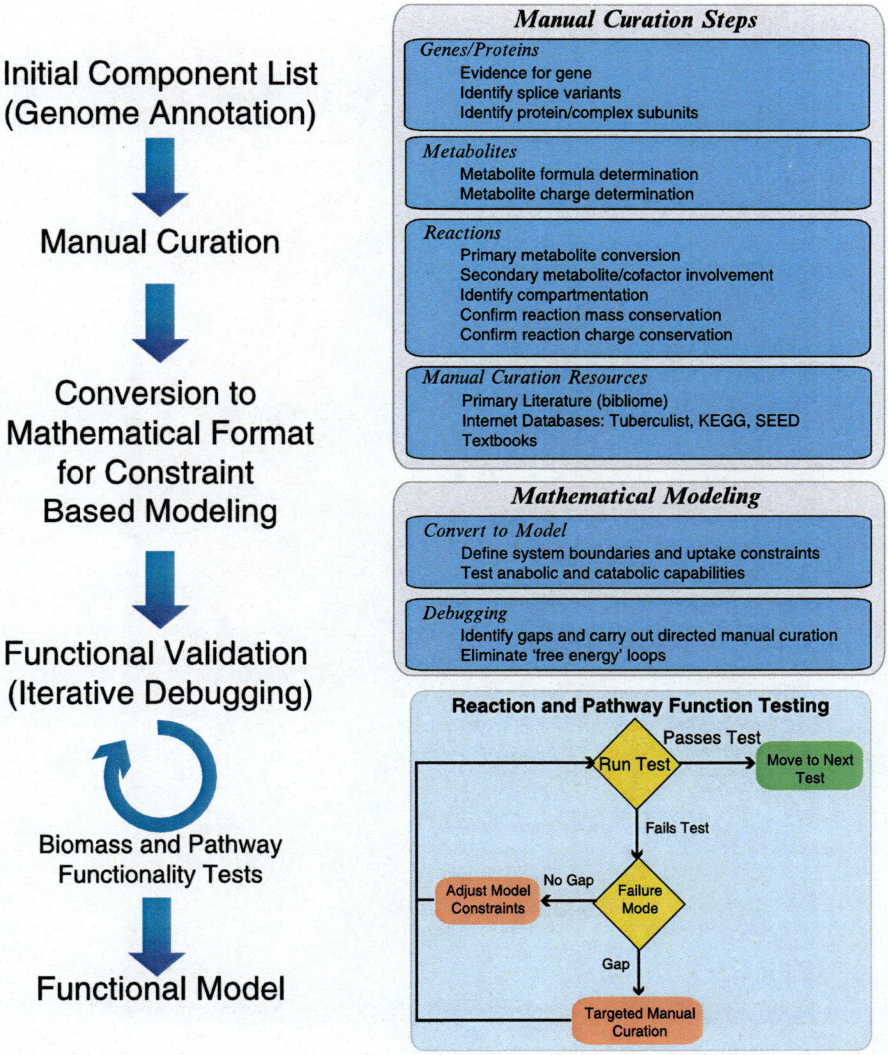

Fig. 1.1 An outline of the general steps for building genome-scale metabolic networks that are amenable to constraint-based modeling. For iNJ661 specifically, the annotated genome for *Mycobacterium tuberculosis* H37Rv was downloaded from The Institute for Genomic Research (TIGR) [68]. Reconstruction content was defined based on the sequence annotation, legacy data, the Tuberculist database [69], ancillary sources such as the Kyoto Encyclopedia of Genes and Genomes (KEGG) and SEED [70]. Debugging was started once the first draft of the reconstruction was completed and functional testing (i.e., flux balance analysis calculations, etc.) were begun. This was carried out in a systematic manner, checking for individual biomass constituents (as products) before analyzing the complete biomass function

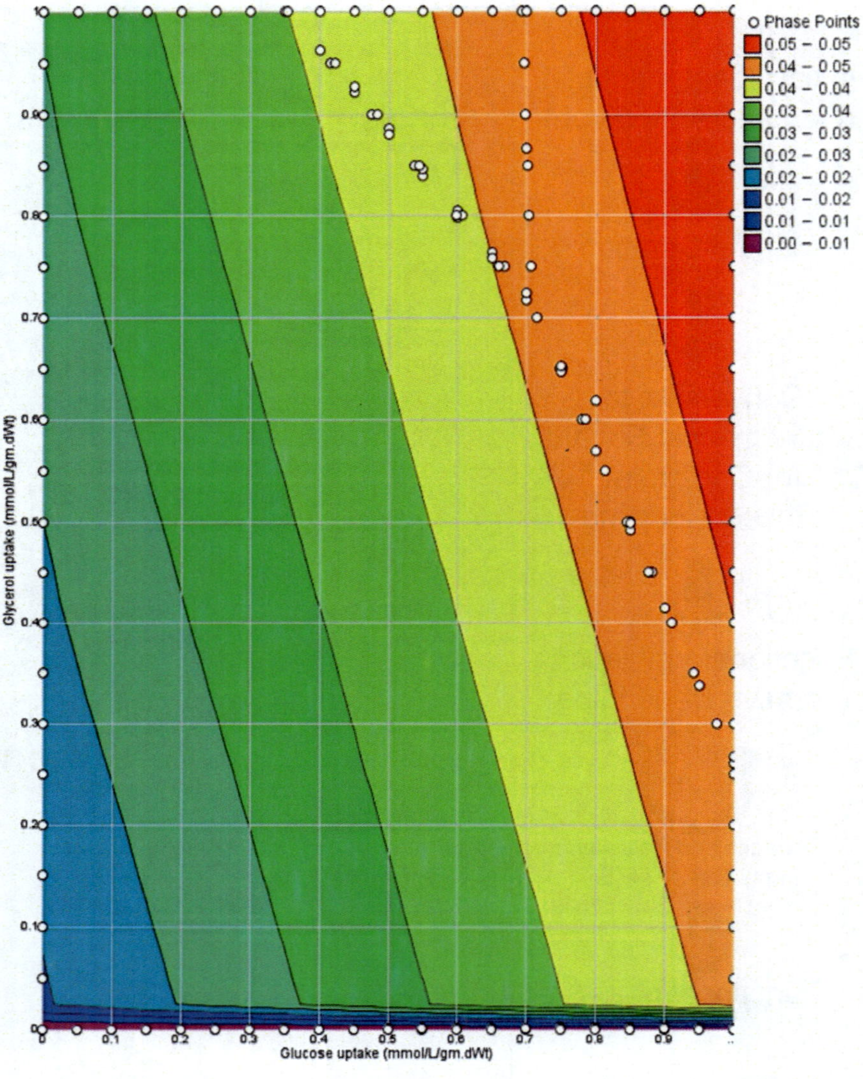

Fig. 1.2 An example of phase-plane diagram for iNJ661 depicting biomass optimization (growth) while varying glycerol and glucose uptake simultaneously. Glycerol is necessary and sufficient for growth, but supplementation with glucose helps achieve higher growth rates. *Open dots* indicate the calculated phase points and the *solid black lines* indicate isoclines

as true positive (verified positive predictions), true negative (verified negative predictions), false positive, and false negative. The specificity of the model is determined by the false negative predictions and the sensitivity is determined by the true positive predictions. Since various processes, notably feedback inhibition and regulation as well as incomplete knowledge of the true metabolic network of the organism, predictions are never expected to be perfect. However, since the model summarizes the current level of knowledge and understanding of the metabolism in the organism, the reasons for the predictions made by the model can be traced back to the network. Thus, the false positive and false negative predictions can serve as hypothesis generating results, and can highlight regions of metabolism that may need further experimental elucidation.

The false positive and false negative rates can be artificially decreased by adjusting the biomass function to include (or exclude) particular metabolites. While this may provide a seemingly better model, such revisions do not actually provide further biological insight. There is more likely information to be gained from further investigation into the possible causes for the incorrect predictions. Alternatively, one can incorporate additional experimental data, such as high-throughput data to make the model more context specific; this approach has been effective in improving the prediction rates for iNJ661 [23].

The concept of correlated reaction sets in functional network models specify groups of reactions that are active in a set; if one reaction in the set has a non-zero flux, then every other reaction in the set will also have a non-zero flux. The converse is also true, if one reaction in the set does not carry a net flux, it implies that every other reaction in the set has no net flux either. There are different ways and different stringencies to defining these correlated sets, that have been discussed elsewhere [2, 20, 24, 25]. Variations among the types of correlated reaction sets include whether the correlations are absolute or partial (i.e., whether they are correlated 85% of the time, 90%, or 100%), qualitatively versus qualitatively correlated, condition dependent or not, etc. [2, 5, 20–22, 24–26].

Antibiotic targets can be analyzed from the perspective of correlated reaction sets. Hard-coupled reaction (HCR) sets are groups of reactions that are always correlated as a result of stoichiometric interactions (so they will be correlated regardless of media condition, although altered expression profiles could result). This concept was applied for tuberculosis to identify drugs that act on single enzymes, but actually knock out complete pathways. The other enzymes in these pathways can serve as equivalent, alternative drug targets.

147 HCR sets were calculated for iNJ661, with the average reaction set involving three reactions. Known *M. tuberculosis* antibiotic targets [27] were mapped to the reactions in the HCR sets, resulting in 25 HCRs with antibiotic targets. A sub-section of the map highlighting some of the druggable HCR sets is illustrated in Fig. 1.3. Since multi-drug treatment regimens are common for tuberculosis treatment [28], it is important to be able to identify different sets of enzyme and reaction targets that are independent of one another, particularly if exploring new drug targets.

The time and resource intensive process of building high-quality, manually curated network reconstructions (Fig. 1.1) has the beneficial result of enabling the analysis of different types of omic data in an integrated manner. Transcriptomic, proteomic, fluxomic, and metabolomic data can all be mapped directly onto a network reconstruction (Fig. 1.4). These data serve to make the organism "context specific," that is representing in vitro growth under different media conditions or data gather from in vivo measurements, with more fidelity. Even growing *M. tuberculosis* in culture in the same condition over the span of weeks has resulted in measurable changes in biomass composition [29], given the slow doubling time of the organism,

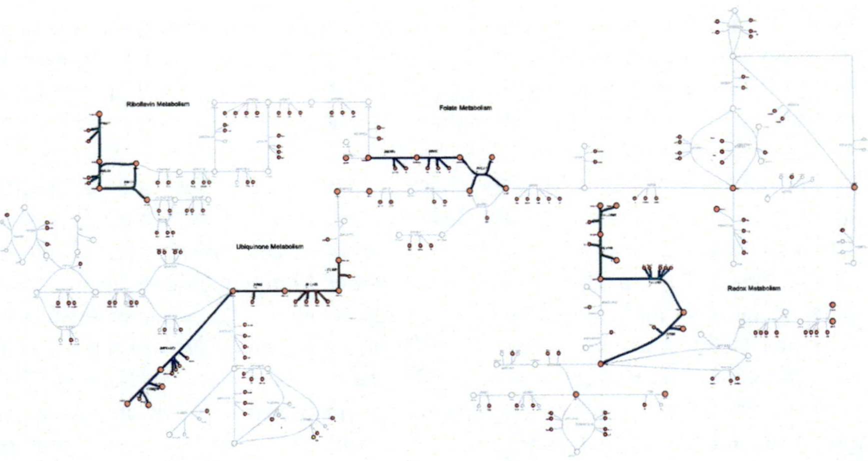

Fig. 1.3 A glimpse of some of the HCR sets that were calculated for iNJ661. These particular HCRs map to particular drug targets (reactions in the same HCR have the same color)

this can occur experimentally and can significantly alter model predictions Furthermore, the integration of omic data into constraint-based models enables predictions to be made, without resorting to any parameter fitting of the models, thus these models serve as data interpretation tools.

Algorithmic approaches for making models context specific [30–32] can also alleviate challenges associated with limitations of existing high-throughput technologies. For example in the field of proteomics, while there have been impressive advancements made in striving towards whole cell coverage [33, 34], current studies in mycobacteria and mammalian host cells are still known to be incomplete. The use of network reconstructions can be used to make sense of and interpret these very large, yet incomplete, datasets. Mass balance constraints link the network together, so gap-filling is not carried out arbitrarily, but is driven by enabling particular biological functions [35, 36].

4 Extending the Model to Account for Host–Pathogen Interactions

Metabolism plays a key role in pathogenesis. As discussed, both simple pathway based and complex network based models have been reconstructed for *M. tuberculosis*. These models have provided great biological insight into pathogenesis, gene essentiality, and potential drug targets. More recently, the metabolic coupling of *M. tuberculosis* to its host has been investigated [37].

Alveolar macrophages play an important role during *M. tuberculosis* infection. *M. tuberculosis* infection begins when droplet nuclei are inhaled, leading to an initial acute infection of the alveoli. Unactivated macrophages phagocytose

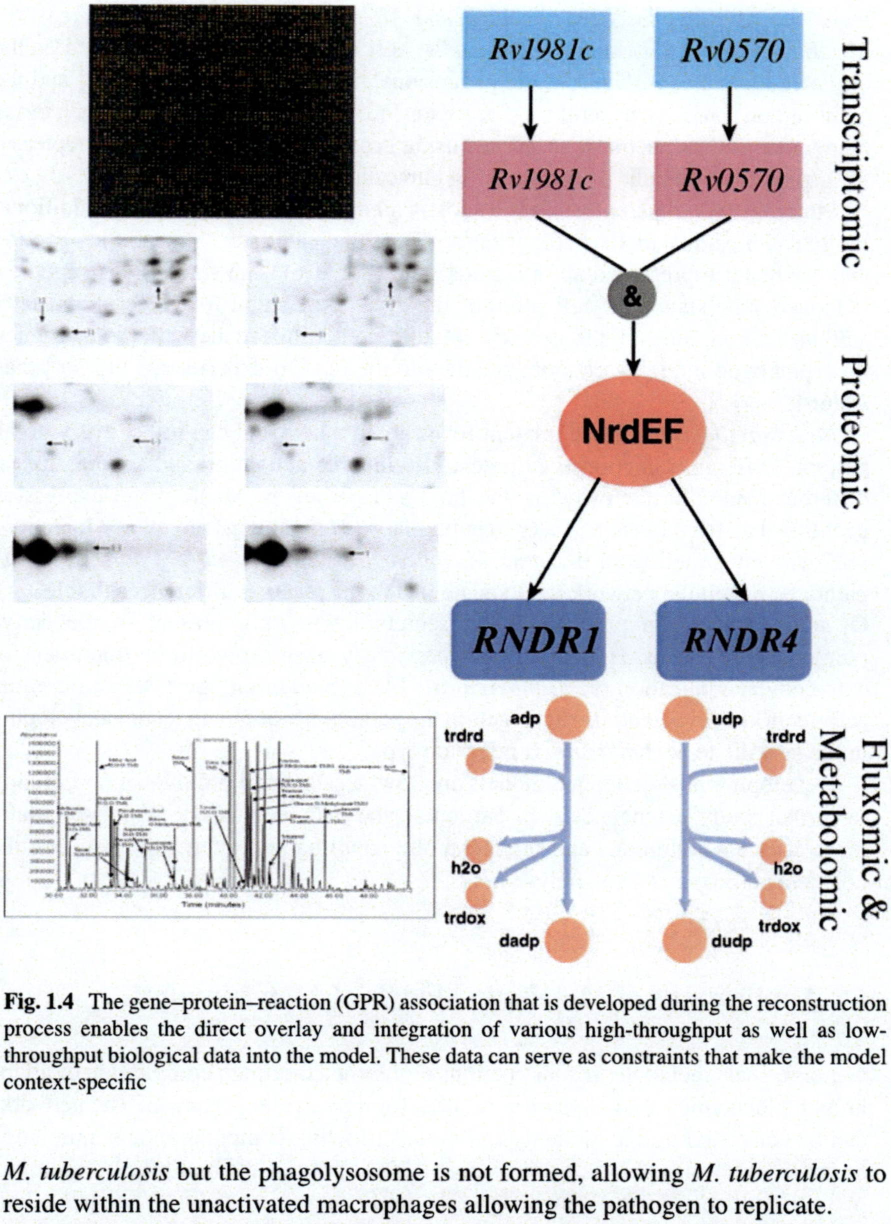

Fig. 1.4 The gene–protein–reaction (GPR) association that is developed during the reconstruction process enables the direct overlay and integration of various high-throughput as well as low-throughput biological data into the model. These data can serve as constraints that make the model context-specific

M. tuberculosis but the phagolysosome is not formed, allowing *M. tuberculosis* to reside within the unactivated macrophages allowing the pathogen to replicate.

Bordbar et al. studied acute *M. tuberculosis* infection by first reconstructing a human alveolar macrophage metabolic network. The global human metabolic network, Recon 1 [38], accounts for the known metabolic genes and reactions of all human cells. Recon 1 was tailored to an alveolar macrophage cell-specific model by integrating transcriptomic [39], proteomic [40, 41], and primary literature articles.

An in silico infection was simulated by integrating the stoichiometric matrices of the macrophage model with *iNJ*661 by constructing a phagosome compartment that allowed metabolic exchange between the host and *M. tuberculosis* cytoplasms (Fig. 1.5). Metabolic exchange constraints between the host and pathogen were set based on the available literature. Particularly, the phagosome environment is hypoxic [42] and the main carbon source is glycerol and fatty acids [43, 44] (Fig. 1.5b). The *M. tuberculosis* biomass objective function was tailored using gene expression data to better represent the pathogen's metabolic demands during intracellular infection.

Studying *M. tuberculosis* in a host–pathogen context provided additional biological insights to its pathogenesis. First, the pathogen's metabolic reaction activity better represented an infectious state (Fig. 1.6). The simulations support a gluconeogenic state with activation of the glyoxylate shunt for glycerol and fatty acid utilization. Second, the gene essentiality predictions made with the integrated host–pathogen model were more accurate to the in vivo experiments [45, 46] than *iNJ*661.

M. tuberculosis can remain latent for years in a host and can infect every organ system. Different infections can potentially involve activation and suppression of different metabolic pathways in the host and pathogen. Macrophage expression profiling data from latent and active (pulmonary and meningeal infections) infections [47] was integrated from different *M. tuberculosis* infection types with the host–pathogen metabolic network to elucidate metabolic reaction differences (Table 1.1). Of note, hyaluronan production and secretion was only present in the active pulmonary infections. Hyaluronan has previously been shown to be important for extracellular replication of *M. tuberculosis* [48]. In addition, there was differential activity in known potential drug targets in the pathogen model, suggesting developing drugs specific to *M. tuberculosis* infection type.

Developing host–pathogen models are now possible and provide an opportunity to further study pathogenesis. In particular, the pathogen can be simulated under more accurate conditions and the metabolic coupling and interaction between the host and pathogen can be analyzed.

5 Applications of Metabolic Models of *M. tuberculosis*

Genome-scale metabolic reconstructions represent a thorough compilation of all the known biochemical and physiological data for a particular organism. The networks can be converted into a mathematical model allowing computation and simulation of different phenotypes. As such, reconstructions provide the biological community of a particular organism a knowledge-base for furthering research. When published, genome-scale metabolic reconstructions are available for public use in research. Since the introduction of *iNJ*661 and GSMN-TB, in 2007, the two *M. tuberculosis* reconstructions have been utilized for many applications including: (1) biological discovery, (2) determining and analyzing potential *M. tuberculosis* drug targets, (3) further model refinement and development, and (4) kinetic modeling of *M. tuberculosis* population dynamics.

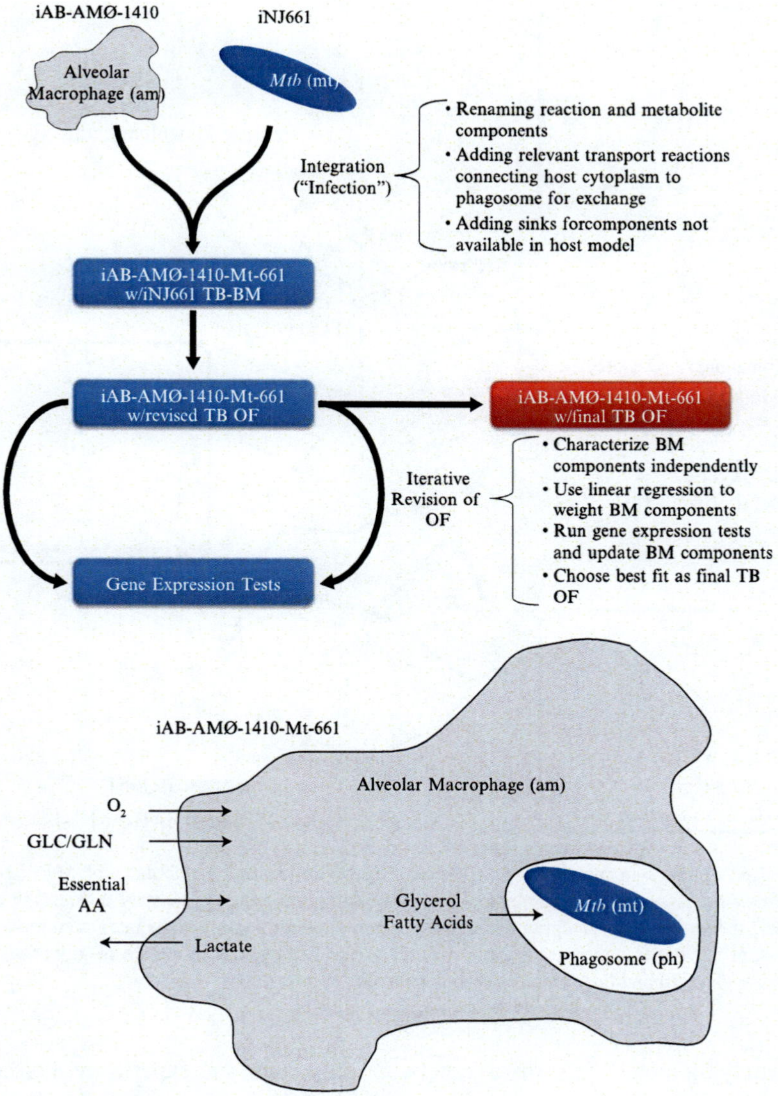

Fig. 1.5 Workflow for building the host–pathogen model. (**a**) Reconstructions of the alveolar macrophage (iAB-AMØ-1410) and *Mycobacterium tuberculosis* (*iNJ*661) were mathematically integrated with physiologically relevant metabolic exchanges. The biomass of *iNJ*661 was updated to better reflect the metabolic demands under infection by utilizing gene expression data. (**b**) Schematic of the metabolic exchanges between the environment, macrophage, and pathogen. Macrophages utilize mainly glucose and glutamine as carbon sources, producing mainly lactate. Glycerol and fatty acids are transported from the host's cytoplasm into the phagosome for *M. tuberculosis* utilization

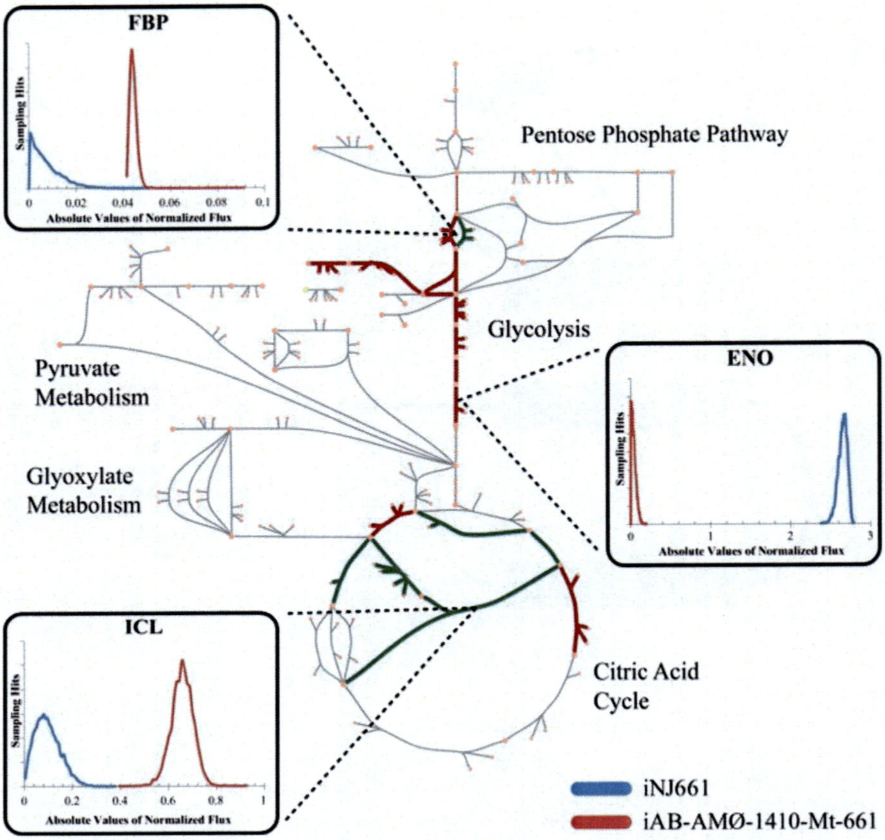

Fig. 1.6 Changes in metabolic reaction activity of the central metabolism of *Mycobacterium tuberculosis*. Randomized sampling was used to determine reaction activity in both the original *iNJ*661 network and the integrated host–pathogen model. Flux states in the integrated model better represent in vivo infection transcription states as the pathogen is in a state of gluconeogenesis, utilizing glycerol and fatty acids as carbon sources

Studying metabolic pathways, network properties, and experimental discrepancies of genome-scale networks can spur biological discovery [49]. Similar studies have been done for *iNJ*661 and GSMN-TB. In 2009, Beste et al. used experimental techniques to determine the essential genes of *Mycobacterium bovis* (BCG) during slow or fast growth [50]. By controlling growth rate, 84 and 256 unique genes were identified as essential under slow and fast in vitro growth on glycerol respectively. The experimentally derived essential genes were compared with simulations of GSMN-TB with 76% accuracy. The functional discrepancies were analyzed and corrected if appropriate in the in silico network.

Recently, high-throughput data was integrated with the network properties of GSMN-TB to determine the metabolic response of *M. tuberculosis* to the host environment. Bonde et al. developed an algorithm called Differential Producibility

Table 1.1 Subsystems of differential metabolic reaction activity in different infection types

Differential subsystem activity		
Latent TB	Pulmonary TB	Meningeal TB
Macrophage		
Pyruvate metabolism	Aminosugar metabolism	Folate metabolism
Thiamine metabolism	Fatty acid metabolism	Nucleotides
Urea cycle/amino group metabolism	Folate metabolism	Pentose phosphate pathway
	Hyaluronan metabolism	Vitamin D metabolism
	Nucleotides	Urea cycle/amino group metabolism
	Pyridine metabolism	
	Thiamine metabolism	
	Val/Leu/Ile metabolism	
	Vitamin D metabolism	
M. tuberculosis		
Alanine and aspartate metabolism		
Glutamate metabolism		Alanine and aspartate metabolism
Pantothenate CoA metabolism		Glutamate metabolism
Polyprenyl metabolism		Pantothenate CoA metabolism

Analysis (DPA) [51]. Essential of reactions and enzymes for metabolite production are determined from the network. High-throughput data are then used to determine the transcriptional and metabolic response to changing conditions. *M. tuberculosis* response to the in vivo environment was studied. Bonde et al. noticed a suppression of central carbon metabolism and an activation of metabolism for cell wall and virulence factors.

Metabolic pathway analysis is a traditional method to study genome-scale networks. Unfortunately, computational limitations do not allow global analysis of large genome-scale networks, including those for *M. tuberculosis*. Ip et al. developed a pathway decomposition method to study *M. tuberculosis* metabolism [52]. Calculated flux distributions can be decomposed into their elementary flux modes for pathway analysis [53]. Ip et al. found though isocitrate lyase is the most efficient path for *M. tuberculosis* biomass and non-growth associated maintenance, it was not the sole metabolic path. If *M. tuberculosis* is provided sufficient levels of octadecenoate, flux can pass through HtrA and malate synthase. Isocitrate lyase has been previously suggested as a potential pharmaceutical target in *M. tuberculosis* [27, 54]. Ip et al. find that *M. tuberculosis* should be able to develop resistance as alternative metabolic pathways are available for the pathogen to utilize carbon for biomass and maintenance.

Second, *M. tuberculosis* reconstructions have been used to identify potential pharmaceutical targets through systems level analysis. In 2008, Raman et al. constructed a database, targetTB, that integrated fluxes derived from genome-scale reconstructions, experimental essentiality data, inference protein–protein interaction networks, the genome sequence of *M. tuberculosis*, and structural properties of

M. tuberculosis proteins [55]. targetTB is a pipeline that defined 451 high-confidence drug targets both applicable broadly to several pathogens and specifically to *M. tuberculosis*.

In a follow-up to targetTB and predicting drug targets, Raman et al. defined a "reactome" network for *M. tuberculosis* using *iNJ*661 [56]. The network was constructed by determining the protein–protein dependencies in *M. tuberculosis* by looking at the "nearness" of proteins with *iNJ*661. A metabolic disruptability index was defined for the reactions of the network, representing the systematic impact of a perturbation on the network. The method was used to determine highly influential proteins and was further tested with FBA to determine potential drug targets.

Metabolic centric approaches to network properties have also been utilized for determining potential pharmaceutical targets. Kim et al. developed a method to determine chokepoints and essential metabolites of four genome-scale networks, including *iNJ*661 [57]. Chokepoints represent enzymes that uniquely produce or consume a particular metabolite, while essential metabolites are required for the network to produce biomass. Kim et al. argue that a metabolic centric approach is more valuable than a reaction/enzyme centric approach for determining pharmaceutical targets as a metabolic centric approach identifies drug combinations of multiple enzyme targets that in principle would lead to less *M. tuberculosis* drug resistance. Enzymes that were both chokepoints and involved in essential metabolites were chosen and compared to the human metabolic network, Recon 1 [38], to ensure that potential drug targets would not affect human metabolism.

Colijn et al. further studied drug interactions with *M. tuberculosis* utilizing expression profiling data and a novel algorithm, E-Flux [58]. Rather than inhibit specific enzymes and reactions that are known to be affected by pharmaceuticals, expression data for a drug-treated case was integrated with GSMN-TB. The upper and lower bounds of the model were adjusted based on the expression levels and mycolic acid production was used as an indicator of *M. tuberculosis* fatty acid biosynthesis. The impact of 75 drugs, their combinations, and nutrient environments were simulated. E-Flux was used to identify potential fatty acid inhibitors of interest in disrupting *M. tuberculosis* fatty acid biosynthesis.

Third, *iNJ*661 has been used for further refinement and development of *M. tuberculosis* metabolic and regulatory models. As previously discussed, Bordbar et al. built a cell-specific metabolic model of the human alveolar macrophage [37]. The network contained a phagosome compartment, which allowed metabolic integration with *M. tuberculosis* (*iNJ*661). The model was utilized to study host–pathogen interaction during acute infection. In another study, Fang et al. modified *iNJ*661 to better represent in vivo condition [59]. Rather than build an integrated network, the metabolic exchanges and biomass components of *iNJ*661 were modified to better represent in vivo infection by comparing the gene-essentiality results with those determined experimentally [46]. The resulting network, *iNJ*661v, highlighted the differential use of metabolic enzymes during infection versus in vitro growth. In addition, *iNJ*661v was used to simulate double gene knockouts for potential synergistic drug targets.

Though genome-scale metabolic reconstructions have shown utility in many areas of *M. tuberculosis* research, the transcriptional regulatory network is not modeled. In 2010, Chandrasekaran and Price developed an automated inference based method to construct a genome-scale regulatory-metabolic network from high-throughput data [23]. Probabilistic regulation of metabolism, PROM, was used to study *E. coli* and *M. tuberculosis*. The constructed *M. tuberculosis* regulatory-metabolic network was validated against existing transcription factor essentiality data and used to determine six new candidate essential genes that can be potential drug targets in *M. tuberculosis*. The orthologs of the six candidate genes have been shown to be essential in *E. coli* and *Bacillus subtilis*.

Finally, Fang et al. constructed a *M. tuberculosis* population growth model utilizing enzyme kinetics, population growth modeling, and *iNJ*661 [60]. The framework involved determining the effect of an inhibitor on the enzyme kinetics of a *M. tuberculosis* enzyme which was then modeled at the genome-scale with *iNJ*661. Initial conditions of substrate and *M. tuberculosis* population are linked with the growth rate of *iNJ*661 and in an iterative fashion the constraint-based model and ordinary differential equations are linked to determine population dynamics. The mathematical framework was used to determine the effect of 3-nitropropionate inhibitor and 5′-O-(*N*-salicylsulfamoyl) adenosine inhibitor. The computational simulations reproduced experimentally derived growth curves.

6 Future Directions and Applications

The continued development of high-throughput data has presented the unique challenge of voluminous datasets to analyze, that will necessarily always be incomplete, which have created the need for methods to interpret data that enable (1) integration of disparate data, (2) can be built on the genome or cell-scale, (3) can incorporate large yet incomplete datasets, (4) have a minimal reliance on parameter fitting. Metabolic network reconstructions can address all of these challenges and consequently represent an ideal tool to help drive the analysis and interpretation of high-throughput datasets in a discovery driven manner.

Our current knowledge of the understanding of *M. tuberculosis* metabolism continues to grow on a daily basis [61–64], thus there is a need to continually update these models. There are undoubtedly many applications that have not yet been conceived; however, given some of the existing areas of progress made in the systems biology of *M. tuberculosis*, there are a number of areas that will benefit from further investment and development. These updates include updating existing content as new experimental data is generated, expanding the scope to include signaling [65] and transcription/translation events [66], reconciling existing models [4, 5, 67], and methods data integration [23, 58] so that ultimately the analysis of in vivo conditions and host–pathogen relationships [37] can be investigated to ultimately develop approaches to identify new drug targets and treatments.

References

1. Orth JD, Thiele I, Palsson BO (2010) What is flux balance analysis? Nat Biotechnol 28(3):245–248
2. Palsson BO (2006) Systems biology: determining the capabilities of reconstructed networks. Cambridge University Press, Cambridge
3. Raman K, Rajagopalan P, Chandra N (2005) Flux balance analysis of mycolic acid pathway: targets for anti-tubercular drugs. PLoS Comput Biol 1(5):e46
4. Beste DJ, Hooper T, Stewart G, Bonde B, Avignone-Rossa C, Bushell ME, Wheeler P, Klamt S, Kierzek AM, McFadden J (2007) GSMN-TB: a web-based genome-scale network model of Mycobacterium tuberculosis metabolism. Genome Biol 8(5):R89
5. Jamshidi N, Palsson BO (2007) Investigating the metabolic capabilities of Mycobacterium tuberculosis H37Rv using the in silico strain iNJ661 and proposing alternative drug targets. BMC Syst Biol 1:26
6. Rom W, Garay S (2004) Tuberculosis, 2nd edn. Lippincott Williams and Wilkins, Philadelphia
7. Thiele I, Palsson BO (2010) A protocol for generating a high-quality genome-scale metabolic reconstruction. Nat Protoc 5(1):93–121
8. Henry CS, DeJongh M, Best AA, Frybarger PM, Linsay B, Stevens RL (2010) High-throughput generation, optimization and analysis of genome-scale metabolic models. Nat Biotechnol 28(9):977–982
9. Kanehisa M (2002) The KEGG database. Novartis Found Symp 247:91–101, discussion 101–103, 119–128, 244–152
10. Lew JM, Kapopoulou A, Jones LM, Cole ST (2011) TubercuList – 10 years after. Tuberculosis (Edinb) 91(1):1–7
11. Overbeek R, Begley T, Butler RM, Choudhuri JV, Chuang HY, Cohoon M, de Crecy-Lagard V, Diaz N, Disz T, Edwards R, Fonstein M, Frank ED, Gerdes S, Glass EM, Goesmann A, Hanson A, Iwata-Reuyl D, Jensen R, Jamshidi N, Krause L, Kubal M, Larsen N, Linke B, McHardy AC, Meyer F, Neuweger H, Olsen G, Olson R, Osterman A, Portnoy V, Pusch GD, Rodionov DA, Ruckert C, Steiner J, Stevens R, Thiele I, Vassieva O, Ye Y, Zagnitko O, Vonstein V (2005) The subsystems approach to genome annotation and its use in the project to annotate 1000 genomes. Nucleic Acids Res 33(17):5691–5702
12. Jamshidi N, Palsson BO (2008) Formulating genome-scale kinetic models in the post-genome era. Mol Syst Biol 4:171
13. Beard DA, Liang SD, Qian H (2002) Energy balance for analysis of complex metabolic networks. Biophys J 83(1):79–86
14. Kummel A, Panke S, Heinemann M (2006) Systematic assignment of thermodynamic constraints in metabolic network models. BMC Bioinformatics 7:512
15. Famili I, Mahadevan R, Palsson BO (2005) k-Cone analysis: determining all candidate values for kinetic parameters on a network scale. Biophys J 88(3):1616–1625
16. Holzhutter HG (2004) The principle of flux minimization and its application to estimate stationary fluxes in metabolic networks. Eur J biochem/FEBS 271(14):2905–2922
17. Jamshidi N, Palsson BO (2010) Mass action stoichiometric simulation models: incorporating kinetics and regulation into stoichiometric models. Biophys J 98(2):175–185
18. Feist AM, Palsson BO (2010) The biomass objective function. Curr Opin Microbiol 13(3):344–349
19. Schellenberger J, Lewis NE, Palsson BO (2011) Elimination of thermodynamically infeasible loops in steady-state metabolic models. Biophys J 100(3):544–553
20. Papin JA, Reed JL, Palsson BO (2004) Hierarchical thinking in network biology: the unbiased modularization of biochemical networks. Trends Biochem Sci 29(12):641–647
21. Price ND, Reed JL, Palsson BO (2004) Genome-scale models of microbial cells: evaluating the consequences of constraints. Nat Rev Microbiol 2(11):886–897
22. Schellenberger J, Palsson BO (2009) Use of randomized sampling for analysis of metabolic networks. J Biol Chem 284(9):5457–5461

23. Chandrasekaran S, Price ND (2010) Probabilistic integrative modeling of genome-scale metabolic and regulatory networks in Escherichia coli and Mycobacterium tuberculosis. Proc Natl Acad Sci USA 107(41):17845–17850
24. Heinrich H, Schuster S (1996) The regulation of cellular systems. Springer, Berlin
25. Jamshidi N, Palsson BO (2006) Systems biology of SNPs. Mol Syst Biol 2:38
26. Burgard AP, Nikolaev EV, Schilling CH, Maranas CD (2004) Flux coupling analysis of genome-scale metabolic network reconstructions. Genome Res 14(2):301–312
27. Mdluli K, Spigelman M (2006) Novel targets for tuberculosis drug discovery. Curr Opin Pharmacol 6(5):459–467
28. Kasper DL, Braunwald E, Fauci AS, Hauser SL, Longo DL, Jameson JL (2005) Harrison's principles of internal medicine, 16th edition, New York, McGraw-Hill
29. Youmans AS, Youmans GP (1968) Ribonucleic acid, deoxyribonucleic acid, and protein content of cells of different ages of Mycobacterium tuberculosis and the relationship to immunogenicity. J Bacteriol 95(2):272–279
30. Becker SA, Palsson BO (2008) Context-specific metabolic networks are consistent with experiments. PLoS Comput Biol 4(5):e1000082
31. Jerby L, Shlomi T, Ruppin E (2010) Computational reconstruction of tissue-specific metabolic models: application to human liver metabolism. Mol Syst Biol 6:401
32. Lewis NE, Schramm G, Bordbar A, Schellenberger J, Andersen MP, Cheng JK, Patel N, Yee A, Lewis RA, Eils R, Konig R, Palsson BO (2010) Large-scale in silico modeling of metabolic interactions between cell types in the human brain. Nat Biotechnol 28(12):1279–1285
33. de Souza GA, Wiker HG (2011) A proteomic view of mycobacteria. Proteomics 11(15):3118–3127
34. Ehebauer MT, Wilmanns M (2011) The progress made in determining the Mycobacterium tuberculosis structural proteome. Proteomics 11(15):3128–3133
35. Reed JL, Patel TR, Chen KH, Joyce AR, Applebee MK, Herring CD, Bui OT, Knight EM, Fong SS, Palsson BO (2006) Systems approach to refining genome annotation. Proc Natl Acad Sci USA 103(46):17480–17484
36. Satish Kumar V, Dasika MS, Maranas CD (2007) Optimization based automated curation of metabolic reconstructions. BMC Bioinformatics 8:212
37. Bordbar A, Lewis NE, Schellenberger J, Palsson BO, Jamshidi N (2010) Insight into human alveolar macrophage and M. tuberculosis interactions via metabolic reconstructions. Mol Syst Biol 6:422
38. Duarte NC, Becker SA, Jamshidi N, Thiele I, Mo ML, Vo TD, Srivas R, Palsson BO (2007) Global reconstruction of the human metabolic network based on genomic and bibliomic data. Proc Natl Acad Sci USA 104(6):1777–1782
39. Kazeros A, Harvey BG, Carolan BJ, Vanni H, Krause A, Crystal RG (2008) Overexpression of apoptotic cell removal receptor MERTK in alveolar macrophages of cigarette smokers. Am J Respir Cell Mol Biol 39(6):747–757
40. Peri S, Navarro JD, Kristiansen TZ, Amanchy R, Surendranath V, Muthusamy B, Gandhi TK, Chandrika KN, Deshpande N, Suresh S, Rashmi BP, Shanker K, Padma N, Niranjan V, Harsha HC, Talreja N, Vrushabendra BM, Ramya MA, Yatish AJ, Joy M, Shivashankar HN, Kavitha MP, Menezes M, Choudhury DR, Ghosh N, Saravana R, Chandran S, Mohan S, Jonnalagadda CK, Prasad CK, Kumar-Sinha C, Deshpande KS, Pandey A (2004) Human protein reference database as a discovery resource for proteomics. Nucleic Acids Res 32(Database issue):D497–D501
41. Uhlen M, Oksvold P, Fagerberg L, Lundberg E, Jonasson K, Forsberg M, Zwahlen M, Kampf C, Wester K, Hober S, Wernerus H, Bjorling L, Ponten F (2011) Towards a knowledge-based Human Protein Atlas. Nat Biotechnol 28(12):1248–1250
42. Honer zu Bentrup K, Russell DG (2001) Mycobacterial persistence: adaptation to a changing environment. Trends Microbiol 9(12):597–605
43. McKinney JD, Honer zu Bentrup K, Munoz-Elias EJ, Miczak A, Chen B, Chan WT, Swenson D, Sacchettini JC, Jacobs WR Jr, Russell DG (2000) Persistence of Mycobacterium tuberculosis

in macrophages and mice requires the glyoxylate shunt enzyme isocitrate lyase. Nature 406(6797):735–738
44. Schnappinger D, Ehrt S, Voskuil MI, Liu Y, Mangan JA, Monahan IM, Dolganov G, Efron B, Butcher PD, Nathan C, Schoolnik GK (2003) Transcriptional adaptation of Mycobacterium tuberculosis within macrophages: insights into the phagosomal environment. J Exp Med 198(5):693–704
45. Sassetti CM, Boyd DH, Rubin EJ (2003) Genes required for mycobacterial growth defined by high density mutagenesis. Mol Microbiol 48(1):77–84
46. Sassetti CM, Rubin EJ (2003) Genetic requirements for mycobacterial survival during infection. Proc Natl Acad Sci USA 100(22):12989–12994
47. Thuong NT, Dunstan SJ, Chau TT, Thorsson V, Simmons CP, Quyen NT, Thwaites GE, Thi Ngoc Lan N, Hibberd M, Teo YY, Seielstad M, Aderem A, Farrar JJ, Hawn TR (2008) Identification of tuberculosis susceptibility genes with human macrophage gene expression profiles. PLoS Pathog 4(12):e1000229
48. Hirayama Y, Yoshimura M, Ozeki Y, Sugawara I, Udagawa T, Mizuno S, Itano N, Kimata K, Tamaru A, Ogura H, Kobayashi K, Matsumoto S (2009) Mycobacteria exploit host hyaluronan for efficient extracellular replication. PLoS Pathog 5(10):e1000643
49. Oberhardt MA, Palsson BO, Papin JA (2009) Applications of genome-scale metabolic reconstructions. Mol Syst Biol 5:320
50. Beste DJ, Espasa M, Bonde B, Kierzek AM, Stewart GR, McFadden J (2009) The genetic requirements for fast and slow growth in mycobacteria. PLoS One 4(4):e5349
51. Bonde BK, Beste DJ, Laing E, Kierzek AM, McFadden J (2011) Differential producibility analysis (DPA) of transcriptomic data with metabolic networks: deconstructing the metabolic response of M. tuberculosis. PLoS Comput Biol 7(6):e1002060
52. Ip K, Colijn C, Lun DS (2011) Analysis of complex metabolic behavior through pathway decomposition. BMC Syst Biol 5:91
53. Schuster S, Dandekar T, Fell DA (1999) Detection of elementary flux modes in biochemical networks: a promising tool for pathway analysis and metabolic engineering. Trends Biotechnol 17(2):53–60
54. Beste DJ, Bonde B, Hawkins N, Ward JL, Beale MH, Noack S, Noh K, Kruger NJ, Ratcliffe RG, McFadden J (2011) ^{13}C metabolic flux analysis identifies an unusual route for pyruvate dissimilation in mycobacteria which requires isocitrate lyase and carbon dioxide fixation. PLoS Pathog 7(7):e1002091
55. Raman K, Yeturu K, Chandra N (2008) targetTB: a target identification pipeline for Mycobacterium tuberculosis through an interactome, reactome and genome-scale structural analysis. BMC Syst Biol 2:109
56. Raman K, Vashisht R, Chandra N (2009) Strategies for efficient disruption of metabolism in Mycobacterium tuberculosis from network analysis. Mol Biosyst 5(12):1740–1751
57. Kim TY, Kim HU, Lee SY (2010) Metabolite-centric approaches for the discovery of antibacterials using genome-scale metabolic networks. Metab Eng 12(2):105–111
58. Colijn C, Brandes A, Zucker J, Lun DS, Weiner B, Farhat MR, Cheng TY, Moody DB, Murray M, Galagan JE (2009) Interpreting expression data with metabolic flux models: predicting Mycobacterium tuberculosis mycolic acid production. PLoS Comput Biol 5(8):e1000489
59. Fang X, Wallqvist A, Reifman J (2010) Development and analysis of an in vivo-compatible metabolic network of Mycobacterium tuberculosis. BMC Syst Biol 4:160
60. Fang X, Wallqvist A, Reifman J (2009) A systems biology framework for modeling metabolic enzyme inhibition of Mycobacterium tuberculosis. BMC Syst Biol 3:92
61. Driscoll MD, McLean KJ, Levy C, Mast N, Pikuleva IA, Lafite P, Rigby SE, Leys D, Munro AW (2010) Structural and biochemical characterization of Mycobacterium tuberculosis CYP142: evidence for multiple cholesterol 27-hydroxylase activities in a human pathogen. J Biol Chem 285(49):38270–38282
62. Elamin AA, Stehr M, Spallek R, Rohde M, Singh M (2011) The Mycobacterium tuberculosis Ag85A is a novel diacylglycerol acyltransferase involved in lipid body formation. Mol Microbiol 81(6):1577–1592

63. Hatzios SK, Bertozzi CR (2011) The regulation of sulfur metabolism in Mycobacterium tuberculosis. PLoS Pathog 7(7):e1002036
64. Ouellet H, Guan S, Johnston JB, Chow ED, Kells PM, Burlingame AL, Cox JS, Podust LM, de Montellano PR (2010) Mycobacterium tuberculosis CYP125A1, a steroid C27 monooxygenase that detoxifies intracellularly generated cholest-4-en-3-one. Mol Microbiol 77(3):730–742
65. Li F, Thiele I, Jamshidi N, Palsson BO (2009) Identification of potential pathway mediation targets in Toll-like receptor signaling. PLoS Comput Biol 5(2):e1000292
66. Thiele I, Jamshidi N, Fleming RM, Palsson BO (2009) Genome-scale reconstruction of Escherichia coli's transcriptional and translational machinery: a knowledge base, its mathematical formulation, and its functional characterization. PLoS Comput Biol 5(3):e1000312
67. Thiele I, Palsson BO (2010) Reconstruction annotation jamborees: a community approach to systems biology. Mol Syst Biol 6:361
68. The Institute for Genomic Research. http://www.tigr.org/
69. Porcelli AM, Ghelli A, Zanna C, Pinton P, Rizzuto R, Rugolo M (2005) pH difference across the outer mitochondrial membrane measured with a green fluorescent protein mutant. Biochem Biophys Res Commun 326(4):799–804
70. Navarro A (2004) Mitochondrial enzyme activities as biochemical markers of aging. Mol Aspects Med 25(1–2):37–48

Chapter 2
Software Platform for Metabolic Network Reconstruction of *Mycobacterium tuberculosis*

Samik Ghosh, Yukiko Matsuoka, Yoshiyuki Asai, Hiroaki Kitano, Anshu Bhardwaj, Vinod Scaria, Rohit Vashisht, Anup Shah, Anupam Kumar Mondal, Priti Vishnoi, Kumari Sonal, Akanksha Jain, Priyanka Priyadarshini, Kausik Bhattacharyya, Vikas Kumar, Anurag Passi, Pratibha Sharma, and Samir Brahmachari

Tuberculosis (TB) is one of the major infectious diseases still prevailing on this planet. Emergence of drug resistant strains and problems of current treatment regimen warrant need for new drugs for TB. At the same time, economic factor plays a

---

Samik Ghosh and Anshu Bhardwaj are Joint first authors.

S. Ghosh
The Systems Biology Institute, Tokyo 108-0071, Japan

Y. Matsuoka
The Systems Biology Institute, Tokyo 108-0071, Japan

ERATO Kawaoka Infection-Induced Host Response Project,
Japan Science and Technology Agency, Tokyo 108-8639, Japan

Y. Asai
Okinawa Institute of Science and Technology, Okinawa 904-0495, Japan

H. Kitano (✉)
The Systems Biology Institute, Tokyo 108-0071, Japan

Okinawa Institute of Science and Technology, Okinawa 904-0495, Japan

Sony Computer Science Laboratories, Inc, Tokyo 141-0022, Japan
e-mail: kitano@sbi.jp

A. Bhardwaj • R. Vashisht • A. Jain • P. Priyadarshini • A. Passi
Council of Scientific and Industrial Research,
2 Rafi Marg, Delhi 110001, India

V. Scaria • A. Shah • A.K. Mondal • P. Vishnoi • K. Sonal • K. Bhattacharyya
V. Kumar • P. Sharma
CSIR-Institute of Genomics and Integrative Biology,
Mall Road, Delhi 110007, India
e-mail: skb@igib.res.in

S. Brahmachari (✉)
Council of Scientific and Industrial Research, 2 Rafi Marg, Delhi 110001, India

CSIR-Institute of Genomics and Integrative Biology, Mall Road, Delhi 110007, India

J. McFadden et al. (eds.), *Systems Biology of Tuberculosis*,
DOI 10.1007/978-1-4614-4966-9_2, © Springer Science+Business Media, LLC 2013

significant role as most patients are in the lowest income bracket of the society. This implies new drugs have to be developed in an innovative manner that allows delivery of drugs at low cost. Drug discovery is in general an expensive and capital-intensive process. A new type of big science is emerging that involves knowledge integration of small sciences as well as coordinating community-based participation. Social dynamics plays critical role in making project successful because open collaboration involves participants with diverse motivations and interests. Thus, proper "social engineering" will play greater role in scientific project planning and management in future. Open Source Drug Discovery (OSDD), initiated by Council for Scientific and Industrial Research (CSIR) of India, is one of such projects aiming at the development of drugs for TB. The fact that drug discovery is a competitive space, bringing in openness and collaboration through e-community-based approach is a challenging task. This article describes the international collaboration among OSDD, the Systems Biology Institute (SBI: Japan), and Okinawa Institute of Science and Technology (OIST: Japan) for reconstruction of a comprehensive and high-precision map of metabolic network of *Mycobacterium tuberculosis* (mTB) through a virtual collaborative space. The fact that OSDD involved large number of non-experts guided by experts in the process further sets it apart from other existing ways of addressing scientific problems of this scale.

1 Issues in Drug Discovery for Tuberculosis

Tuberculosis (TB) is still a major killer in developing countries as 9.4 million new patients are reported in 2009 globally and significant percentage of them are multi-drug resistant TB and some are extensively drug resistant [1], but only a handful of drug discovery projects exist due to economic affordability of patients mismatching against possible investment for the development of such drugs. Unless, cost effective drug development can be achieved, those who suffer from neglected diseases will have no hope for new treatment and fast recovery. Developing technologies to significantly mitigate these problems is socially significant.

However, we are facing the reality that there are diseases where effective cure is not generally available for extreme drug resistance strain of mTB. Even if the drug is developed, the cost of drug discovery may be too high to make it widely available for poorest segment of patient population. Cost of developing drug is too high not only due to high failure rate and long duration from discovery stage to approval, but also the R&D cost of each product is high (an estimated average cost of $454 million per product [2]).

Considerable efforts have to be made to significantly improve drug discovery productivity and to deliver drugs at affordable cost. It is critical for R&D productivity that in-depth understanding of complexity of biological systems and means to predict potential outcome of candidate compounds when used in cells, model animals, and patients [3, 4] to be better established. Proper introduction of systems biology approach to drug discovery is expected to rectify the situation by providing better understanding of biology behind diseases at system-level and ultimately by enabling us to use precision computational models of cells, organs, and patients.

Introduction of precision modeling may open up improved productivity for combinatorial drug design including re-purposing of existing drugs. An apparent problem of combinatorial drugs is huge search space intrinsic from combinatorial nature of such an approach. Brute force approaches will be too expensive for combination of more than three components. Enabling systematic computational identification of possible combination at early stage is critical for productivity improvement. This is critically important in addressing the urgent medical needs where potential opportunities and early cases have been reported [5–8]. National Cancer Institute is carrying out a large-scale systematic study on combinatorial drugs and initial results seem to be promising [9]. Many of CNS related diseases and drug resistant infectious diseases might require multiple points of interventions as well [10]. Improved efficiency of combinatorial drug discovery pipeline, possibly enabled by precision modeling approach, provides exciting opportunities.

An interesting opportunity in combinatorial drug is that it may be possible to discover novel combinations of existing drugs for new indications. For example, there is an interesting study that demonstrated the combined use of chlorpromazin (antipsycotics agent) and pentamidine (antiparasitic agent) can be equally or more effective than paclitaxel for a specific cancer [6]. It is interesting to note that the cost of paclitaxel in Japan is about 44,000 Yen (about 450USD) for 100 mg, whereas cost for chlorpromazine is about 9 Yen and pentamidine is about 2,800 Japanese Yen. The point is not whether this specific combination is effective or not at the end. The point is there will be numbers of such combinations that can create drugs at significantly lower cost. Combined price is less than 3,000 Yen that is almost 1/10 of paclitaxel. This opens up tremendous opportunities not only for industrial countries, but also for developing countries where cost of drug is a critical issue. It should be noted that patents of many drugs expire sooner or later that implies options for combinations at lower cost will continue to increase in future [11].

Furthermore, such an approach may open up a new opportunity to reuse compounds that were failed in clinical and preclinical trials as one of the compounds in combinatorial drugs. It is possible compounds that have not been as effective as expected as a mono-therapy drug may be re-purposed in the context of combination. With new criteria for combinatorial drug context, it is possible that the entire compound library may have to be revisited.

Although combinatorial drug is a promising approach, the issue is how to discover effective combination at practical efficiency. This is potentially a combinatorial explosion process, and without innovative scheme it would only result in a low efficiency hit-and-error process. Random screening is too inefficient for this approach and may not capture interesting synergetic combinations. An exhaustive screening has been tried for the two-component combination, but it was limited to combination among 1,200 candidate compounds and difficult to be scaled for multiple component combinations [6]. Most likely such an undertaking would be far beyond the capability of a single pharmaceutical company or publicly funded projects. Thus, it is essential that computational approach to be established at the practical level that can predict possible combinations for further study.

2 Knowledge Integration: A Challenge in Precision Modeling

The challenge of developing precision biological models is that it requires integration of knowledge and data at all levels from genomics and proteomics to imaging and physiology. While various data from high-throughput experiments provide us with genome-wide characteristics, understanding detailed mechanisms has to depend on individual "small sciences." It is unfeasible to obtain such knowledge by a single large-scale project due to both financial and sociological reasons. Each researcher is interested in a specific aspect of biology using organisms that they think are most suitable for the study. Due to diversity of biological systems, choice is diverse and researchers make a choice for good reasons. Even if one obtained a large-scale funding, it is not practical to force sufficiently large numbers of researchers to put their systems aside and work on a new species and biological problems.

At the same time, using existing resources such as pathway databases for modeling is not a solution either. Although these pathway databases are developed mainly based on manual curation of publications, it does not mean they are well covered or accurate. Since pathway databases have to cover broad range of pathways, each pathway is curated and represented with limited coverage, depth, and accuracy. Current "Gold Standard" is manually curated models carefully build based on the literature and various data resources by a small group of people who spend months on the same pathway to the extent that they acquire in-depth insights on the pathway [12]. This is what we call "deep curation" as exemplified by a series of comprehensive molecular interaction maps [13, 14]. However, the deep curation of large-scale network maps from the literature is extremely labor-intensive and stressful work. Also, it is very difficult to maintain motivations to continuously up-dating maps and models to keep up with the new discoveries for many years. Automated literature mining has been extensively investigated, but nowhere near the stage to replace human curators. At the same time, quality control dependent on the individual groups, and updating and fixing errors can be slowed by this centralization. How we can solve this problem impacts productivity and practicality of computational approach for drug discovery for wider targets.

3 Needs for Virtual Big Science

Scientific projects with large funding to achieve defined mission are often called Big Science. Successful big science projects shall have clearly defined goals, possible means to achieve it, and strong social justifications to support such endeavor through public funding. At the same time, what type of project can be supported widely beyond scientific community depends on social needs at that time. While most biology has been and continue to be small science, the human genome project and a series of large-scale genome projects can be considered big science in biology. A typical characteristic of big science is a project with large-scale engineering to support a specific scientific discovery. Big science is not a new phenomenon. The legendary Manhattan Project, and more recently the Large Hadron Collider the Human Genome Project connected experts by bringing them together physically. These projects are essentially equipment driven data acquisition projects, and such project will continue to provide us with new findings by improving measurement equipment. There are desires to obtain comprehensive understanding of specific cellular systems and biological processes using high-throughput measurements so that comprehensive picture of biological systems can be observed from a specific aspect. Emergence of systems biology as mainstream biology is accelerating this tendency as it often requires measurements and analysis of various large-scale and multi-faceted data. At the same time, the reality is that new knowledge critical for in-depth and precise understanding is often derived from small science. This means that a new type of big science is needed that consolidates data and knowledge not only from large-scale projects, but also from discoveries by small science. Thus, it is inevitable to form a "virtual big science" by connecting large numbers of researchers around the globe to attain large-scale knowledge integration in an emergent manner. The implication of this is that the initiative needs to have a widely acceptable objectives, leadership, and proper sociological design to make it sustainable.

Some of the more recent genome annotation jamborees have also followed a similar approach. However, the growth of Internet brought in a paradigm shift in implementing large collaborative projects. Galaxy Zoo (http://www.galaxyzoo.org/) is one of the pioneers in using Internet for launching the first and largest ever citizen-science experiment involving non-experts in classifying galaxies. This collaborative approach has contributed mightily to the outcome of this project. Translating this concept to IPR intensive areas to solve challenging scientific problems seems to be a promising path towards speedy and affordable solutions. The OSDD project of the Council of Scientific and Industrial Research (CSIR), India, has taken a similar approach for solving challenges in drug discovery process [15]. Similar experiments, involving larger e-community, are being done with challenging problems like protein folding (http://www.ncbi.nlm.nih.gov/pubmed/20686574). However, drug discovery is a very competitive space making it even more challenging to open it for global participation through virtual communities. The following discussion elaborates more on the process and the framework designed to achieve these goals.

4 Open Source Drug Discovery for Tuberculosis

Drug discovery for neglected diseases can be a successful emergent collaborative project. It carries a good cause, socially appealing, needs collective efforts, and participants will have a sense of pride for their contribution. Towards this OSDD project was initiated by CSIR of Indian government [16]. OSDD targets drug discovery for *Tuberculosis* through open collaboration realizing that confidentiality and IP rights slows down the drug discovery process and makes it extremely expensive due to large-scale failure of discovery effort going from hit to lead to preclinical and clinical trials. The OSDD project was conceived and designed to make drug discovery cost effective by distributed co-creation in an open source mode. As one of the many projects, OSDD launched the Connect to Decode Project (http://c2d.osdd.net) as an open call to comprehensively re-annotate the genome of *M. tuberculosis* to facilitate its systems level understanding. A large number of students and researchers registered for the project, pan India as well as overseas, and contributed on a voluntary basis. Over 830 researchers and students participated in five components of the project. More than 400 students curated literature (>10,000 published literature) for Mtb genome across five components, namely, Interactome/Pathway Annotation (IPW), TB Gene Ontology (TBGO), Glycomics, Structural/fold annotation, and Immunome. One of these components, IPW, was designed to achieve two goals, the first was to create a protein–protein functional interaction network and the second to reconstruct a comprehensive and detailed metabolic map of Mtb discussed here. Due to the nature of the project, Indian students involved are highly passionate and motivated in contributing for the common goal. The desire to learn and excel in resource limited setting further fuels the motivation for contributing to the project. Unlike past efforts for collective pathway reconstruction and curation such as yeast metabolic map, any volunteer researchers and students can join the effort. Thus this project has distinctive open-endedness in both quantity and quality of participation. This is a novel model where motivation was the key as opposed to incentive and sets an example towards a novel social engineering model for involving large communities in solving challenging problems. However, the best contributors from all five components were shortlisted based on their contributions and were awarded a net-book for their contribution (India 800 foundation). Thus the OSDD project has become an emotional enterprise rather than a professional enterprise.

This project satisfies some of the criteria for successful emerging projects such as clear and appealing goal, motivations of participants, financial support, and willingness to address a global issue. The distributed and collective genome re-annotation was indeed a social experiment. With this experiment, a large-scale distributive reconstruction of biological networks was shown to be possible with a proper software platform, well-defined workflow, and project management *when* the objectives of the project designed to motivate potential participants (Fig. 2.1).

5 OSDD Mtb Metabolome Challenge

The OSDD Mtb Metabolome challenge involved manually mining literature on Mtb research, specifically on the metabolic reactions involved towards understanding the function of enzymatic proteins at the global scale. While similar efforts have been reported in the past, the OSDD Mtb Metabolome challenge engendered a unique open source community collaborative platform involving researchers and student volunteers from across India and worldwide who worked to mine knowledge buried in the experimental studies and unlock relevant information for each enzyme and metabolites systematically. The process involved well-defined protocols

Fig. 2.1 A scene from Connect2Decode onsite session at New Delhi, India. Over 200 participants get together for a week for the final assembly of the metabolic map. Each student is given a laptop and hands-on training for assembling the metabolic reactions into pathways using cell designer (http://c2d.osdd.net; http://osdd-c2d.blogspot.com/). Detailed workflow is in Figure 2.1

and work-packages for knowledge acquisition, representation, peer-review by students under active guidance from experts in the field. The overall roadmap of the challenge is schematically elucidated in Fig. 2.2.

While the collection of individual metabolic reactions and enzymes for the knowledge aggregation phase of map reconstruction, their integration in a global context is the key towards understanding the network dynamics at a systems level. Towards this direction, metabolome challenge project, in collaboration with The Systems Biology Institute, Japan and Okinawa Institute of Science and Technology (OIST), Japan (http://www.oist.jp), employed a systems biology computational platform for reconstructing a large-scale, standard compliant metabolic map.

The integrated platform, as outlined in Fig. 2.3, involved two key computational tools—*CellDesigner* (developed by Kitano's group under JST, NEDO, and MEXT funding) is a graphical molecular network editing and analysis software suite that complies with Systems Biology Markup Language (SBML) and Systems Biology Graphical Notation (SBGN) standards. *CellDesigner* has been successfully used to create large-scale molecular interaction maps based on literature curation (http://www.celldesigner.org/models.html).

The need to extend the computational tools to an online, community collaboration paradigm motivated the use of *Payao*, by Kitano group and OIST. *Payao is* a web-based biological pathway sharing and tagging service (http://www.payaologue.org).

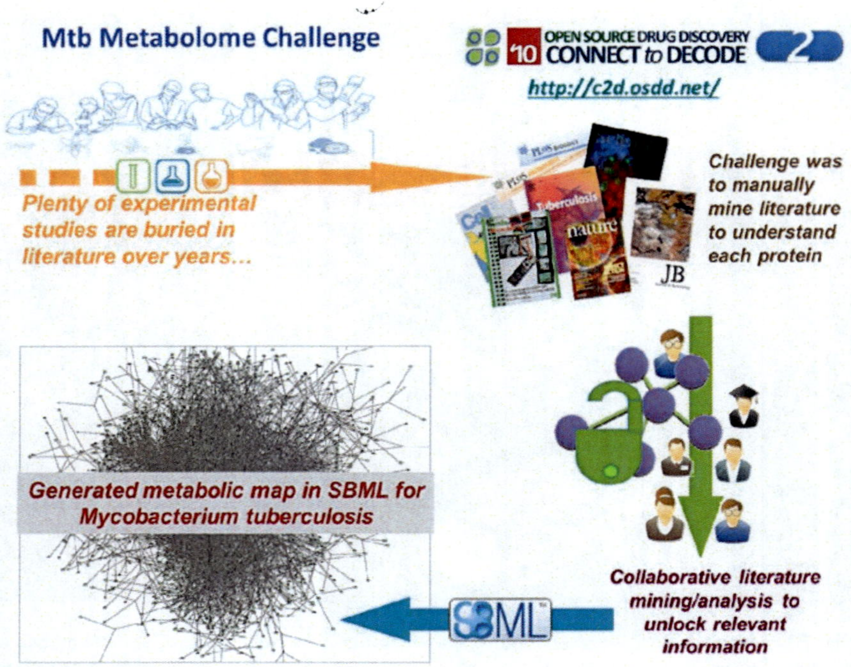

Fig. 2.2 The Mtb Metabolome Challenge (http://c2d.osdd.net)

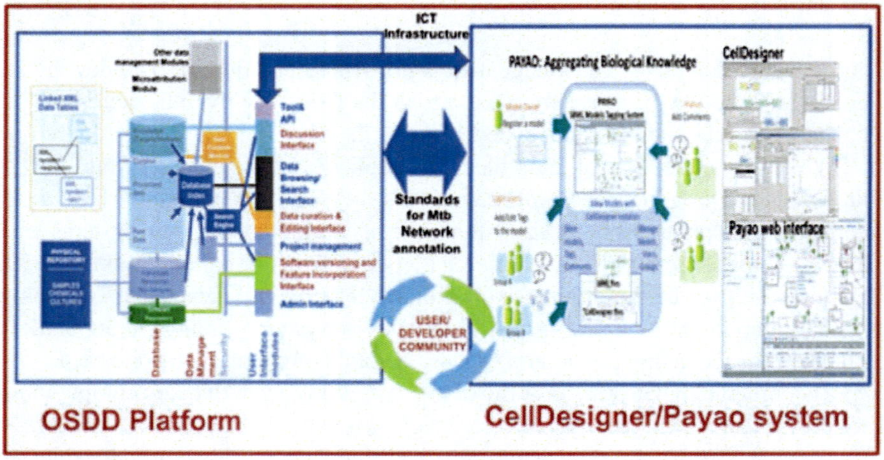

Fig. 2.3 Integrated map curation platform deployed for the Mtb Metabolome Challenge

It aims to provide a Google Maps (http://www.maps.google.com) equivalent for biological pathways, wherein researchers can share large-scale, curated, and annotated network maps using software like *CellDesigner* and publish it to the community online. Both tools employ systems biology standards for knowledge representation and exchange, namely, The SBML (http://sbml.org), a set of standards

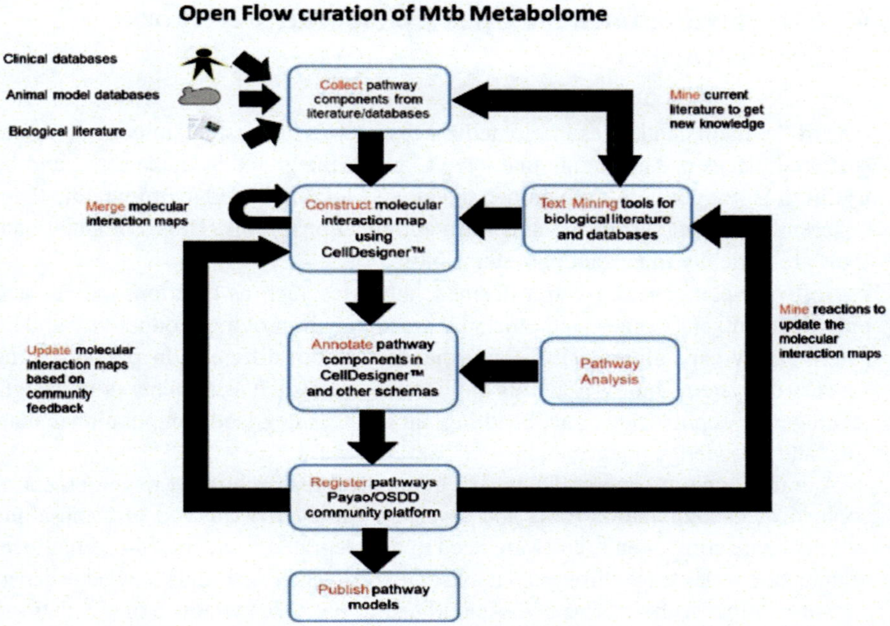

Fig. 2.4 Distributed pathway reconstruction workflow. Republished with permission from Nature Publishing Group: Kitano, H., Ghosh, S., and Matsuoka, Y., Social engineering for virtual "big science" in systems biology, Nature Chemical Biology, 7 (6) 323–326, 2011 [17]

developed to facilitate effective and efficient sharing of models defined as a set of biochemical reactions, and The SBGN (http://sbgn.org), a common graphical representation standard in the life sciences.

Based on the integrated curation platform, systematic workflow was employed in close global partnership between India and Japan teams, to *collect, mine, construct, annotate,* and *publish* the metabolic map, as outlined in Fig. 2.4.

A project-wide distributed reconstruction has been tested using *CellDesigner* network editing software [18] that was provided by the Systems Biology Institute, Tokyo, Japan, as a part of agreement with CSIR.

Participants used Google documents to collect and curate data on the metabolic reactions of Mtb following standard operating protocols and manuals in data format amenable for programmatic processing. This data is then converted into SBML files using the Connect to Decode plug in developed by SBI. Different SBML files for each pathway were drawn and then merged using *CellDesigner* to generate the complete map. Thus, the interactions are merged and re-layed out in *CellDesigner*. Multiple iterations of pathway construction and integration took place. After a few month of distributed curation session, everyone from Indian side and Tokyo side got together in Delhi for a week for the final assembly of the entire network. Final draft network covers the TB metabolic network with around 1,394 genes in 13 meta-pathways. The data curated at each level is shared with the scientific community using the OSDD portal.

6 Software Platform for Open Collaborative Network Reconstruction

One of the main challenges in biomedical research is the vast quantity of data, and scattered pieces of knowledge that have to be all integrated to make sense and be useful. It is not possible for a human to extract useful knowledge or integrate them coherently without systematic aids from computational tools. Thus, computational tools are critically important in systems biology.

Software platforms have transformed industries such as aviation, movies and entertainment, electronics, and others by drastically improving productivity and by offering new capabilities. Biological sciences are not different. In particular, the success of systems biology, and its application areas such as systems drug design, leverages on sophisticated data handling, modeling, integrated computational analysis, and knowledge integration.

A cornerstone in open collaborative biomedical research is the development of sophisticated computation tools and services. Particularly, instead of stand-alone and disparate components, software need to be integrated in an end-to-end platform architecture to leverage different databases, experimental data, and knowledge generated at multiple scales of research. As outlined in Fig. 2.5, systems biology platform needs to build on community-driven standards, pathway and network modeling, together with community collaborative platform which empower social engineering in virtual big science projects.

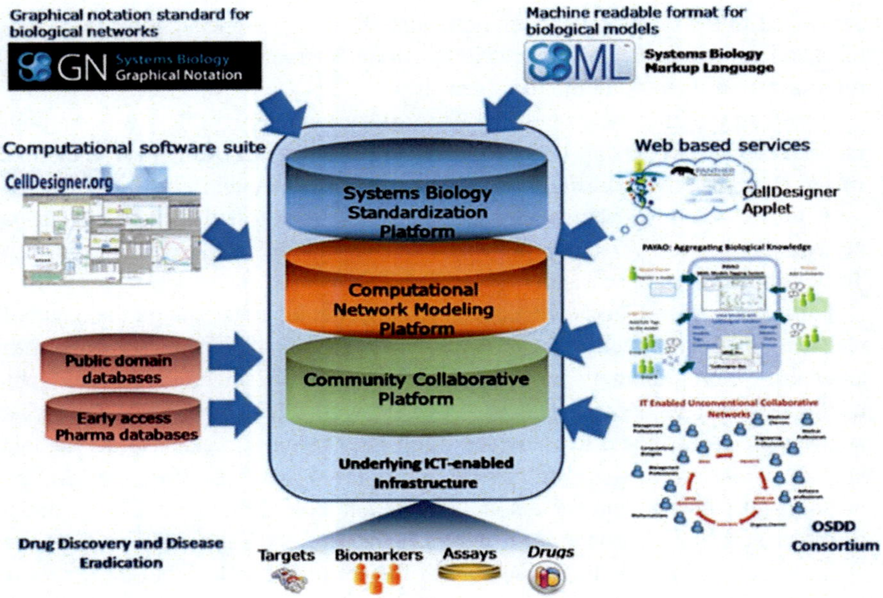

Fig. 2.5 Software platforms for systems biology

While several reviews have focused on standardization and software platforms to drive large-scale systems biology research, in this article, we delve into community collaborative platforms, particularly in the context of open innovations for drug discovery in neglected areas like tuberculosis.

As explained earlier, creating an extensive model of gene-regulatory and biochemical networks with the latest data is a painstaking task. Curation is essential to create an accurate model. Yet as science and technology advances rapidly, once curated models soon become out-of-date and need to be revised constantly. Many pathways and networks are now available online via pathway database, such as Reactome, BioModels.net, Panther Pathways, and many pathway editors are available [12]. What is needed is a framework to facilitate tracking and update mechanism for modelers and researchers in the community to contribute to the collaborative model building and curation process.

WikiPathways [19] is an effort for such a collaborative platform in the Wiki style. While the Wiki system has its strength in collaborative editing and version tracking, it does not provide access control or explicit community tagging mechanisms. In a community-driven model enrichment environment, it is effective to differentiate privileges to special interest group (SIG) members for curation activities—commenting on existing tags, adding tags to models, annotating individual component inside a model, and validating the annotations. In view of the complexity of biological pathways and the expertise of biologists in different areas, a community platform for biology requires an exquisite balance of federated resource sharing and quality control of information by a SIG of experts in the particular pathway or process. An access-control privilege system allows the community to share and disseminate the knowledge, while enabling a dedicated SIG to maintain high-quality, curated information.

Payao, developed jointly by The Systems Biology Institute, Tokyo, Japan and OIST, Japan, is a community-driven molecular pathway curation framework. The system is named after a fish aggregating device, an artificial floating raft where fish congregate, and popular in Okinawa/Philippine area. Payao aims to become a biological knowledge aggregating system, which enables a community to work on the same models simultaneously, insert tags as pop-up balloon to the parts of the model, exchange comments, record the discussions, and eventually update the models accurately and concurrently.

Payao serves for enrichment phase of the curation. It is a web-based platform, providing an interface for adding tags and comments to the components (such as Species, Reactions, and specified area) of the model, as well as community management functionality. The information on the users and tag data are stored in a relational database on the server.

Payao adopts community standards, accepting SBML [20] format models and displays them in SBGN [21] compliant *CellDesigner* [18] graphical notation. As Payao accepts pathway models stored in SBML format and uses *CellDesigner* APIs (Application Program Interface) for visualization, the most suitable SBML editor for Payao is *CellDesigner*. In SBML format, models can capture details of biochemical process descriptions, not only protein–protein interactions. Adopting SBML format enables the models to be easily used as the base of computational data analysis or simulation of dynamic behaviors. The Payao platform enriches the model curation process by providing a host of features for user management, tagging, and model updates.

Forming a community is an important step for curation. Different expert groups can contribute variety of information to the model. As Web-based Payao can be accessible from all physical locations, it enables experts across the world to communicate in a collaborative curation effort.

Community is formed around a pathway model. It is the model owner who sets access control over the registered model. In the Payao system, access controls can be set by specifying the privileges to individuals as well as to user-categories, such as guest, login user, and model user (who are invited to access the model by the model owner). This enables a user to stage the curation process; initiate the curation within a small group (e.g. SIG), then switch the access control of the model for public viewing.

The tagging on the visually represented pathways is a characteristic of Payao, which makes the curators easy to grasp the nature of the pathway while discussing on the specific component of the pathways. Like Google Maps, tags are displayed in a bubble form attached to the items (Species, Reactions, or any specified area), and click to expand and display the content of the information in the tag. Tags can be specific keywords, links, PubMed IDs as well as free text.

The Mtb metabolic map curated using Google docs and merged using Cell Designer was converted into SBML and is being made available through the Payao platform for larger community access and tagging.

7 Sustainability of Large-Scale Interaction Map Development

The challenge remains how to design and generate platform that enable continuous updates of the new data into the system and subsequent quality checks towards better annotation and data analysis.

We should carefully look at reasons why projects like Wikipedia and Linux took off and keep flying. In case of Linux, there was a hacker culture that support open source and sharing of knowledge as signified by Free Software Foundation by Richard Stallman where contribution to the community at large was the pride of the hackers. At the same time, there was practical need to develop open source operating systems as opposed to closed commercial systems such as Microsoft Windows. Among the several efforts for open source operating systems such as FreeBSD, Linux survived mainly because it hit the right moment and had more applications and publications than other initiatives. Thus, if it was not Linux, then it could have been FreeBSD or other initiative that filled this space. Wikipedia essentially inherit similar culture. The goal that is widely shared and exciting, and a sense of participation have been key factors driving the community-based initiative. While these motivations are sustainable over time is yet to be seen, it was indeed effective to get things started and matured enough.

The biological community too is getting infected with the Web 2.0 concept as it is being generally realized that biological problems are too complex to be solved and translated into public good by any one individual or organization. A formalized recognition system which may include micro-impact factors or micro-attribution for contributions may be a very effective way of encouraging more participation as done in Sysborg2.0, the OSDD collaborative platform [12]. Such index or credit should also become part of the merit system as is the case with Citation index which is a widely used measure of scientific contributions. For micro-attribution to be accepted in the merit system, it has to acquire universality as the currency in scientific evaluation. Simply asking for contribution and assigning micro-attributions assume that people are motivated by individual benefits. It is generally true to assume that, but it is also weak motivation factors to be a successful project. In the most successful project, people are driven by the vision, passion, and dedication aligned with individual aspiration to future. In OSDD there is a serious attempt towards addressing these issues by giving due credit to significant contributions through Sysborg2.0 by including the contributors as authors in publications also. We have followed this approach in various publications that have originated from community-driven projects [13, 22]. However, there are critical issues in community building including reaching a critical mass of active members to ensure sustainability and community-driven feedback loops. OSDD Community as of now stands at more than 5,000 registered users from more than 130 countries indicating development of a self-sustaining group. At any given time more than 10 % of the participants are actively involved.

8 Can We Scale-Up?

A related question to sustainability of open collaboration is the issue of scale. The same skepticism of motivation and culture may apply. However, if the project frames the mission in a socially appealing and attractive way, there may be a chance that broader participation can be expected. Due to social significance of finding effective cure for drug resistant TB, it may attract those who are willing to contribute even without personal or professional benefits, and they may simply be pride of being a part of it.

It should be noted that such collaborations are possible now due to development of various standards and software that comply with such standards. Standards like SBML [20], SBGN [21], and BioPAX [22] and tools and platforms like TBrowse [23] and Sysborg2.0 [15] ensure a certain level of interoperability. However, technology alone cannot make things work, particularly when projects have to involve large numbers of interested parties with varying motivations, carrier aspirations, and opinions. The OSDD Community is a large group of researchers and students with heterogeneous expertise and interests. The OSDD portal provides a common platform for facilitating interactions among the members which enables identification of complementary skills and interests for fruitful collaborations and quick outcomes. The successful implementation of the Connect to Decode Phase I project has lead to launch of its second phase (http://c2d.osdd.net) where models for predicting anti-tuberculosis property are already published [24]. Thus, broader social consideration can be a major "key for success" when launching increasingly complex projects [25]. Social engineering will be recognized as an indispensable part of research activity in coming years for large-scale and complex big sciences because it is the people who do science, not technology or machines.

Acknowledgments The Indian team is fully supported by CSIR/OSDD, India. The Japanese team is, in part, supported by funding from the HD-Physiology Project of the Japan Society for the Promotion of Science (JSPS) to the Okinawa Institute of Science and Technology (OIST), and the International Strategic Collaborative Research Program (BBSRC-JST) of the Japan Science and Technology Agency (JST), the Exploratory Research for Advanced Technology (ERATO) programme of JST to the Systems Biology Institute (SBI).

References

1. Donald PR, van Helden PD (2009) The global burden of tuberculosis—combating drug resistance in difficult times. N Engl J Med 360:2393–2395
2. PricewaterhouseCoopers (2007b) Pharma 2020: Virtual R&D—which path will you take?
3. FDA (2004) Challenge and opportunity on the critical path to new medical products, http://www.fda.gov/ScienceResearch/SpecialTopics/CriticalPathInitiative/CriticalPathOpportunitiesReports/ucm077262.htm
4. PricewaterhouseCoopers (2007a) Pharma 2020: the vision—which path will you take?
5. Apsel B, Blair JA, Gonzalez B, Nazif TM, Feldman ME, Aizenstein B, Hoffman R, Williams RL, Shokat KM, Knight ZA (2008) Targeted polypharmacology: discovery of dual inhibitors of tyrosine and phosphoinositide kinases. Nat Chem Biol 4:691–699
6. Borisy AA, Elliott PJ, Hurst NW, Lee MS, Lehar J, Price ER, Serbedzija G, Zimmermann GR, Foley MA, Stockwell BR et al (2003) Systematic discovery of multicomponent therapeutics. Proc Natl Acad Sci USA 100:7977–7982
7. Hopkins AL (2008) Network pharmacology: the next paradigm in drug discovery. Nat Chem Biol 4:682–690
8. Kitano H (2007) A robustness-based approach to systems-oriented drug design. Nat Rev Drug Discov 6:202–210
9. Kummar S, Chen HX, Wright J, Holbeck S, Millin MD, Tomaszewski J, Zweibel J, Collins J, Doroshow JH (2010) Utilizing targeted cancer therapeutic agents in combination: novel approaches and urgent requirements. Nat Rev Drug Discov 9:843–856
10. Lim J, Hao T, Shaw C, Patel AJ, Szabo G, Rual JF, Fisk CJ, Li N, Smolyar A, Hill DE et al (2006) A protein-protein interaction network for human inherited ataxias and disorders of Purkinje cell degeneration. Cell 125:801–814
11. Taneja B, Yadav J, Chakraborty TK, Brahmachari SK (2009) An Indian effort towards affordable drugs: "generic to designer drugs". Biotechnol J 4:348–360
12. Bauer-Mehren A, Furlong LI, Sanz F (2009) Pathway databases and tools for their exploitation: benefits, current limitations and challenges. Mol Syst Biol 5:290
13. Caron E, Ghosh S, Matsuoka Y, Ashton-Beaucage D, Therrien M, Lemieux S, Perreault C, Roux PP, Kitano H (2010) A comprehensive map of the mTOR signaling network. Mol Syst Biol 6:453

14. Oda K, Kitano H (2006) A comprehensive map of the toll-like receptor signaling network. Mol Syst Biol 2(2006):0015
15. Bhardwaj A, Scaria V, Raghava GP, Lynn AM, Chandra N, Banerjee S, Raghunandanan MV, Pandey V, Taneja B, Yadav J et al (2011) Open source drug discovery—a new paradigm of collaborative research in tuberculosis drug development. Tuberculosis (Edinb) 91:479–486
16. Singh S (2008) India takes an open source approach to drug discovery. Cell 133:201–203
17. Kitano H, Ghosh S, Matsuoka Y (2011) Social engineering for virtual 'big science' in systems biology. Nat Chem Biol 7:323–326
18. Funahashi A, Matsuoka Y, Jouraku A, Morohashi M, Kikuchi N, Kitano H (2008) Cell Designer 3.5: a versatile modeling tool for biochemical networks. Proc IEEE 96:1254–1265
19. Pico AR, Kelder T, van Iersel MP, Hanspers K, Conklin BR, et al (2008) WikiPathways: Pathway Editing for the People. PLoS Biol 6(7): e184.doi:10.1371/journal.pbio.0060184
20. Hucka M, Finney A, Sauro HM, Bolouri H, Doyle JC, Kitano H, Arkin AP, Bornstein BJ, Bray D, Cornish-Bowden A et al (2003) The systems biology markup language (SBML): a medium for representation and exchange of biochemical network models. Bioinformatics 19:524–531
21. Le Novere N, Hucka M, Mi H, Moodie S, Schreiber F, Sorokin A, Demir E, Wegner K, Aladjem MI, Wimalaratne SM et al (2009) The Systems Biology Graphical Notation. Nat Biotechnol 27:735–741
22. Demir E, Cary MP, Paley S, Fukuda K, Lemer C, Vastrik I, Wu G, D'Eustachio P, Schaefer C, Luciano J et al (2010) The BioPAX community standard for pathway data sharing. Nat Biotechnol 28:935–942
23. Bhardwaj A, Bhartiya D, Kumar N, Scaria V (2009) TBrowse: an integrative genomics map of Mycobacterium tuberculosis. Tuberculosis (Edinb) 89:386–387
24. Periwal V, Rajappan JK, Jaleel AU, Scaria V (2011) Predictive models for anti-tubercular molecules using machine learning on high-throughput biological screening datasets. BMC Res Note 4:504
25. Hill C (2007) The post-scientific society, Issues in Science and Technology. Fall:78–84

Chapter 3
Probing Gene Regulatory Networks to Decipher Host–Pathogen Interactions

Kanury V.S. Rao, Dhiraj Kumar, and Shekhar C. Mande

Abstract Given the pivotal role that macrophages play in determining the outcome of infection, it is now becoming apparent that a better understanding of the molecular interplay between the pathogen and this host cell type will be crucial for developing more effective strategies for tuberculosis therapy. In this context the need to capture the dynamic features of this crosstalk, so as to dissect the evolving stages of host–pathogen equilibration, is also beginning to be appreciated. A promising way to probe the ongoing dialog between the macrophage and the pathogen is through gene expression profiling. An analysis of the gene expression pattern of the infected host cell on the one hand, and that of the infecting pathogen on the other, provides a coarse grained insight into the nature and dynamics of interactions between these two entities. While much more work needs to be done in this direction, initial studies are beginning to shed light on the mechanisms by which the pathogen equilibrates within the host intracellular environment. However, an important goal here would be to extract the gene regulatory networks that emerge within the pathogen and the host cell, and to then precisely map the interface between these two networks. In addition to yielding important information on crosstalk mechanisms, such mapping should also help to identify novel targets for drug development.

K.V.S. Rao (✉) • D. Kumar
International Centre for Genetic Engineering and Biotechnology,
New Delhi, India
e-mail: kanury@icgeb.res.in

S.C. Mande (✉)
National Centre for Cell Science, Pune, India
e-mail: shekhar@nccs.res.in

J. McFadden et al. (eds.), *Systems Biology of Tuberculosis*,
DOI 10.1007/978-1-4614-4966-9_3,

1 Introduction

Despite the availability of effective drugs, tuberculosis (TB) persists as a disease of high burden in the human population. At least to a large extent this is a reflection of the stable manner in which the pathogen *Mycobacterium tuberculosis* (MTB) entrenches itself within the host and then successfully evades the myriad of antimicrobial responses that it encounters. Infection by MTB is initiated by its penetration into the alveoli of the lung, wherein it is then taken up by the macrophages. Interestingly, while macrophages are phagocytes that provide the first line of defense against microbial pathogens, they are nonetheless successfully coopted by the mycobacteria to serve as the niche for survival and persistence of the pathogen. This is achieved through interference with the intracellular trafficking events that follow uptake, allowing the bacteria to occupy an immature phagosomal compartment. Encapsulation within immature phagosomes insulates the pathogen from several of the potent antimicrobial pathways and also provides a window for the bacilli to replicate during the early stages of infection. Over time, however, the bacteria arrest their replication and enter into a dormant state. This state is clinically termed as "latency" and is supported by the acquisition of phenotypic tolerance as a result of metabolic adaptation or quiescence. Intracellular persistence also requires that the pathogen engage with an additional array of cellular mechanisms. Some of these include autophagy, lipid turnover, and pathways that influence the redox equilibrium. Finally, the ability to control the activation of apoptotic pathways and induce necrotic cell death is key to successful dissemination of the pathogen within the host.

It is, therefore, becoming increasingly clear that a better resolution of the crosstalk between the intracellular mycobacteria host cellular machinery is central to understanding disease biology as well as for developing new methods for treatment.

2 Delineation of Mycobacterial Genes That Mediate Adaptation and Survival in Macrophages

Any attempt at deciphering the host–pathogen dialog first requires a description of the key molecular players, on the side of both the pathogen and the host, that mediate these interactions. A seminal study published in 2005 yielded a systematic identification of the MTB genes that were required for survival and growth in macrophages [1]. This study exploited transposon site hybridization, a microarray-based technique that was previously developed to comprehensively identify genes from large pools of transposon mutants that were essential for MTB growth under a variety of conditions [2]. Based on this method a screen was devised to identify the MTB transposon mutants that were unable to survive in murine macrophages that were either unactivated or activated with IFN-α either prior to or subsequent to infection. The rationale behind this experimental design was to simulate the in vivo conditions of initial and/or latent infection, as well as those wherein the infected macrophages were under pressure from an ongoing immune response. A total of 126 MTB genes were identified in this study and a clustering of these genes into functional groups identified several putative operons that were each required for survival in macrophages. The functional attributes of these putative operons included transport functions, constituents of the ESX-1 secretion apparatus, and components of the lipid metabolism machinery.

A particularly striking aspect of these results was that little or no correlation was found between the essentiality of a gene for intra-macrophage survival and its expression. While it remains to be seen whether this feature also holds true in human macrophages these findings, nonetheless, caution about over-interpreting gene expression data when elucidating intracellular survival mechanisms of MTB. It would seem that host–pathogen interactions in MTB infections are far more complex than they are for other bacterial pathogens, and that gene expression profiling may at best reveal short-term adaptations to changing host environments [1].

3 Host Factors That Support Intracellular MTB

Complementing the studies identifying MTB genes that are essential for intracellular survival of the pathogen are those aimed at defining the host factors that are coopted by the bacteria to facilitate their survival. This represents the other side of the coin in terms of describing the molecular interface that mediates the engagement between the host and the pathogen. An intracellular pathogen such as MTB may be considered to integrate as a dominant hub within the host cellular network, interacting with and simultaneously influencing several of the constituent nodes of the host cell. It would, therefore, not be unreasonable to posit that this regulation is central to defining the intracellular survival capabilities of the pathogen. RNA interference screens have proved extremely useful at distinguishing the macrophage-specific factors that are targeted by the intracellular MTB. The goal in such experiments is to identify those proteins whose "knockdown" yields a significant effect on the intracellular bacillary load. An early study employed a human kinome-specific screen to identify host kinases that were required for intracellular growth of *Salmonella typhimurium* [3]. Several kinases were identified which, interestingly, clustered in a single network that was centered on AKT1 (or, PKB), a serine-threonine protein kinase that has been implicated in regulating a wide range of cellular functions including proliferation and death. Of particular interest was the fact that inhibitors of AKT1 successfully prevented intracellular growth of various bacteria including MTB. A similar, kinome-targeted screen that was performed specifically against murine macrophages infected with MTB identified several additional kinases that were implicated in regulating a diverse range of cellular activities [4].

A more comprehensive description of the host factors required for supporting survival/persistence on intracellular MTB was obtained through a recent genome-wide RNAi screen performed in the human macrophage-like THP-1 cells [5]. In the first part of this study, the screen was performed against cells infected with H37Rv, a virulent laboratory strain of MTB. This experiment implicated a total of 274 host factors of which 269 were those where gene silencing caused a marked reduction in the intracellular pathogen load. This list was subsequently filtered in a second round where the effect of knockdown of these 274 proteins was evaluated in cells that had independently been infected with a panel of diverse clinical isolates of MTB that differed at the level of both genotype and the phenotypic properties such as drug sensitivity and in vivo replication rates. The hallmark of this study was the extraction of a core list of 74 host factors that were important for supporting the intracellular survival of all of the MTB strains tested.

Surprisingly, an analysis of the known functions of these molecules revealed that at least a significant proportion of them did not derive from what could classically be defined as antimicrobial pathways. Rather, the functional annotations of these molecules encompassed several of the core cellular functions involved in maintaining overall homeostasis. These included metabolism, signaling, transcription regulation, and the stress response [5]. In other words, these findings revealed the pervasive nature of the influence that the pathogen exerts on the host cell, emphasizing the intimate manner in which the pathogen integrates within the host cellular milieu. Importantly, in similarity with observations for the MTB genes that are essential for intracellular survival, only a subset of the genes for these 74 host proteins were found to be regulated, at the expression level, upon MTB infection [6]. That is, not all features related to the expression of mycobacterial virulence are reflected at the level of alterations in gene expression and this aspect needs to be considered when attempting to interrogate the host–pathogen crosstalk.

4 Capturing the Host Cellular Responses to MTB Infection Through Genome-Wide Expression Profiling

The cooption of diverse host cellular functions by the invading pathogen is, at least in part, likely to be induced through targeted effects on the transcriptional machinery. These perturbations can be efficiently captured by monitoring changes in the gene expression profile of the infected host cell. Rationalized extrapolations of the results from such experiments can then also provide insights into how such effects translate into modifying cellular functions.

Several studies have been conducted to date where whole genome expression profiling of cells infected with MTB was examined in order to dissect cellular response to MTB infections. One of the earliest study of this kind compared gene expression profile of macrophage activation in response to different bacterial species including *E. coli*, *Salmonella*, *Staphylococcus*, *Listeria*, and *Mycobacteria* representing Gram-positive, Gram-negative, and Mycobacterial species [7]. An interesting observation here was identification of several genes that were commonly regulated across the various bacterial species tested. The functional class enrichment of genes that showed significant regulation suggested that genes belonging to pro-inflammatory immune function including cytokines and chemokines were highly perturbed upon infection. In addition, expression levels of genes belonging to functional classes such as receptors, signaling molecules, transcription, cell adhesion, tissue remodeling, enzyme function, and apoptosis regulators were also significantly influenced.

In the context of macrophages infected with MTB, one of the earliest analyses was performed by Ehrt et al. [8]. Using a selected set of nearly 10,000 probe sets, they compared alterations induced in the macrophage gene expression profile following either MTB infection, or activation by IFN-α treatment, or in response to the combined stimuli of both MTB and IFN-α-dependent activation. From a detailed analysis of the results, this group was able to derive several significant inferences. For example they observed that, following IFN-α stimulation, many more genes were downregulated in their expression levels as opposed to those that were upregulated. These results provided an early glimpse of the complex nature of effects that even simple stimulation conditions can cause in the target cell. In addition though, they also highlighted that understanding the regulatory networks for such switch-like responses across the genome could be extremely valuable for an overall understanding of host responses. The intricate role played by signaling molecules in mediating this transcriptional reprogramming could also then be established when many of the expression patterns were found to be reversed in the absence of genes such as inducible nitric oxide synthase and phagocyte oxidases [8].

Some of the subsequent studies concentrated on probing MTB-induced effects on expression of only a select set of genes, which were mostly those involved in immune regulation [9, 10]. These investigations revealed a diverse range of perturbation of several genes that included IL8, osteopontin, MCP1, MIP1α, RANTES, MPIF-1, and GRO-β among many others. Further, expression of genes involved in cell migration such as MMP-9, VEGF, CCR3, VCAM1, and integrin was also upregulated [9]. In a separate study, Shi et al. showed that MyD88 signaling was crucial in mediating the IFN-α-dependent transcriptional activation program of macrophages. Interestingly, they further observed that while MTB infection also targeted several of the genes that were induced by IFN-α its effects, however, were independent of MyD88 signaling [11]. This latter observation provides an interesting perspective on how MTB exploits the inherent plasticity of gene regulatory networks so as to modulate macrophage function, and yet keep activation of an inflammatory immune response under check.

Over the subsequent years, several groups have reported microarray profiling of gene expression in macrophages infected with MTB. For instance, Wang et al. [12] showed infection-induced regulation of many cytokine and chemokine genes as well as the interferon-response gene Stat1. These findings were also supported by studies from other groups [13]. Using macrophages derived from mouse strains that were either susceptible or resistant to TB infection, Keller et al. [14] identified genes that were both commonly and differentially regulated by MTB. This comparison revealed an interesting pattern wherein susceptibility was found to be associated with activation of tissue damage pathways whereas microbial elimination pathways were more prominently activated in macrophages from the resistant mouse strain [14]. Many other studies have also addressed specific aspects of MTB-induced transcriptional reprogramming including time-dependent changes, role of specific mycobacterial factors, the role of host signaling molecules in this process, and the diversity of macrophage responses to different strains of MTB [15–23].

In a provocative study, Thuong et al. [24] used an ex vivo infection model of macrophages from human subjects with different clinical phenotypes and identified CCL1 as a host gene responsible for susceptibility to TB.

A particularly relevant study in terms of exploring reciprocal regulation at the level of both host and pathogen was that of Tailleux et al. [25]. Here, gene expression profiles of both host and the mycobacteria were monitored at different times in two different host cell types: macrophages and dendritic cells (DCs) [25]. It was previously shown that environment within DCs was more restrictive to mycobacteria as compared to that within macrophages [26]. They characterized the expression profiles of macrophages and DCs from the same donors infected with MTB, as well as that of MTB in both these cell types. The functional analysis of genes that were commonly or differentially regulated in each of the cell types clearly provided signatures of cell-type specific responses. More interestingly, the expression profile of mycobacteria from macrophages and DCs showed marked differences. Whereas MTB within the DC phagosomes displayed an expression profile that was vastly indicative of a stress response, those from macrophage phagosomes were reflective of a replicative phenotype in terms of the gene expression pattern displayed [25]. This is so far the only study where both adaptive responses of host and the pathogen in two different host cell types have been simultaneously characterized.

All of the above studies revolved around identifying those host cell genes whose expression levels were significantly altered in response to mycobacterial infections. In terms of drawing more mechanistic inferences, however, these studies would have greatly benefited from a more structured analysis wherein the whole transcriptional data is classified into clusters of genes showing similar expression pattern. The strength of such pattern-based clusters lies in their ability to establish associations among hitherto unrelated genes, simply based on the co-expression phenotype. These clusters can then subsequently be analyzed for enrichment of specific gene ontologies (GO, functional classes) to establish relationship between gene function and their regulation. Such data driven analysis of gene expression data helps to understand the gross response pattern of cells to stimuli. They also permit retrospective assignment of at least some of the genes to one or more specific mechanisms.

Although several microarray experiments have been performed to date, the inferences drawn have been largely empirical in nature. Nonetheless, they have provided preliminary glimpses into the complexity of host responses, in terms of reprogramming of the transcriptional machinery, to mycobacterial infection. To gain a deeper understanding, however, novel analytical techniques will need to be applied to further interrogate these microarray datasets. Such techniques generally rely on high quality computational and statistical analytical tools and, although have yet to be adopted for studying of host–MTB interactions, have contributed significantly towards better mechanistic understanding in the case of cancer biology, immunology, and other such disease situations. The next few sections will discuss how some of these tools can potentially be applied for probing the host–pathogen dynamics in MTB infections. It is our view that the future utilization of such methodologies can yield unprecedented levels of insights into the interplay that occurs during the various stages of the infection process.

5 Analyzing Transcript Profiles Within a Dynamical Framework

The above approaches for analyzing microarray data describe cellular perturbations that can be interpreted in the context of existing information on biological function of molecules. In order to gain mechanistic insights into the response machinery, however, it is important to analyze the expression data in a manner that accounts for functions of the differentially expressed genes, as well the regulatory relationships that exist between them. Since microarray experiments provide gene expression data as the output this, in a eukaryotic system, represents the culmination of events involving transcription factor activation and interaction of TFs with regulatory elements of specific genes. Eukaryotic genomes with thousands of genes pose a serious challenge in terms of extracting any mechanistic information from the transcriptome profile. In this regard, time course expression data becomes extremely important for two reasons. One of these is that temporal profiles capture the dynamic pattern of gene expression, whereas the second aspect is that time-dimensional data facilitate discernment of causality since past events have causal links with both present and future events [27]. Several approaches are now available for recreating a gene regulatory network from time series data with higher accuracy and better scalability [27]. These approaches can be broadly classified into two categories: model-based approaches and machine learning-based approaches. Network identification by regression, singular value decomposition and regression analysis, mode of action by network identification, time series network identification, and Inferator are some of the model-based approaches used extensively in understanding time series gene expression data [28–32]. Some of the machine learning techniques used to understand gene regulatory networks are partial correlation, graphic Gaussian models, Dynamic Bayesian Network analysis, state space models, Granger causality, etc. [33–37]. Other strategies for such analysis also include linear models, Bayesian network [38, 39], neural networks [40], differential equations [41], and stochastic models [42]. It needs to be cautioned though that all modeling approaches are associated with both advantages and disadvantages, and these intrinsic limitations need to be kept in mind when drawing any mechanistic conclusions [43, 44].

An interesting model class in this context is the Boolean network model, which was originally introduced by Kauffman [45, 46]. Boolean network models are reviewed in detail by [47]. In this model, the state of a gene is represented by a Boolean variable (ON or OFF) and interactions between the genes are represented by Boolean functions. The latter describe the state of a gene on the basis of the states of a few other genes. The model converges typically to an attractor state for given input states of different transcription factors, which could represent a combination of several intermediate states. The attractor state thus captures effect of simultaneous triggering ON or OFF states of several transcription factors on the phenotypic outcome. Although these are relatively simplistic models they, nonetheless, efficiently capture much of the complexity of the gene regulatory network [48].

There are only two allowed states for a given entity (gene) in the Boolean framework: expressed or not expressed. The level of each entity is updated on the basis of several other entities at the previous step through a Boolean function. Therefore, the level of each gene is determined according to the level of its regulators at the previous time step and the Boolean function deriving the regulation [49]. For cases where rules for Boolean function could not be established due to lack of information, another variant of Boolean model was described. This is called the probabilistic Boolean model where the Boolean function for a given entity can have several possibilities, each with a certain probability that is derived based on experimental data [50]. Subsequent analysis then tests feasibility of the rules governing the Boolean function, eventually yielding mechanistic insights.

The genome-wide expression data captures the complex dynamics of the cellular response machinery. That is, changes in activity of certain pathways—in response to either stimuli or perturbations—lead to specific changes in the gene expression profile, which can then be considered as an expression signature of the inducing pathways [51]. Even pathways primarily operating through posttranslational modification like phosphorylation have been shown to imprint recognizable expression signatures [52–54]. Further, the complexity of a given pathway is also reflected in the complexity of expression data that its perturbation generates [55]. To decipher pathway complexity therefore, one useful approach is to deconstruct the pathway into underlying modules based on structures observed in the gene expression profile [52, 56]. A statistical method called Bayesian Factor Regression Modeling was devised accordingly, which could elucidate modules of the signaling network [55, 57, 58]. Such kinds of analyses could prove useful for extracting the functional relevance of specific perturbations induced by MTB in the host cell signaling modules.

6 *Cis*-Regulatory Map of the Human Genome

The approaches discussed earlier are feasible because of the availability of genome sequence data, where upstream regulatory regions of genes (promoters, enhancers) can be identified and possible trans-regulators (transcription factors) that may further modulate the gene expression through these *cis*-elements can also be defined [59]. Different strategies such as linear, Bayesian, or Boolean approaches employ different sets of assumptions for interpreting the biochemical basis of transcriptional regulation [31, 60–62]. In some landmark studies, sequence preferences for nucleosome and transcription factors were used to predict gene expression in yeast [63], whereas Drosophila segmentation was predicted by calculating probabilities of all possible configuration of trans factors on *cis*-regulatory elements [64]. The *cis*-regulatory map has been extensively developed and analyzed in models such as Drosophila and Sea Urchins [59, 65, 66].

Availability of such a regulatory map for the human genome, although a difficult task considering the size and complexity, could nonetheless be immensely beneficial for understanding of how the response machinery is perturbed under a variety of conditions including MTB infection. The map though would need to be extensive and also include information on transcription factor binding, *cis*-elements, histone modifications, epigenetic factors like methylation and acetylation, and posttranslational modification of transcription factors [59]. Such a map could then not only provide us details of the cellular response machinery to infection but also allow us to test various tools and regions of the regulatory map to be targeted in order to achieve effective intervention against infection-induced perturbations.

7 Gene Expression Profiling Can Also Unravel Perturbations in the Host Molecular Interaction Network

It is now recognized that the molecular components of a cell interact in an intricate manner to form a network of tightly linked interactions. This network organization of bio-molecules is now believed to be responsible for imparting to the cell the ability to respond to diverse kinds of biological, chemical, and physical stimuli. Interestingly, the networks of bio-molecules follow the basic principles of the graph theory, a well-studied tenet in physics and mathematics. This fundamental finding has eventually led to the characterization of several biological networks such as the metabolic network, the signaling network, the gene regulatory network, the protein–protein interaction network, and the cell-to-cell interaction network among many others. Most of these networks were largely built through compilations from experimental datasets. For example, a protein–protein interaction network would integrate large datasets obtained through various experiments such as co-immunoprecipitation followed by mass spectrometry, or co-localization experiments stored across several databases like STRING, BIND, MINT, etc. Such molecular networks, however, represent only a static view since they do not incorporate any temporal data on either gene expression levels or modulations in function of their protein products. Consequently then, gene expression profiling through time-series microarray profiling could provide the requisite information needed to incorporate dynamical features into the network [67].

A good example of this is the earlier described study involving a genome-wide RNAi screen for host factors that were involved in supporting MTB infection in human macrophages [5]. As mentioned, a total of 274 genes could be identified for their role in regulating infection with the laboratory virulent strain H37Rv. Interestingly, a search in the global interaction databases revealed that these 274 genes interacted together in a large clustered network. To then add a temporal dimension to this clustered network, the gene expression data obtained from these infected macrophages—at various times postinfection—was superimposed onto the protein interaction network. The consequent identification of many highly connected modules that were co-regulated across the time course of the experiment demonstrated the utility of this exercise. From a functional perspective, this integration of two independent analytical methodologies provided insights into how the various host cellular functions were modulated by the pathogen, over the course of infection. That is, the resulting dataset serves as a rich information source for further investigations on the host–pathogen interaction dynamics.

An interesting extension of this approach was recently developed as a graph theoretic based tool termed as Express Path Analysis [6]. In this the gene expression data was first generated from macrophages infected with either a virulent or an avirulent strain of MTB. Genes that were differentially regulated under these two conditions were then integrated with the molecular association network. The subsequent application of graph theory parameters such as betweenness, stress, and degree centrality [68] then enabled identification of the key molecules that regulated divergent responses to the virulent and the attenuated strains [6]. The host factors that served as mediators of MTB virulence could then be defined, with Src kinase proving to be a key player in this regard. Significantly, pharmacological inhibition of this kinase in infected macrophages resulted in increased killing of the intracellular bacilli [6], thus demonstrating the potential utility of this approach.

Recently, Bonde et al. [69] devised yet another novel analytical technique which was termed as Differential Producibility Analysis (DPA). This methodology was applied to the analysis of microarray data in order to deduce the metabolic state of MTB in infected macrophages. An ingenious incorporation of Flux Balance Analysis to interpret the microarray datasets facilitated the delineation of a specific switch in the transcriptional program of MTB, which was activated upon entry into the host macrophage [69]. These results revealed a novel facet of the MTB response to the hostile macrophage environment wherein expression of genes influencing the central metabolism pathways was downregulated, while those directing synthesis of cell wall components and virulence factors were induced. The latter response presumably reflects a defense mechanism to protect against antimicrobial pathways. It has been suggested that the DPA may prove useful in unraveling the mechanisms of virulence and persistence of MTB within the host [69].

An interesting approach for inferring large-scale gene regulatory networks has been the context likelihood relatedness, based on several gene expression studies [70]. In this approach, transcription profiling of an organism over many different experimental conditions is used to infer regulatory circuits. Moreover, this method can also be effectively applied to define community structures in large networks [71]. One may thus obtain information on the genome-scale regulation of genes and the cross talk between different circuitry. This information may further be used for modeling the host response to a pathogen for defined conditions.

The MTB genome encodes approximately 200 transcription factors, 13 sigma factors, 11 eukaryotic-like Ser/Thr kinases, and more than 150 transcription factors [72]. Among these, the SigmaA is the housekeeping sigma factor, whereas others are involved in potential sensing of environmental signals. It is therefore understood that the regulatory circuit in MTB is likely to be highly complex, with extensive and redundant mechanisms of sensing environmental signals. Towards this, many large-scale gene expression studies have been carried out; moreover, several ChIP-chip experiments have also started becoming available in recent times [73]. Integration and analysis of many expression studies have led to the reconstruction of a gene regulatory network of MTB consisting of 783 nodes in the regulatory network with

45 transcription factors and 937 edges [74]. Interestingly, this network revealed significant difference in the distribution of in- and out-degree, with the out degree not following power-law distribution. The network also revealed a four-layer architecture, with 15 transcription factors being unregulated, and 19 being regulated exclusively by feedback loops. Such network properties are characteristic of prokaryotic regulatory circuits, thus suggesting simplistic models of sensing signal processing.

8 Modeling Drug Treatments and Mycobacterial Persistence

The success of MTB as a pathogen hinges on its ability to adapt to fluctuating environments such as changes in pH, oxygen tension, and nutrient availability. From a more general perspective, one of the two contrasting views on adaptation suggests that in a population each individual organism is anticipated to adapt to changes in environment. The alternate theory posits that population diversity allows adaptation by subsets of organisms to different environments. Although Darwinian arguments favor the latter explanation, certain observations on adaptation are paradoxical. Persistence exhibited by many bacterial species in response to antibiotic stress, by switching between slow growing population and the population that grows faster when the antibiotic stress has been removed, is one such phenomenon. Such a phenomenon has understandably been evolved for fitness against environmental stress [75]. Persistence has therefore been widely believed to be an epigenetic trait, as the persisters in a population are genotypically identical to those that do not survive.

Long-term persistence of MTB within the human population relies on its ability to enter into a dormant phase within the macrophages. Two models for this have been proposed. One of these suggests that, under in vitro stress conditions, bacteria are metabolically reprogrammed, but in a manner wherein they are still culturable. This situation is reminiscent of stationary phase cultures, where an inoculum from a long stationary phase is able to grow in vitro. However, in the in vivo murine model, evidence suggests that bacteria remain nonculturable, which can only be activated by an external signal [76]. The relevance of either of these models for the disease state is currently under debate. The understanding of the dormant phase has been addressed by various gene expression studies using well-established models of latency. A central role of the transcription factor DevR (also referred to as DosR) has been suggested. DevR regulates ~50 genes mediated via primary and secondary promoter binding sites [77]. Most of these genes are involved in adaptation to different environmental cues including hypoxia, nitric oxide, carbon monoxide, and ascorbic acid. Thus, by modulating the expression of these genes, DevR is believed to play an important role in the onset of dormancy.

Inherent difficulties in understanding drug susceptibility and the role of dormancy in the disease have prompted the development of many novel approaches in recent times. Exciting new results indicate, for example, that mutational rates enhance during latency, and thereby increase chances of drug resistance [78]. The theory of reactivation of dormant bacteria to cause the disease is under severe test following demonstration that bacteria actively divide during infection, possibly at a reduced rate [79]. Similarly, a mathematical model of latency has suggested that the balance between population of dormant and actively dividing bacilli is modulated by the strength of the host immune response [80]. Thus, the future holds promise for a better understanding of the development of drug resistance, and its link with Mtb latency.

9 Extracting Protein Interaction Networks from Gene Expression Data to Understand Latency

Another interesting way to view gene expression data is from the standpoint of interactions between the protein products of the expressed genes. This is especially true in the case of prokaryotic systems where genes associated with a given physiological function or pathways are proximally located in an operon, with each of them showing similar regulatory behavior in any given context. Consequently, it is possible to construct a functional association matrix between proteins based on their corresponding gene expression data. Genome-wide maps of protein:protein interactions can thus be generated which can then be exploited to not only yield information on missing functional pathways in cells, but can also potentially help in understanding the phenotypic outcome of cells to perturbations. In view of this, a few studies have been addressed towards obtaining protein:protein interaction maps in *M. tuberculosis*. One of the first attempts to map subcellular localization of proteins in *M. tuberculosis* also led to identification of some of the missing enzymes by analysis of functional interaction networks [81]. Similarly, attempts have been made to map protein:protein interactions in MTB by experimental and computational approaches [74, 82–85]. The role of genome-wide protein:protein interactions in latency has recently been addressed with the objective of resolving the communication channels between genes that are upregulated and those that are downregulated [86]. All these studies, explored at the systems level, have provided valuable insights into functionalities of different proteins and understanding of mycobacterial latency.

A transcription regulatory network based on time course microarray data indicated that DevR (DosR) is an important initiator of dormancy. However, other transcription factors such as NadR, SigE, SigC, and FurB may also be important in maintaining late stage of dormancy [74]. Involvement of SigD, HrcA, and Rv0494 in the early phase of dormancy was also hypothesized. Thus, DevR appears to be capable of receiving diverse stress signals such as hypoxia, nutrient deprivation, and stationary phase conditions in order to differentially regulate large numbers of genes and, thereby, trigger the onset of dormancy. DevR being a member of the two component regulatory system, the obvious partner to impart signal to DevR is DevS (DosS). Among the critical genes that DevR is known to upregulate during dormancy is Rv2623, which is an ATP-binding protein. It has therefore been suggested that Rv2623 may play an important role in dormancy by mediating ATP-dependent signaling [87]. Another recent analysis of interaction networks and Boolean modeling has indicated, as was anticipated, the central role of DevR is dormancy signals [86, 87]. Among the 50 genes that are commonly found to be upregulated in different models of dormancy, and the 34 genes that are downregulated, many are under the direct or indirect transcriptional control of DevR. Interestingly, 34 downregulated genes exhibit better evolutionary conservation than the 50 upregulated genes. Moreover, these 84 genes appear to form a tightly regulated gene cluster, as indicated by their expression correlations from several microarray studies. It is therefore likely that these 84 genes form a regulon-like structure and play a vital role in dormancy signals.

DevR is likely to receive signals from DosT, DosS, or Rv0845. As a consequence of this signal, expression levels of a large number of genes are perturbed. Among the many genes that DevR upregulates is the transcription factor Rv0081. Expression of Rv0081–Rv0088 locus has been demonstrated to be complex, where Rv0081 downregulates its own expression in contrast to DevR upregulating its expression [88]. In the functional interaction network, Rv0081 is linked to Rv0082, which in turn interacts with the NADH ubiquinone oxidoreductase complex, which is interestingly downregulated during dormancy. Thus, an intricate network of interactions is linked to the switching off of the respiratory system, which is under the central control of the DevR transcriptional regulator. The Boolean model reveals a fascinating crosstalk among the up- and downregulated genes of dormancy. The model converges to an attractor cycle, where only four transcription factors can control expression of 92 % of the genes involved in dormancy. The four genes, namely, DosS (Rv3132c), DosR (Rv3133c), Rv0081, and CRP (Rv3676) are themselves capable of controlling expression of a large number of genes involved during dormancy. Therefore, these four proteins appear to be the core regulators for initiating and maintaining dormancy signals. Thus, through a variety of experiments and computational studies, addressed at the systems level, interesting insights have been obtained in mycobacterial latency. Whether these will eventually lead to a better understanding of drug-induced persistence and/or provide insights into drug resistance as in other bacteria, however, remains to be seen.

10 Future Perspective

Profiling gene expression in both MTB and the host cell has clearly provided us with interesting new insights into the nature of perturbations that are caused when these two entities interact with each other. At least from the standpoint of the pathogen, such studies have been well complemented by parallel analyses dissecting MTB gene regulatory networks under a variety of situations that mimic the in vivo conditions. These latter studies have provided useful information on the reprogramming of these networks, in response to diverse environmental stresses. Delineation of the host cellular responses to MTB infection, however, is still in the early stages largely because of the sheer complexity of the underlying processes involved. Moreover, because of the extensive contributions from posttranscriptional mechanisms, the transcriptome profile often shows little correlation with changes at the level of the corresponding proteome in mammalian cells. This poses a severe limitation to extracting functional interpretations solely on the basis of microarray data. The future, therefore, presumably lies in generating multidimensional data that combines microarray with proteomic approaches and, perhaps also, other high-throughput techniques such as ChIP-seq and metabolomics. The challenge, however, will be to develop innovative new tools, some of which have been discussed here, to integrate such multivariate datasets so as to provide quantitative insights into the molecular crosstalk between the host cell and the pathogen. In addition to providing a deeper understanding of the infection process, such information could well reveal new strategies for therapeutic intervention.

References

1. Rengarajan J, Bloom BR, Rubin EJ (2005) Genome-wide requirements for Mycobacterium tuberculosis adaptation and survival in macrophages. Proc Natl Acad Sci USA 102(23):8327–8332
2. Sassetti CM, Boyd DH, Rubin EJ (2001) Comprehensive identification of conditionally essential genes in mycobacteria. Proc Natl Acad Sci USA 98(22):12712–12717
3. Kuijl C, Savage ND, Marsman M, Tuin AW, Janssen L, Egan DA, Ketema M, van den Nieuwendijk R, van den Eeden SJ, Geluk A, Poot A, van der Marel G, Beijersbergen RL, Overkleeft H, Ottenhoff TH, Neefjes J (2007) Intracellular bacterial growth is controlled by a kinase network around PKB/AKT1. Nature 450(7170):725–730
4. Jayaswal S, Kamal MA, Dua R, Gupta S, Majumdar T, Das G, Kumar D, Rao KV (2010) Identification of host-dependent survival factors for intracellular Mycobacterium tuberculosis through an siRNA screen. PLoS Pathog 6(4):e1000839
5. Kumar D, Nath L, Kamal MA, Varshney A, Jain A, Singh S, Rao KV (2010) Genome-wide analysis of the host intracellular network that regulates survival of Mycobacterium tuberculosis. Cell 140(5):731–743
6. Karim AF, Chandra P, Chopra A, Siddiqui Z, Bhaskar A, Singh A, Kumar D (2011) Express path analysis identifies a tyrosine kinase Src-centric network regulating divergent host responses to Mycobacterium tuberculosis infection. J Biol Chem 286(46):40307–40319
7. Nau GJ, Richmond JF, Schlesinger A, Jennings EG, Lander ES, Young RA (2002) Human macrophage activation programs induced by bacterial pathogens. Proc Natl Acad Sci USA 99(3):1503–1508
8. Ehrt S, Schnappinger D, Bekiranov S, Drenkow J, Shi S, Gingeras TR, Gaasterland T, Schoolnik G, Nathan C (2001) Reprogramming of the macrophage transcriptome in response to interferon-gamma and Mycobacterium tuberculosis: signaling roles of nitric oxide synthase-2 and phagocyte oxidase. J Exp Med 194(8):1123–1140
9. Ragno S, Romano M, Howell S, Pappin DJ, Jenner PJ, Colston MJ (2001) Changes in gene expression in macrophages infected with Mycobacterium tuberculosis: a combined transcriptomic and proteomic approach. Immunology 104(1):99–108
10. Volpe E, Cappelli G, Grassi M, Martino A, Serafino A, Colizzi V, Sanarico N, Mariani F (2006) Gene expression profiling of human macrophages at late time of infection with Mycobacterium tuberculosis. Immunology 118(4):449–460
11. Shi S, Nathan C, Schnappinger D, Drenkow J, Fuortes M, Block E, Ding A, Gingeras TR, Schoolnik G, Akira S, Takeda K, Ehrt S (2003) MyD88 primes macrophages for full-scale

activation by interferon-gamma yet mediates few responses to Mycobacterium tuberculosis. J Exp Med 198(7):987–997
12. Wang JP, Rought SE, Corbeil J, Guiney DG (2003) Gene expression profiling detects patterns of human macrophage responses following Mycobacterium tuberculosis infection. FEMS Immunol Med Microbiol 39(2):163–172
13. Hu CH, Xie JP, Li Y, Yue J, Xu YZ, Wang HH (2004) Differential expression of apoptosis-related gene induced by clinical and laboratory Mycobacterium tuberculosis strain in macrophages U937 revealed by oligonucleotide microarray. Yi Chuan Xue Bao 31(3):231–235
14. Keller C, Lauber J, Blumenthal A, Buer J, Ehlers S (2004) Resistance and susceptibility to tuberculosis analysed at the transcriptome level: lessons from mouse macrophages. Tuberculosis (Edinb) 84(3–4):144–158
15. Blumenthal A, Lauber J, Hoffmann R, Ernst M, Keller C, Buer J, Ehlers S, Reiling N (2005) Common and unique gene expression signatures of human macrophages in response to four strains of Mycobacterium avium that differ in their growth and persistence characteristics. Infect Immun 73(6):3330–3341
16. Calamita H, Ko C, Tyagi S, Yoshimatsu T, Morrison NE, Bishai WR (2005) The Mycobacterium tuberculosis SigD sigma factor controls the expression of ribosome-associated gene products in stationary phase and is required for full virulence. Cell Microbiol 7(2):233–244
17. Khajoee V, Saito M, Takada H, Nomura A, Kusuhara K, Yoshida SI, Yoshikai Y, Hara T (2006) Novel roles of osteopontin and CXC chemokine ligand 7 in the defence against mycobacterial infection. Clin Exp Immunol 143(2):260–268
18. Pai RK, Pennini ME, Tobian AA, Canaday DH, Boom WH, Harding CV (2004) Prolonged toll-like receptor signaling by Mycobacterium tuberculosis and its 19-kilodalton lipoprotein inhibits gamma interferon-induced regulation of selected genes in macrophages. Infect Immun 72(11):6603–6614
19. Raju B, Hoshino Y, Belitskaya-Levy I, Dawson R, Ress S, Gold JA, Condos R, Pine R, Brown S, Nolan A, Rom WN, Weiden MD (2008) Gene expression profiles of bronchoalveolar cells in pulmonary TB. Tuberculosis (Edinb) 88(1):39–51
20. Schreiber T, Ehlers S, Heitmann L, Rausch A, Mages J, Murray PJ, Lang R, Holscher C (2009) Autocrine IL-10 induces hallmarks of alternative activation in macrophages and suppresses antituberculosis effector mechanisms without compromising T cell immunity. J Immunol 183(2):1301–1312
21. Talaat AM, Lyons R, Howard ST, Johnston SA (2004) The temporal expression profile of Mycobacterium tuberculosis infection in mice. Proc Natl Acad Sci USA 101(13):4602–4607
22. Weiss DJ, Evanson OA, Deng M, Abrahamsen MS (2004) Sequential patterns of gene expression by bovine monocyte-derived macrophages associated with ingestion of mycobacterial organisms. Microb Pathog 37(4):215–224
23. Widdison S, Watson M, Piercy J, Howard C, Coffey TJ (2008) Granulocyte chemotactic properties of M. tuberculosis versus M. bovis-infected bovine alveolar macrophages. Mol Immunol 45(3):740–749
24. Thuong NT, Dunstan SJ, Chau TT, Thorsson V, Simmons CP, Quyen NT, Thwaites GE, Thi Ngoc Lan N, Hibberd M, Teo YY, Seielstad M, Aderem A, Farrar JJ, Hawn TR (2008) Identification of tuberculosis susceptibility genes with human macrophage gene expression profiles. PLoS Pathog 4(12):e1000229
25. Tailleux L, Waddell SJ, Pelizzola M, Mortellaro A, Withers M, Tanne A, Castagnoli PR, Gicquel B, Stoker NG, Butcher PD, Foti M, Neyrolles O (2008) Probing host pathogen cross-talk by transcriptional profiling of both Mycobacterium tuberculosis and infected human dendritic cells and macrophages. PLoS One 3(1):e1403
26. Tailleux L, Neyrolles O, Honore-Bouakline S, Perret E, Sanchez F, Abastado JP, Lagrange PH, Gluckman JC, Rosenzwajg M, Herrmann JL (2003) Constrained intracellular survival of Mycobacterium tuberculosis in human dendritic cells. J Immunol 170(4):1939–1948
27. Li Z, Li P, Krishnan A, Liu J (2011) Large-scale dynamic gene regulatory network inference combining differential equation models with local dynamic Bayesian network analysis. Bioinformatics 27(19):2686–2691

28. Bonneau R, Reiss DJ, Shannon P, Facciotti M, Hood L, Baliga NS, Thorsson V (2006) The Inferelator: an algorithm for learning parsimonious regulatory networks from systems-biology data sets de novo. Genome Biol 7(5):R36
29. Cantone I, Marucci L, Iorio F, Ricci MA, Belcastro V, Bansal M, Santini S, di Bernardo M, di Bernardo D, Cosma MP (2009) A yeast synthetic network for in vivo assessment of reverse-engineering and modeling approaches. Cell 137(1):172–181
30. di Bernardo D, Thompson MJ, Gardner TS, Chobot SE, Eastwood EL, Wojtovich AP, Elliott SJ, Schaus SE, Collins JJ (2005) Chemogenomic profiling on a genome-wide scale using reverse-engineered gene networks. Nat Biotechnol 23(3):377–383
31. Gardner TS, di Bernardo D, Lorenz D, Collins JJ (2003) Inferring genetic networks and identifying compound mode of action via expression profiling. Science 301(5629):102–105
32. Yeung MK, Tegner J, Collins JJ (2002) Reverse engineering gene networks using singular value decomposition and robust regression. Proc Natl Acad Sci USA 99(9):6163–6168
33. Li X, Rao S, Jiang W, Li C, Xiao Y, Guo Z, Zhang Q, Wang L, Du L, Li J, Li L, Zhang T, Wang QK (2006) Discovery of time-delayed gene regulatory networks based on temporal gene expression profiling. BMC Bioinformatics 7:26
34. Mukhopadhyay ND, Chatterjee S (2007) Causality and pathway search in microarray time series experiment. Bioinformatics 23(4):442–449
35. Schafer J, Strimmer K (2005) An empirical Bayes approach to inferring large-scale gene association networks. Bioinformatics 21(6):754–764
36. Yu J, Smith VA, Wang PP, Hartemink AJ, Jarvis ED (2004) Advances to Bayesian network inference for generating causal networks from observational biological data. Bioinformatics 20(18):3594–3603
37. Zou M, Conzen SD (2005) A new dynamic Bayesian network (DBN) approach for identifying gene regulatory networks from time course microarray data. Bioinformatics 21(1):71–79
38. Friedman N, Linial M, Nachman I, Pe'er D (2000) Using Bayesian networks to analyze expression data. J Comput Biol 7(3–4):601–620
39. Hartemink AJ, Gifford DK, Jaakkola TS, Young RA (2001) Using graphical models and genomic expression data to statistically validate models of genetic regulatory networks. Pac Symp Biocomput 6:422–433
40. Vohradsky J (2001) Neural model of the genetic network. J Biol Chem 276(39): 36168–36173
41. Mestl T, Plahte E, Omholt SW (1995) A mathematical framework for describing and analysing gene regulatory networks. J Theor Biol 176(2):291–300
42. McAdams HH, Arkin A (1997) Stochastic mechanisms in gene expression. Proc Natl Acad Sci USA 94(3):814–819
43. Hasty J, McMillen D, Isaacs F, Collins JJ (2001) Computational studies of gene regulatory networks: in numero molecular biology. Nat Rev Genet 2(4):268–279
44. Smolen P, Baxter DA, Byrne JH (2000) Mathematical modeling of gene networks. Neuron 26(3):567–580
45. Glass L, Kauffman SA (1973) The logical analysis of continuous, non-linear biochemical control networks. J Theor Biol 39(1):103–129
46. Kauffman S (1969) Homeostasis and differentiation in random genetic control networks. Nature 224(5215):177–178
47. Huang S (1999) Gene expression profiling, genetic networks, and cellular states: an integrating concept for tumorigenesis and drug discovery. J Mol Med (Berl) 77(6):469–480
48. Shmulevich I, Lahdesmaki H, Dougherty ER, Astola J, Zhang W (2003) The role of certain Post classes in Boolean network models of genetic networks. Proc Natl Acad Sci USA 100(19):10734–10739
49. Karlebach G, Shamir R (2008) Modelling and analysis of gene regulatory networks. Nat Rev Mol Cell Biol 9(10):770–780
50. Shmulevich I, Dougherty ER, Kim S, Zhang W (2002) Probabilistic Boolean networks: a rule-based uncertainty model for gene regulatory networks. Bioinformatics 18(2):261–274

51. Kumar D, Srikanth R, Ahlfors H, Lahesmaa R, Rao KV (2007) Capturing cell-fate decisions from the molecular signatures of a receptor-dependent signaling response. Mol Syst Biol 3:150
52. Bild AH, Yao G, Chang JT, Wang Q, Potti A, Chasse D, Joshi MB, Harpole D, Lancaster JM, Berchuck A, Olson JA Jr, Marks JR, Dressman HK, West M, Nevins JR (2006) Oncogenic pathway signatures in human cancers as a guide to targeted therapies. Nature 439(7074):353–357
53. Huang E, Ishida S, Pittman J, Dressman H, Bild A, Kloos M, D'Amico M, Pestell RG, West M, Nevins JR (2003) Gene expression phenotypic models that predict the activity of oncogenic pathways. Nat Genet 34(2):226–230
54. Sweet-Cordero A, Mukherjee S, Subramanian A, You H, Roix JJ, Ladd-Acosta C, Mesirov J, Golub TR, Jacks T (2005) An oncogenic KRAS2 expression signature identified by cross-species gene-expression analysis. Nat Genet 37(1):48–55
55. Chang JT, Carvalho C, Mori S, Bild AH, Gatza ML, Wang Q, Lucas JE, Potti A, Febbo PG, West M, Nevins JR (2009) A genomic strategy to elucidate modules of oncogenic pathway signaling networks. Mol Cell 34(1):104–114
56. Lamb J, Crawford ED, Peck D, Modell JW, Blat IC, Wrobel MJ, Lerner J, Brunet JP, Subramanian A, Ross KN, Reich M, Hieronymus H, Wei G, Armstrong SA, Haggarty SJ, Clemons PA, Wei R, Carr SA, Lander ES, Golub TR (2006) The Connectivity Map: using gene-expression signatures to connect small molecules, genes, and disease. Science 313(5795):1929–1935
57. Brunet JP, Tamayo P, Golub TR, Mesirov JP (2004) Metagenes and molecular pattern discovery using matrix factorization. Proc Natl Acad Sci USA 101(12):4164–4169
58. Carvalho CM, Chang J, Lucas JE, Nevins JR, Wang Q, West M (2008) High-dimensional sparse factor modeling: applications in gene expression genomics. J Am Stat Assoc 103(484):1438–1456
59. Kim HD, Shay T, O'Shea EK, Regev A (2009) Transcriptional regulatory circuits: predicting numbers from alphabets. Science 325(5939):429–432
60. Beer MA, Tavazoie S (2004) Predicting gene expression from sequence. Cell 117(2):185–198
61. Das D, Nahle Z, Zhang MQ (2006) Adaptively inferring human transcriptional subnetworks. Mol Syst Biol 2(2006):0029
62. Nguyen DH, D'Haeseleer P (2006) Deciphering principles of transcription regulation in eukaryotic genomes. Mol Syst Biol 2(2006):0012
63. Raveh-Sadka T, Levo M, Segal E (2009) Incorporating nucleosomes into thermodynamic models of transcription regulation. Genome Res 19(8):1480–1496
64. Segal E, Raveh-Sadka T, Schroeder M, Unnerstall U, Gaul U (2008) Predicting expression patterns from regulatory sequence in Drosophila segmentation. Nature 451(7178):535–540
65. Negre N, Brown CD, Ma L, Bristow CA, Miller SW, Wagner U, Kheradpour P, Eaton ML, Loriaux P, Sealfon R, Li Z, Ishii H, Spokony RF, Chen J, Hwang L, Cheng C, Auburn RP, Davis MB, Domanus M, Shah PK, Morrison CA, Zieba J, Suchy S, Senderowicz L, Victorsen A, Bild NA, Grundstad AJ, Hanley D, MacAlpine DM, Mannervik M, Venken K, Bellen H, White R, Gerstein M, Russell S, Grossman RL, Ren B, Posakony JW, Kellis M, White KP (2011) A cis-regulatory map of the Drosophila genome. Nature 471(7339):527–531
66. Oliveri P, Tu Q, Davidson EH (2008) Global regulatory logic for specification of an embryonic cell lineage. Proc Natl Acad Sci USA 105(16):5955–5962
67. Hegde SR, Manimaran P, Mande SC (2008) Dynamic changes in protein functional linkage networks revealed by integration with gene expression data. PLoS Comput Biol 4(11):e1000237
68. Barabasi AL, Oltvai ZN (2004) Network biology: understanding the cell's functional organization. Nat Rev Genet 5(2):101–113
69. Bonde BK, Beste DJ, Laing E, Kierzek AM, McFadden J (2011) Differential producibility analysis (DPA) of transcriptomic data with metabolic networks: deconstructing the metabolic response of M. tuberculosis. PLoS Comput Biol 7(6):e1002060

70. Faith JJ, Hayete B, Thaden JT, Mogno I, Wierzbowski J, Cottarel G, Kasif S, Collins JJ, Gardner TS (2007) Large-scale mapping and validation of Escherichia coli transcriptional regulation from a compendium of expression profiles. PLoS Biol 5(1):e8
71. Trevino S 3rd, Sun Y, Cooper TF, Bassler KE (2012) Robust detection of hierarchical communities from Escherichia coli gene expression data. PLoS Comput Biol 8(2):e1002391
72. Manganelli R, Provvedi R, Rodrigue S, Beaucher J, Gaudreau L, Smith I (2004) Sigma factors and global gene regulation in Mycobacterium tuberculosis. J Bacteriol 186(4):895–902
73. Hartkoorn RC, Sala C, Uplekar S, Busso P, Rougemont J, Cole ST (2012) Genome-wide definition of the SigF regulon in Mycobacterium tuberculosis. J Bacteriol 194(8):2001–2009
74. Balazsi G, Heath AP, Shi L, Gennaro ML (2008) The temporal response of the Mycobacterium tuberculosis gene regulatory network during growth arrest. Mol Syst Biol 4:225
75. Kussell E, Leibler S (2005) Phenotypic diversity, population growth, and information in fluctuating environments. Science 309(5743):2075–2078
76. Chao MC, Rubin EJ (2010) Letting sleeping dos lie: does dormancy play a role in tuberculosis? Annu Rev Microbiol 64:293–311
77. Chauhan S, Sharma D, Singh A, Surolia A, Tyagi JS (2011) Comprehensive insights into Mycobacterium tuberculosis DevR (DosR) regulon activation switch. Nucleic Acids Res 39(17):7400–7414
78. Ford CB, Lin PL, Chase MR, Shah RR, Iartchouk O, Galagan J, Mohaideen N, Ioerger TR, Sacchettini JC, Lipsitch M, Flynn JL, Fortune SM (2011) Use of whole genome sequencing to estimate the mutation rate of Mycobacterium tuberculosis during latent infection. Nat Genet 43(5):482–486
79. Gill WP, Harik NS, Whiddon MR, Liao RP, Mittler JE, Sherman DR (2009) A replication clock for Mycobacterium tuberculosis. Nat Med 15(2):211–214
80. Magombedze G, Mulder N (2012) A mathematical representation of the development of Mycobacterium tuberculosis active, latent and dormant stages. J Theor Biol 292:44–59
81. Mawuenyega KG, Forst CV, Dobos KM, Belisle JT, Chen J, Bradbury EM, Bradbury AR, Chen X (2005) Mycobacterium tuberculosis functional network analysis by global subcellular protein profiling. Mol Biol Cell 16(1):396–404
82. Sanz J, Navarro J, Arbues A, Martin C, Marijuan PC, Moreno Y (2011) The transcriptional regulatory network of Mycobacterium tuberculosis. PLoS One 6(7):e22178
83. Strong M, Graeber TG, Beeby M, Pellegrini M, Thompson MJ, Yeates TO, Eisenberg D (2003) Visualization and interpretation of protein networks in Mycobacterium tuberculosis based on hierarchical clustering of genome-wide functional linkage maps. Nucleic Acids Res 31(24):7099–7109
84. Venancio TM, Aravind L (2009) Reconstructing prokaryotic transcriptional regulatory networks: lessons from actinobacteria. J Biol 8(3):29
85. Wang Y, Cui T, Zhang C, Yang M, Huang Y, Li W, Zhang L, Gao C, He Y, Li Y, Huang F, Zeng J, Huang C, Yang Q, Tian Y, Zhao C, Chen H, Zhang H, He ZG (2010) Global protein-protein interaction network in the human pathogen Mycobacterium tuberculosis H37Rv. J Proteome Res 9(12):6665–6677
86. Hegde SR, Rajasingh H, Das C, Mande SS, Mande SC (2012) Understanding communication signals during Mycobacterial latency through predicted genome-wide protein interactions and Boolean modeling. PLoS One 7(3):e33893
87. Drumm JE, Mi K, Bilder P, Sun M, Lim J, Bielefeldt-Ohmann H, Basaraba R, So M, Zhu G, Tufariello JM, Izzo AA, Orme IM, Almo SC, Leyh TS, Chan J (2009) Mycobacterium tuberculosis universal stress protein Rv2623 regulates bacillary growth by ATP-Binding: requirement for establishing chronic persistent infection. PLoS Pathog 5(5):e1000460
88. He H, Bretl DJ, Penoske RM, Anderson DM, Zahrt TC (2011) Components of the Rv0081-Rv0088 locus, which encodes a predicted formate hydrogenlyase complex, are coregulated by Rv0081, MprA, and DosR in Mycobacterium tuberculosis. J Bacteriol 193(19):5105–5118

Chapter 4
Metabolism of *Mycobacterium tuberculosis*

Dany J.V. Beste and Johnjoe McFadden

Abstract Despite decades of research many aspects of the biology of *Mycobacterium tuberculosis* remain unclear and this is reflected in the antiquated tools available to treat and prevent tuberculosis. Consequently, this disease remains a serious public health problem responsible for 2–3 million deaths each year. Important discoveries linking *M. tuberculosis* metabolism and pathogenesis have renewed interest in the metabolic underpinning of the interaction between the pathogen and its host. Whereas, previous experimental studies tended to focus on the role of single genes, antigens or enzymes, the central paradigm of systems biology is that the role of any gene cannot be determined in isolation from its context. Therefore, systems approaches examine the role of genes and proteins embedded within a network of interactions. We here examine the application of this approach to studying metabolism of *M. tuberculosis*. Recent advances in high-throughput experimental technologies, such as functional genomics and metabolomics, provide datasets that can be analysed with computational tools such as flux balance analysis. These new approaches allow metabolism to be studied on a genome scale and have already been applied to gain insights into the metabolic pathways utilised by *M. tuberculosis* in vitro and identify potential drug targets. The information from these studies will fundamentally change our approach to tuberculosis research and lead to new targets for therapeutic drugs and vaccines.

1 Introduction

Tuberculosis (TB) is a disease that plagued ancient Egyptians and still remains a major threat to human health thousands of years later. The control of tuberculosis has been significantly hindered by the limited resources available for both the

D.J.V. Beste (✉) • J. McFadden, Ph.D. (✉)
Faculty of Health and Medical Sciences, University of Surrey, Guildford GU2 7XH, UK
e-mail: d.beste@surrey.ac.uk; j.mcfadden@surrey.ac.uk

J. McFadden et al. (eds.), *Systems Biology of Tuberculosis*,
DOI 10.1007/978-1-4614-4966-9_4, © Springer Science+Business Media, LLC 2013

prevention and treatment of tuberculosis. A truly effective vaccine is lacking as the 90-year-old *Mycobacterium bovis* bacillus Calmette–Guerin live attenuated vaccine is not universally protective and does not produce immunity against re-infection or reactivation. Lengthy (6–9 months) and complex (three or more different drugs) treatment is required using currently available anti-TB drugs. The economic and logistic burden of administering these drug regimens in industrially undeveloped countries where TB is most prevalent is enormous and combined with poor patient compliance are important factors in the emergence of drug-resistant TB isolates that are causing ongoing epidemics. These factors underscore the urgent need for the development of novel and effective therapeutics and vaccines and new approaches will be required to achieve these goals.

Mycobacterium tuberculosis is an unusual bacterial pathogen, which has the remarkable ability to cause both acute life-threatening disease and also clinically latent infections which can persist for the lifetime of the human host. Unlike many pathogens *M. tuberculosis* does not rely on the production of specific toxins to cause disease but rather the secret of this bacterium's great success seems to be the ability to adapt and survive within the changing and adverse environment provided by the human host during the course of an infection. It is becoming apparent that key to this adaptation is the metabolic reprogramming of *M. tuberculosis* during both the acute and chronic phase of TB disease and therefore a more complete understanding of mycobacterial metabolism remains a major goal of TB research.

Whilst recent increases in research funding have progressed our understanding of the basic biology of *M. tuberculosis* this has not yet impacted on the global TB trends which remain at staggering levels. A possible reason why it has been difficult to translate basic research into effective strategies for combating tuberculosis is that TB research has until recently focused on studying individual parameters in isolation which can consequently result in an overestimation of the importance of these factors. This effect may be particularly profound for a persistent pathogen such as *M. tuberculosis* that lacks classical virulence factors. The systems biology framework, which investigates the dynamic interactions of many components, provides an alternative and complementary strategy to the more traditional reductionist approaches to TB research. This methodology has started to be applied to the metabolism of *M. tuberculosis* on a genome scale and promises to drive biological discovery in the TB research field by providing scaffolding for the interpretation of "omic" scale datasets, directing hypothesis driven discovery and also assisting in the identification of novel drug targets.

2 Central Metabolism of *M. tuberculosis*

Application of metabolic modelling approaches to *M. tuberculosis* is aided by the fact that metabolism is a reasonably well-studied system even in mycobacteria. Moreover, metabolism has been shown to be involved in the virulence of *M. tuberculosis*, playing a key role in the development and maintenance of both acute and persistent TB infections [1–7]. It is perhaps not surprising therefore that several modelling efforts in tuberculosis have focused on metabolism.

Much of what is known about metabolism in *M. tuberculosis* has been gleaned from conventional biochemical and molecular studies over many decades. The pathogen appears typical of bacteria of the Actinomycetales order, with a predominantly aerobic metabolism that is able to catabolise a wide range of substrates to generate biomass and energy. The genome encodes all the enzymes of the Embden–Meyerhof–Parnas pathway (EMP) and pentose phosphate pathway (PPP) and has a complete, or nearly complete tricarboxylic acid (TCA) cycle (see below). The pathogen also encodes a functional glyoxylate shunt as well as several enzymes connecting the TCA cycle and glycolysis that may be used for either anaplerosis or gluconeogenesis.

There are, however, several features of central metabolism in *M. tuberculosis* that appear to be unusual. Although the link between glycolysis and the TCA cycle is complete in *M. tuberculosis*, the closely related pathogen *M. bovis* lacks a functional pyruvate kinase and is therefore unable to deliver sugars from glycolysis to the TCA cycle. It is thus unable to utilise carbohydrates as the sole carbon source [8]. This function is therefore unnecessary in vivo, as this pathogen causes very similar disease in humans to *M. tuberculosis.* The role of isocitrate lyase has been intensively studied since the demonstration that both of the isocitrate lyase genes encoded by this pathogen, *icl*1 and *icl*2 (although some strains only have *icl*1) play an essential role in virulence [1, 2]. This finding has been generally interpreted to be due to this enzyme's role in the glyoxylate shunt and a metabolic shift in the principal carbon source from carbohydrates to fat in the host. However, the role of the isocitrate lyases maybe more complex than just fat catabolism, as these enzymes also function as methyl citrate lyases in the methyl citrate cycle [9], which is used to catabolise propionate, derived from the oxidation of odd-numbered chain and branched chain fatty acids. ICL has also been shown to be essential for intracellular ATP level reduction in a nutrient starvation model of persistence [10] and the glyoxylate shunt has been shown to operate concurrently with an oxidative TCA cycle which is completed by an anaerobic α-ketoglutarate ferredoxin oxidoreductase [11]. More recently, we have demonstrated an essential role for ICL during slow growth rate on glycerol, a substrate that would be expected to be catabolised via glycolysis and the TCA cycle [12, 13].

It was reported that the TCA cycle was atypical in *M. tuberculosis* as the pathogen was proposed to lack α-ketoglutarate dehydrogenase (KDH) activity and thereby the standard connection between α-ketoglutarate and succinate via succinyl CoA [14]. These findings prompted the proposal that *M. tuberculosis* operates an alternative route (the SSA shunt) between α-ketogluterate and succinate via the enzyme α-ketoglutarate decarboxylase (KGD, putatively encoded by Rv1248c, to produce succinic semialdehyde which could be converted to succinate by succinic semialdehyde dehydrogenase (SSADH encoded by *gabD1/D2*) [15]. It was also pointed out [15] that *M. tuberculosis* has all the enzymes required for a GABA shunt capable of converted α-ketoglutarate to succinic semialdehyde (and then on to succinate) via glutamate and 4-aminobutyrate (GABA). However, neither of these

SSA-based shunts accounts for the synthesis of succinyl CoA, which is an essential precursor of both heme and branched fatty acids. Recently, the enzyme encoded by Rv1248c was shown to be a carboligase with 2-hydroxy-3-oxoadipate synthase (HOA synthase) activity capable of condensing α-ketogluterate with glyoxylate to yield 2-hydroxy-3-oxoadipate (HOA) which decomposes to 5-hydroxylevulinate (HLA) [16], undermining evidence for a SSA shunt in *M. tuberculosis*. However, the enzyme does appear to form SSA in the absence of glyoxylate [16, 17] so it may be the SSA shunt functions when levels of glyoxylate in the cell are low. Indeed recent work demonstrated that Rv1248c appears to be multifunctional enzyme with classical succinyl-transferring KDH activity, but also KDG and carboligase activity [17]. Additionally, an alternative route to succinyl CoA from α-ketoglutarate has also recently been shown to be active involving a CoA-dependent ferredoxin oxidoreductase (KOR), which operates preferentially under anaerobic conditions [11]. Recent evidence has also emerged that, under anaerobic conditions, *M. tuberculosis* operates a reverse TCA cycle involving the reduction of fumarate to succinate (which is then secreted) by fumarate reductase, possibly as a means of generating redox potential and maintaining the membrane potential in the absence of oxygen [18]. It therefore seems that *M. tuberculosis* encodes a number of alternative pathways that could operate around the TCA cycle, although the significance of most of them in vivo remains to be determined. Figure 4.1 illustrates the central metabolic pathways of *M. tuberculosis*, as understood in 2012.

3 Experimental Systems for Systems Biology

Systems biology is an iterative procedure of experimental data acquisition, model building, hypothesis generation and experimental verification. One of the constraints upon this approach surrounds the experimental basis of this work. Models should be developed and validated with accurate and reproducible data. Moreover, the mathematical underpinning of many modelling approaches such as ^{13}C-MFA have an absolute requirement for the cultivation of the organism under steady-state conditions where metabolite concentrations are maintained at constant levels. This makes it very difficult to apply these approaches directly to pathogens such as *M. tuberculosis* growing in vivo, as such steady states are not attainable in mammalian cells. However, a standard approach in systems biology is to initially study systems in highly controlled experimental environments that allow models to be parameterised before their subsequent application in real life situations. One of the pioneers of systems biology, Hiroaki Kitano [19], uses an example from racing car design to illustrate this approach. Cars are initially designed using a computer and then tested in a wind _tunnel before being deployed in the actual race. By controlling airflow, wind tunnels transform a highly dynamic unsolvable system into steady state that is amenable to mathematical modelling. Kitano argues that systems biologist's similarly needs biological "wind tunnels" to develop their models. We here describe the application of one of the few biological wind tunnels: the chemostat.

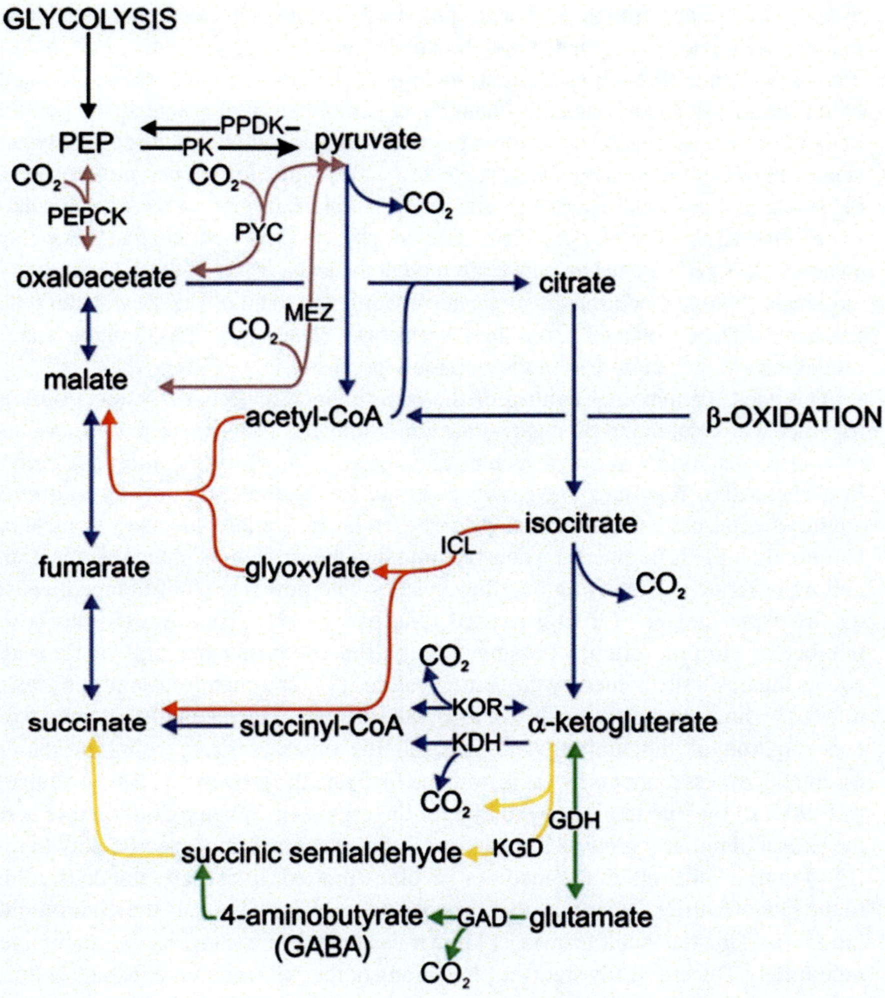

Fig. 4.1 Central metabolism in *M. tuberculosis*. The standard TCA cycle is shown in *blue* with the variant (SSA) pathway in *yellow* and GABA pathway in *green*. Anaplerotic/gluconeogenic reactions are shown in *purple* with the glyoxylate shunt in *red*. Only enzymes mentioned in the text are indicated, including pyruvate kinase (PK), pyruvate phosphate dikinase (PPDK), KOR (α-ketoglutarate ferredoxin oxidoreductase), KGD (α-ketoglutarate decarboxylase), GabD1/D2 (succinic semialdehyde dehydrogenase), GDH (glutamate dehydrogenase), GAD (glutamate decarboxylase), ICL (isocitrate lyase) and MEZ (malic enzyme (malate dehydogenase, decarboxylating)), PEPCK (phosphoenolpyruvate carboxykinase), PK (pyruvate kinase) and PYC (pyruvate carboxylase)

Traditional batch cultivation remains the standard for most microbiological investigations. Typically, the microbe is inoculated into a stirred vessel filled with rich media. The organism will grow at close to maximal rate (logarithmic phase) until either nutrients (including oxygen) become limiting or inhibitory products accumulate to levels that retard growth (stationary phase). Although convenient and suitable for many microbiological, genetic and functional genomic applications, this culture method is unsuitable for most systems biology applications, because (a) it is dynamic with cells adapting to a constantly changing environment; (b) it is not usually possible to monitor rates of substrate utilisation or product accumulation; (c) the culture system is usually uncontrolled and thereby subject to wide fluctuations in parameters such as pH or oxygen concentration; (d) several microenvironments exist in most batch culture vessels that allow microbial growth in different physiological states so that the average value of measured parameters may not represent the actual value of those parameters in any single cell (the mode and the mean are very different so no cell exists with the actual parameter values obtained from measurement). This latter consideration makes modelling of batch culture systems extremely problematic.

The need for maximal control of the experimental aspects of systems biology together with attainment of steady-state conditions has led to resurgence in the use of continuous culture systems such as chemostats [20]. During continuous culture in a chemostat, microbes are grown at a rate set by the experimenter and other environmental parameters such as pH, oxygen levels are also precisely controlled. Culture medium is pumped at a constant rate into the vessel whilst the volume of the culture is kept constant by an overflow system. The flow rate (f) of the media is set by the experimenter to give a desired dilution rate (D). The dilution rate is the number of culture volumes passing through the chemostat per unit of time and equals the flow rate divided by the culture volume (V). The chemostat controls growth rate (μ) by limiting the availability of a growth substrate. The medium contains a fixed concentration of the limiting substrate, all the other nutrients being present in essentially excess amounts. By adjusting the feed rate the growth rate can be adjusted to 1–90% of the maximum growth rate for the organism. When a dilution rate is set the cells will initially grow as in a batch culture at the maximum specific growth rate (μ_{max}) until a substrate in the medium becomes limited. Eventually the cells adjust to the rate of nutrient supply so that the specific growth rate equals the dilution rate, i.e. $D=\mu$. This balanced growth is known as steady state and may be maintained indefinitely. During steady state the physiology of the cells remains constant, cellular processes being controlled by the concentration of the limiting substrate.

The chemostat therefore enables growth in a tightly regulated steady-state environment and thereby eliminates the inherent variability and dynamics of constantly changing batch cultures. The chemostat system is thereby analogous to the aerodynamic wind tunnel. It effectively freezes the dynamics of microbial growth to attain a steady-state system that is amenable to constraint-based modelling approaches, such as flux balance analysis (FBA), which critically depend on the assumption that concentrations of internal metabolites are held constant during the experiment. Data from chemostat cultivations is therefore more precise, reproducible and statistically significant than those obtained from batch cultivations [21–23]. Moreover, because the cultures are relatively homogenous, the mean value of measured parameters in samples removed from the chemostat is likely to correspond to the mode value of those parameters in the bulk population; allowing application of these values for model parameterisation.

The slow growth rate of pathogenic mycobacteria, problems associated with clumping of bacilli and safety considerations have all provided obstacles for researchers attempting continuous cultures of *M. tuberculosis*. James and colleagues [24] were the first to successfully cultivate *M. tuberculosis* in a chemostat at a growth rate of 24 h in a complex nutrient-rich medium and have used their system successfully to investigate the responses of *M. tuberculosis* to oxygen [25] and iron limitation [26] and also mutation rates at different pH [27]. These studies demonstrated that the chemostat provides a reliable and reproducible environment for culturing mycobacteria and is therefore a very useful tool for "omic" scale analysis such as DNA microarrays. It has been demonstrated that gene expression data from organisms, including *M. tuberculosis*, grown in the chemostat is significantly more reproducible than batch culture DNA-array data [23, 25, 28].

Our group has developed a system for growing mycobacteria in a carbon limited chemically defined minimal medium which can be used as a reproducible platform for systems biology studies [12, 29–31]. Initial studies using this experimental system to grow *M. bovis* BCG (a non-pathogenic surrogate for *M. tuberculosis*) provided vital information on the biomass composition of the tubercle bacillus [29]. Studies prior to this are limited and were performed in poorly defined batch cultivations. For genome-scale metabolic models, the equations defining the biomass synthesis are very important and can impact on the predictive accuracy of the model. For two different growth rates ($D=0.03$; $t_d=23.1$ or $D=0.01$; $t_d=69.3$), the elemental and macromolecular composition of the biomass was measured and shown to change as a function of the growth rate. This study demonstrated that more than half of the dry mass of the mycobacterial cell was comprised of carbohydrates and lipids with only a quarter of the dry weight consisting of protein and RNA, but that these proportions change, depending on the growth rate. This data allowed a stoichiometric composition model for *M. bovis* BCG to be reconstructed, which is an important first step in the development of a metabolic network [29].

4 Metabolic Model Building

The ultimate goal of system biology approaches to studying TB is to construct a complete model of infection incorporating both the pathogen and host, but this is currently infeasible as the information about the different components to be included in the model is lacking. Studies with other organisms have demonstrated that metabolism is, by far, the best understood cellular network and is thereby an excellent starting point for a systems-based approach [32–34].

However, metabolism is complex. Even the simplest organisms synthesise many hundreds of metabolites connected by a similar number of enzyme-catalysed reactions. Each reaction is described by a set of kinetic parameters (e.g. K_m, V_{max}) which, in combination with substrate/product concentrations, determine its rate. Although K_m values are constants (for a particular substrate/product combination) and may be determined experimentally, intracellular concentrations of substrate, products and enzyme (influencing V_{max}) vary over wide ranges and are not easily measured. Even a single enzyme reaction is therefore a highly dynamic system; and systems of just a few reactions steps are usually mathematically described by a set of ordinary differential equation with a large number of parameters and variables whose values are extremely challenging to measure experimentally. Kinetic models have therefore only been applied to the dynamics of small well-defined systems, such as glycolysis in *Escherichia coli* [35] that are very far from being genome scale.

However, it is relatively straightforward to generate a metabolic network that describes the biochemical reactions that an organism is predicted to be capable of performing, in terms of stoichiometric formulas (see Chap. 1). It is therefore possible to build a model consisting of all the stoichiometric reactions predicted by the annotation and link these pathways and networks connected by flux values between each reaction. These models can be interrogated with tools such as FBA and metabolic flux analysis (MFA) to gain insight into the underlying structure of the network, identify essential genes and pathways and simulate experiments. However, because metabolic networks contain multiple branch points and parallel pathways there is not a unique solution but a vast space of possible solutions (the system is underdetermined). It is therefore necessary to apply constraint-based approaches, which reduce the solution space and thereby predict metabolic capabilities or internal fluxes [36–41]. FBA uses the procedure of optimisation to reduce the solution space (Chap. 1) optimising some parameter, which might be biomass production rate (and thereby growth rate), ATP synthesis, substrate consumption, product production or any other parameter of the model. Clearly there is a strong assumption in FBA that the cell applies a similar optimisation strategy and thereby grows at its optimal growth rate, ATP production rate or rate of other optimised parameter. If that assumption is correct then FBA will find the correct solution—the one that the cell finds—and the FBA solution will correspond to the biological reality. It is of course an open question how often and in what circumstances microbial cells such as *M. tuberculosis* do actually optimise simplistic parameters such as growth rate, particularly during in vivo growth. MFA applies an alternative approach to reducing the solution space: applying additional measurements as constraints [42, 43]. These may be measurements of intracellular metabolites, enzyme activities or indeed any additional measurement constraints, but the most powerful method currently available is ^{13}C-MFA, which derives solutions for the intracellular fluxes from the distribution of ^{13}C from a substrate into central metabolites and the amino acid products derived from central metabolites.

There are of course limitations to these approaches such as the requirement for steady or quasi-steady state conditions. Also, since no consideration is made of either transcriptional, translational, metabolic regulation or enzyme kinetics the predictive capabilities of constraint-based models are limited to situations when these factors are not significantly influencing reaction rates [34]. Nevertheless, these approaches have been successfully applied to predict the metabolic capabilities of many different cellular systems [44–49]. The application of both of FBA and MFA to *M. tuberculosis* will be discussed below.

5 Metabolic Models of *M. tuberculosis*

The first *M. tuberculosis* constraint-based model was constructed by Raman et al. and consisted of all the reactions in mycolic acid synthesis [50]. This sub-model of metabolism was composed of 219 reactions that involved 197 metabolites, catalysed by 28 enzymes. FBA was used to simulate mycolic acid metabolism and to identify potential drug targets in these pathways. The study illustrates the importance of optimisation in FBA. As already discussed, FBA reduces the solution space by optimising a parameter, usually known as the objective function, so choice of the choice of the parameter to be used as the objective function has considerable influence on the solutions obtained. Popular objective functions include maximisation or minimisation of ATP production; maximisation of redox potential; maximisation of the rate of synthesis of a particular product, or minimisation of nutrient uptake, but the most commonly used parameter is maximisation of growth rate which has been successfully applied in many systems including nutrient limited chemostat culture of *E. coli* [51]. Its use is more problematic for slow growing pathogens, such as *M. tuberculosis*, since it has not been established that these organisms do actually maximise their growth rate. The study used two objective functions that optimised the production of mycolic acids. The first, termed C1, optimised production of only the most abundant mycolate, whereas the objective function C2 maintained the known ratios of different mycolates. To test the predictive accuracy of these objective functions in silico deletions were performed and compared to transposon site hybridisation (TraSH) mutagenesis data. The highest predictive accuracy was obtained with the objective function C2 with an 82% correlation with experimental data. FBA identified 16 essential genes in this study and this primary list was then filtered to remove any genes encoding proteins that were complemented by homologues and also those with close homologues in the human proteome. This feasibility analysis identified seven potential drugs targets for anti-TB drug design (discussed below).

Although targeting a small sub-system such as mycolic acid synthesis can yield valuable information on specific pathways, it has limited value in elucidating the metabolic capability of *M. tuberculosis*. This latter objective is best approached by constructing a genome-scale network of metabolism [12, 52]. The first published genome-scale network was built using *Streptomyces coelicolor* as a starting model [12]. Orthology relationships were mapped between the related species using the KEGG databases and this preliminary model was further supplemented with data from the BioCyc database. This automatic process, however, accounted for only 57% of the final model. The remaining model was reconstructed by labour intensive manual curation based upon primary research publications, textbooks and review articles, and also by picking the brains of experts in the field. The final model utilised two biomass formulations which were derived from published data of cell composition obtained from a variety of sources, including our own chemostat-derived data obtained from fast and slow-growing BCG. BIOMASS 1 reflects the actual macromolecular composition of in vitro-grown *M. tuberculosis*, whereas BIOMASSe consisted of only those cellular components, such as DNA, RNA, protein, co-factors and the cell wall skeleton, that were considered to be essential for in vitro growth. The advantage of having these two biomass formulations is that the model could be used to predict gene essentiality both in vitro (with the minimal BIOMASSe as the objective function) and in vivo (with the more complete BIOMASS 1 including many virulence factors as the objective function).

The final functional genome-scale metabolic network of *M. tuberculosis* (GSMN-TB) consisted of 739 metabolites participating in 849 reactions and involves 726 genes. The model is freely available as both an excel file or in sbml format, and is accessible via a user-friendly web tool for constraint-based simulations (http://sysbio.sbs.surrey.ac.uk/tb/). FBA-based predictions of in vitro gene essentiality using BIOMASSe as the objective function correlated well with predictions of gene essentiality obtained by global transposon mutagenesis [53], with an overall predictive accuracy of 78% [12]. Quantitative validation of the model was also performed using data from continuous culture chemostat experiments [29]. The model predicted a lower rate of glycerol consumption than the experimentally determined values. A plausible explanation for the discrepancy was that, in addition to consumption of glycerol, the tubercle bacillus also utilised oleic acid released from hydrolysis of the Tween 80 dispersal agent present in the media. Opening an additional oleic acid transport flux corrected this discrepancy and we have recently confirmed that Tween 80 is indeed being consumed in these experiments [13]. A second genome-scale reconstruction of *M. tuberculosis*, *iNj661*, was published by Jamshidi and Palsson [52], as described in Chap. 1.

The mycolic acid synthesis sub-model and the two genome-scale network reconstructions available for *M. tuberculosis* illustrate the different approaches which can be applied to reconstructing, validating and applying metabolic models. They also provide a reference for future metabolic reconstructions. The next challenge is to combine these three models and build upon them by integrating any new experimental data in order to expand and refine the reconstructions in an iterative cycle. In this way the model can serve as an up-to-date representation of the cumulative knowledge of *M. tuberculosis*'s metabolic capabilities. For comparison, the *E. coli* genome-scale network has undergone six different successive reconstructions over the last 18 years, each one contributing positively to a large number of different studies [54]. A well-curated reconstruction is a perquisite for all systems biology approaches to studying *M. tuberculosis*.

6 Metabolic Models of Host–Pathogen Systems

M. tuberculosis is an intracellular pathogen that replicates primarily in the phagosome compartment of macrophages so its biology is intimately connected to that of its host cell. To simulate the combined system, Bordbar et al. [57] built a novel metabolic model that integrated the iNj66I model of *M. tuberculosis* with a cell-specific alveolar macrophage model, iAB-AMØ-1410 (based on the global human metabolic reconstruction, Recon 1 [37]) to build an integrated host–pathogen genome-scale

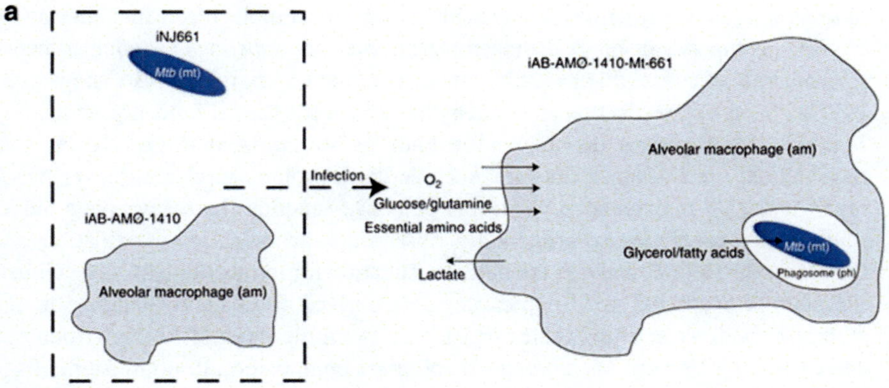

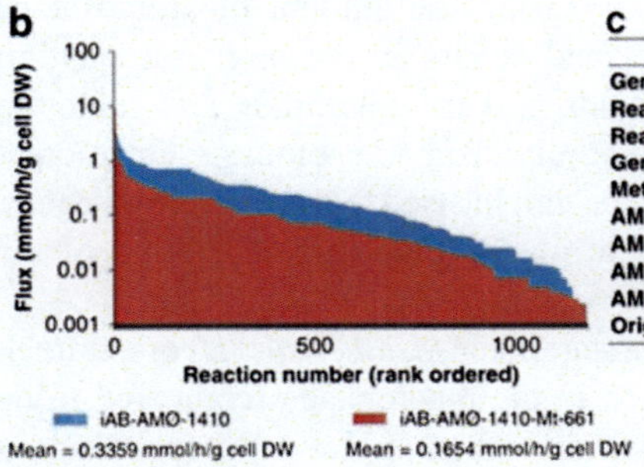

c

	iAB-AMØ-1410	iNJ661	iAB-AMØ-1410-Mt-661
Genes	1410	661	2071
Reactions (intracellular)	3012	939	3951
Reactions (exchange)	382	88	538
Gene associated reactions	69.3%	77%	71.1%
Metabolites	2572	828	3400
AMØ max biomass	0.0270	–	0.0238
AMØ max ATP production	0.6843	–	0.1703
AMØ max NO production	0.0359	–	0.0163
AMØ max NADH production	1.7806	–	0.5054
Original Mtb max biomass	–	0.0522	0.0021

Fig. 4.2 Results obtained by integration of the alveolar macrophage (iAB-AMØ-1410) and *M. tuberculosis* (iNJ661) reconstructions. (**a**) Metabolic links between the extracellular space (e), alveolar macrophage (am), phagosome (ph) and *M. tuberculosis* (Mtb) in iAB-AMØ-1410-Mt-661. The model is compartmentalised using the abbreviations as shown. In the model, the major carbon sources of the alveolar macrophage were glucose and glutamine. The macrophage compartment was also aerobic and requires the essential amino acids. Despite its use of oxygen, the macrophage exhibits anaerobic respiration and produces excess lactate. In the *M. tuberculosis* compartment of the model, the major carbon sources available in the phagosome environment were glycerol and fatty acids. The phagosome environment was also functionally hypoxic. (**b**) The flux span of iAB-AMØ-1410-Mt-661 was significantly reduced (51%) compared with its progenitor macrophage model, iAB-AMØ-1410. This shows a stricter definition of the alveolar macrophage solution space without adding additional constraints on the alveolar macrophage portion of the network. (**c**) Reaction, metabolite and gene properties of the three reconstructions. Maximum production rates of ATP, nitric oxide, redox potential (NADH) and biomass are shown. From [57]

reconstruction, iAB-AMØ-1410-Mt-661 (Chap. 1). The combined model was essentially composed of three compartments representing the macrophage, the phagosome and the pathogen residing within the phagosome. These were connected via metabolite and gas exchange reactions that allowed the *M. tuberculosis* compartment to uptake substrates and excrete waste products into the phagosome compartment (Fig. 4.2). The exchanges do of course instantiate several assumptions regarding the infectious state. The macrophage was assumed to be consuming glucose, glutamine and essential amino acids and excreting lactate. The phagosome environment that provided resources for *M. tuberculosis* replication was assumed to be depleted in glucose and rich in glycerol and fatty acids. A key aspect of the reconstruction was the biomass composition of both macrophage and *M. tuberculosis* compartments, remembering of course that biomass composition plays a key role in FBA and is often used as the objective function and thereby has a very substantial influence on the flux solutions. Macrophages do not readily multiply so the iAB-AMØ-1410 biomass reflected only maintenance function, such as lipid, protein, mRNA turnover, DNA repair and ATP maintenance. With this objective function the macrophage model successfully predicted experimentally observed rates of glucose oxidation and lactate production. To provide the biomass equation for intracellular *M. tuberculosis* the authors examined gene expression data derived from in vivo mouse model studies as well as in vitro studies that aimed to mimic the infection environment. They then adjusted the *M. tuberculosis* biomass composition to optimise the fit to the gene expression data. This involved increasing the amount of amino acids, mycolic acids, mycobactin, mycocerosates and sugars in the biomass equation; reducing ATP maintenance, DNA, fatty acids and phospholipids and removing peptidoglycans and phenolic glycolipids entirely from the biomass equation to construct a new objective function. It should be emphasised that the resulting behaviour of the reconstituted model is dependent on the precise composition of this adjusted biomass.

The authors then examined changes in flux state of the *M. tuberculosis* compartment as a consequence of its simulated replication in the macrophage (compared to the iNj66I model). The simulation predicted a shift in carbon uptake with suppression of glycolysis and up-regulation of gluconeogenesis, together with production of acetyl-CoA coming from macrophage-derived fatty acids via the glyoxylate shunt. Concomitant with the utilisation of fatty acids as carbon source was up-regulation of fatty acid oxidation pathways. There was a shift toward mycobactin and mycolic acid synthesis with reduced flux through nucleotide, peptidoglycan and phenolic glycolipid pathways. Many of these changes are likely to be a consequence of the altered biomass composition.

The accuracy of the model was tested by comparison of gene essentiality predictions of the *M. tuberculosis* component of the model with genes identified to be conditionally essential for infection in a mouse lung model (but not essential in vitro) by TraSH [55]. A total of 374 genes investigated by TraSH were in the model. Of these, the in silico analysis predicted that only 9 genes were conditionally essential in the *M. tuberculosis* compartment of the integrated model. Of those nine in silico-predicted essential genes, only two genes were also essential experimentally by TraSH. Many of the discrepant results are likely to be due to differences between the simulated macrophage system and the mouse model that was used to obtain the TraSH data. Four of the nine genes were components of nitrate reductase, which has been shown to play a role in in vitro models of infection [56] and was thereby incorporated into the model when the gene expression data was used in the fitting exercise. However, even discrepant model predictions can be informative. Systems biology models are essentially a mathematical instantiation of biological hypotheses. In this case, one of the hypotheses embedded in the iAB-AMØ-1410-Mt-661 model was that nitrate reductase was required for survival of *M. tuberculosis* in the mouse. This hypothesis was tested by comparison of model predictions with TraSH data [55] and shown to be discrepant [57]. However, the situation is of course more complicated as nitrate reductase, although apparently non-essential in the mouse lung model [55], has been shown to contribute to the virulence of BCG in immunodeficient mice [58]. Therefore, alternative models will have to be constructed for different host: pathogen combinations. Testing in silico predictions experimentally is an essential step in the refinement and improvement of systems biology models in an iterative cycle of model→prediction→experimental test→model.

6.1 Applications of the Models

6.1.1 Using Models to Interrogate Genome Annotation

Genome-scale networks are usually constructed initially from genome annotation and are thereby subject to errors in that annotation. However, the metabolic model scrutinises the metabolic component of genome annotation at a system level for functionality and can thereby be used to find pathway holes or inconsistencies in the annotation. There are several "orphan reactions" in GSMN-TB, that is, reactions that are required for network functionality but for which there is no annotated *M. tuberculosis* gene predicted to perform that function. For example, sulfolipid synthesis in *M. tuberculosis* generates the metabolite adenosine 3′,5′-bisphosphate (PAP in the model) which will accumulate and thereby become toxic (unbalanced in the model) if it is not catabolised. The model is therefore infeasible unless the reaction catalysed by the enzyme 3′,5′-bisphosphate nucleotidase (which converts the metabolite to AMP and inorganic phosphate) is included in the network, as an orphan reaction. Examining model feasibility thereby generates clues to incomplete or incorrect genome annotation and may even provide novel drug targets that are not apparent in the genome annotation.

In silico models also allow genome annotation to be scrutinised by systems-based experimental data. For example, the route for glycerol utilisation is generally assumed to proceed via glycerol kinase (encoded by *glpK*) followed by dehydrogenation; however, the genome annotation of *M. tuberculosis* includes several alcohol dehydrogenases that could be involved in an alternative uptake pathway whereby glycerol is first oxidised by glycerol dehydrogenase before being phosphorylated (this pathway is annotated in the KEGG *M. tuberculosis* pathway map). However, incorporation of this pathway into the initial GSMN-TB model led to the prediction that the gene *glpK* is dispensable for growth on media with glycerol as sole carbon source. Global mutagenesis data demonstrated that *glpK* was in fact essential for growth on glycerol, which was confirmed by construction of a single *glp*K knock-out mutant [30]. This information was incorporated into a refined GSMN-TB model in which the annotated alcohol dehydrogenases do not provide an alternative glycerol uptake pathway.

Other systems-based insights into the metabolism of *M. tuberculosis* can be obtained by simply performing growth simulations with the model. For instance, it is often claimed that *M. tuberculosis* requires operation of the glyoxylate shunt for growth on lipids. However, FBA-based simulation of GSMN-TB indicated that although the isocitrate lyase reaction of the glyoxylate shunt is predicted to be essential for growth on simple fatty acids such as acetate, it was not predicted to be essential for growth on complex lipids, such as phospholipids. The reason is apparent on examination of the flux solution: catabolism of phosopholipids yields glycerol as well as acetate, which can be used for gluconeogenesis without operation of the shunt.

6.1.2 Interpretation of Experimental Data

The functional genomics revolution has provided the means to generate high-throughput datasets but integration and interpretation of vast numbers of data points to generate new hypotheses remains a formidable challenge. Computational models can serve to bridge the gap between data and hypothesis driven research providing a framework for integration of high-throughput data that can lead to model revisions (to resolve discrepancies between model predictions and experiment) and semi-automatic generation of new hypotheses [59, 60]. Even simply overlaying "omic" data onto genome-scale metabolic models provides a metabolic context to interpret this data and can also highlight incomplete or incorrect knowledge [59].

Gene Essentiality Data

One of the most straightforward applications of genome-scale models is to predict essential genes that can then be compared to experimental data. For example, using a high-throughput TraSH screen we identified the genes essential for the growth of *M. bovis* BCG on glycerol and compared this with gene essentiality predictions using GSMN-TB [30]. Whilst there was a good correlation between the GSMN-TB and the experimentally observed gene essentialities (76.66%) the analysis demonstrated how the model could be used to highlight gaps in our knowledge of TB's metabolism. Some of the discrepancies can be attributed to an undefined level of inaccuracy in global mutagenesis assays but may also be due to gene regulation of isoenzymes. For instance, both menaquinol oxidase systems (the aa3-type and bd-type) are predicted to be non-essential as they are functionally redundant in the model. This contradicts the global mutagenesis data, which indicated that the aa3-type cytochrome *c* oxidase is in fact essential and likely to be the main electron transport system operating under aerobic conditions.

Transcriptome Data

Whereas it is relatively simple to obtain multiple measurements to define the physiological or metabolic state of bacteria in vitro, only limited information can be obtained for bacteria in vivo. In particular, it is very challenging to perform metabolomic, proteomic, biochemical, physiological or structural studies with the small numbers of organisms obtained from infected animals. However, it is possible to perform transcriptomic studies on in or ex vivo grown organisms and these methods have been applied to the TB bacillus to obtain transcriptomic profiles of bacteria growing in cultured macrophages, mouse models and in human lesions [61–65]. The transcriptional profile of a cell (via translation, enzyme activity, etc.) determines most aspects of the physiological state; therefore, it should be possible to predict a physiological state from knowledge of the complete cellular transcriptome. However, the mapping between messenger RNA levels and physiological state is highly complex and non-linear depending on many unknown factors such as mRNA stability, translation efficiency and post-translational modification of proteins. Traditional approaches to defining metabolic responses from transcriptome data have generally relied on examining expression levels of key (rate-controlling) genes in metabolic pathways (for instance, [66]). However, metabolic control analysis has demonstrated that control is distributed throughout the entire metabolic network, such that the flux through any particular pathway is controlled globally [67, 68] rather than by a particular enzymatic step. This makes a simple mapping of differentially expressed genes onto metabolic pathways an unrealistic strategy for successful predictions of global metabolic state changes.

Several system-level approaches have been proposed to extract metabolic information from gene expression profiles. In the reporter metabolites approach [69], the local connectivity of a metabolite in the bi-partite, substance/reaction graph is used to identify a set of genes associated with each metabolite. Subsequently, for each of the metabolites, the distribution of the microarray-derived signal of genes associated with the metabolite is compared with the background distribution of the microarray-derived signal for all genes, resulting in the identification of the transcription regulation focal points of metabolism: network nodes that are directly affected by clusters of differentially expressed genes. In another approach, Shlomi [70] used Mixed Integer Linear programming to minimise the discrepancy between the internal metabolic flux distribution and the transcriptional profile of genes encoding metabolic enzymes. Their approach identifies flux distributions, which are consistent with the stoichiometric constraints of the genome-scale metabolic reaction network and at the same time maximise the number of active metabolic fluxes associated with up-regulated genes and the number of non-active metabolic fluxes associated with down-regulated genes. Yet another approach, E-flux, was recently developed and used to examine *M. tuberculosis* microarray data in the context of both the genome-scale metabolic reaction network, by constraining upper bounds of metabolic reactions to values proportional to the microarray signals of genes associated with these reactions [71]. Two models were used for the analysis: the Raman et al. model of mycolic acid pathways [50] and the GSMN-TB genome-scale metabolic model [12]. E-flux was applied to microarray data obtained from a large study that investigated the response of *M. tuberculosis* to 75 different drugs, drug combinations and nutrient conditions [72]. Eight of the tested drugs target mycolic synthesis, and this was correctly predicted by E-flux analysis of the microarray data, indicating that the method may be useful for target identification of novel inhibitors.

Shi et al. [73] applied a similar in silico method to interrogate quantitative PCR (qPCR) transcriptome data obtained from a model of respiratory infection of mice in which *M. tuberculosis* replicates in the lung for approximately 20 days, followed by stabilisation of bacterial numbers due to expression of acquired cell-mediated immunity. The qPCR data was not genome scale but focused on a set of genes predominantly involved in the pathways of central metabolism and lipid synthesis. This data was first interpreted qualitatively. Observed changes in mRNA abundance suggested that as tubercle bacilli stop replicating in the mouse lung and respond to the decreased demand for energy and biosynthetic precursors by down-regulating glycolysis, PPP and the TCA cycle. The main function of central metabolism appears to shift from providing energy and biosynthetic precursors for bacterial growth to accumulating the storage compounds such as triacyl glycerides (TAG) and glutamate. To gain a genome-scale insight into the underlying metabolic changes, two in silico cells were constructed by adjusting biomass composition of the GSMN-TB model. One cell represented growing *M. tuberculosis*, while the other represented the more minimal cell composition predicted for non-growing *M. tuberculosis* in the mouse lung. FBA was then used to predict the consequent changes in flux distribution in the cell. The resulting flux distributions were broadly consistent with the gene expression data and the hypothesis that growth arrest in the mouse lung is associated with a re-routing of carbon flow in central metabolism from metabolic pathways generating energy and biosynthetic precursors to pathways for storage compounds, such as TAG and glutamate.

Our own laboratory developed an alternative method, differential producibility analysis (DPA). The method [74] relies on FBA to link genes with metabolites on a system-wide level. A gene essentiality scan is first performed on every gene but instead of using biomass as the objective function, each metabolite is, in turn, used as the objective function. It is thereby possible to generate a mapping that identifies, for each metabolite, the genes that are required for its synthesis (the *producibility plot*). In the next step of DPA, the experimental data is interrogated. Gene expression signals for a particular experiment are assigned onto each gene which, using the producibility plot, are mapped onto each metabolite. Metabolites are then ranked to identify the metabolites that are *most affected* by genes that are up-regulated in the target experiment and (separately) are ranked to metabolites that are *most affected* by genes that are down-regulated in the same experiment. The whole procedure of DPA effectively transforms a gene-based transcriptome signal into a metabolite-based metabolome signal.

The DPA method was first tested with *E. coli* transcriptome data and shown to successfully identify metabolic responses to environmental perturbation (shift to anaerobic growth) and gene knock-out. This method was then applied to several *M. tuberculosis* in vitro transcriptome datasets and was able to identify metabolic responses. Applying DPA to transcriptomic data obtained from *M. tuberculosis* replicating in mice-derived macrophage [63] revealed a previously unrecognised feature of the response of *M. tuberculosis* to the macrophage environment [74]: a down-regulation of genes influencing metabolites in central metabolism and concomitant up-regulation of genes that influence synthesis of cell wall components

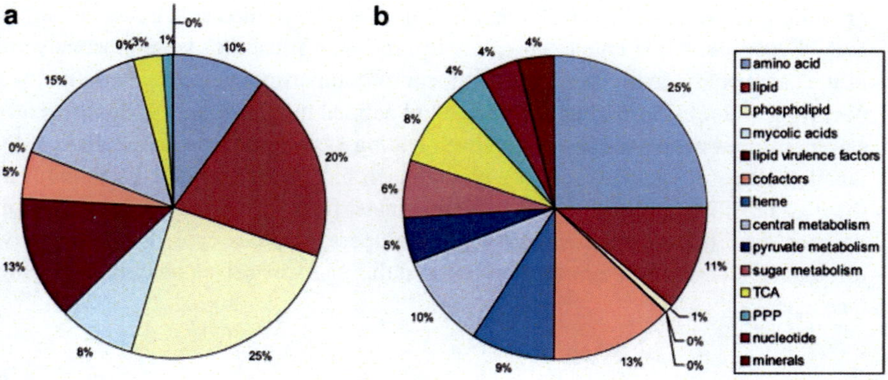

Fig. 4.3 Pi chart illustrating the role of *M. tuberculosis* metabolites in macrophages. Pi chart illustrating the role of metabolites associated by DPA with up-regulated (**a**) or down-regulated, (**b**) genes in the mouse macrophage [63]

and virulence factors (Fig. 4.3). DPA suggests that a significant feature of the response of the tubercle bacillus to the intracellular environment is a channelling of resources towards remodelling of its cell envelope, possibly in preparation for attack by host defences. Interestingly, application of DPA [74] to transcriptome data obtained from *M. tuberculosis* bacilli recovered from human sputum [75] generated a very different metabolic signature to the mouse macrophage data. DPA can thereby be used to unravel the mechanisms of virulence and persistence of *M. tuberculosis* and other pathogens and may have general application for extracting metabolic signals from other "-omics" data.

Stable Isotope Metabolite Profiling

Seminal studies performed many decades ago, mostly in *E. coli*, established the major pathways for carbon substrate utilisation in bacteria through metabolite tracer analysis. More recently, stable isotope studies are being increasingly used to monitor metabolism. The usual approach is to feed the microbe a ^{13}C-labelled carbon substrate (uniformly and/or positionally labelled) and then measure the labelling profile using nuclear magnetic resonance (NMR) and/or Mass Spectrometry (MS). NMR can provide positional information but it is less sensitive than MS and, crucially, it is soften difficult to identify metabolites responsible for a particular NMR signals. MS is being increasingly used for metabolite profiling since it combines high mass accuracy (providing accurate metabolite identification) with high sensitivity. However, it measures only mass so (unlike NMR) it does not distinguish between isotopomers labelled at different positions but with the same mass.

Stable isotope profiling has been applied in a several studies of *M. tuberculosis* metabolism [76–78]. For example, ^{13}C labelled substrates were used to demonstrate that phosphoenolpyruvate carboxykinase (PCK) predominantly catalyses the conversion of oxaloacetate to phosphoenolypyruvate (PEP) when *M. tuberculosis* is growing on acetate [78]. Cavalho et al. [76] used ^{13}C labelled substrates to confirm that *M. tuberculosis* is able to co-catabolise multiple substrates simultaneously and in the same study demonstrated that a form of compartmentalised metabolism was occurring whereby individual substrates had defined metabolic fates. More recently, stable isotopes were used in a study investigating *M. tuberculosis* metabolism of ^{13}C labelled glucose, aspartate and CO_2 under anaerobic conditions [18]. By monitoring the ^{13}C labelling profile of secreted succinate from these cultures the authors demonstrated that *M. tuberculosis* is able to operate a TCA cycle in the reductive direction under anaerobic conditions and that this pathway drives succinate secretion (see also 4.2).

^{13}C Metabolic Flux Analysis

The goal of metabolic analysis is to understand the metabolic pathways that are being utilised under particular conditions. As described earlier, it is possible to obtain estimates of the ranges of fluxes that are compatible with substrate inputs and outputs in a system using FBA. However, there are usually a great number of possible flux solutions that are compatible with the data so FBA utilises the method of optimisation to determine the flux distribution that optimises some parameters, such as growth rate. This method has been very successfully applied in fast-growing organisms but its application to the slow-growing pathogen, *M. tuberculosis* is uncertain. A more direct means of establishing the intracellular fluxes is through ^{13}C-MFA. This powerful technique has been successfully applied to identify functional flux states in various microbes ([79] provides a recent review of the technique and its application) and has enormous potential for studying the metabolism of *M. tuberculosis*.

During a ^{13}C-MFA experiment an organism in metabolic steady state (usually cultivated in a chemostat) is grown in the presence of ^{13}C-labelled carbon substrate (uniformly and/or positionally labelled). For isotopic steady-state experiments, mixtures of unlabelled and ^{13}C-labelled substrates are used as otherwise all the metabolites would eventually become labelled to the same degree as the substrate and the experiment will be uninformative. The positional labelling patterns (which carbon atoms are labelled) of the amino acids and/or metabolites (as determined by either MS and/or NMR) are then used as additional constraints in MFA to solve the internal fluxes and thereby reconstruct the paths through central metabolism that the carbon took inside the cells.

Our earlier work used FBA to predict flux distribution in fast and slow-growing *M. tuberculosis* with glycerol as sole carbon source and predicted an increased flux through the glyoxylate shunt at slow growth rate [12]. This was surprising as the shunt is usually considered to be used solely for growth on two carbon compounds such as acetate and the essentiality of this enzyme for in vivo growth has been widely interpreted to indicate that the pathogen consumes host lipids (yielding acetate on beta oxidation) during growth inside the macrophage [1, 2, 80]. We confirmed that activity of the enzyme isocitrate lyase (ICL, the key enzyme of the glyoxylate shunt)

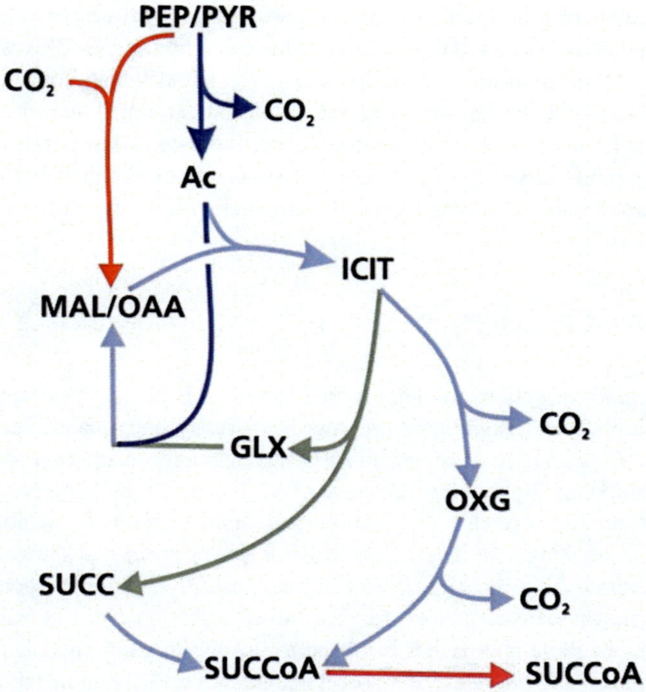

Fig. 4.4 Schematic of the GAS pathway which is characterised by flux through the glyoxylate shunt and anaplerotic reactions for oxidation of pyruvate and succinyl CoA synthetase for the generation of succinyl CoA. Metabolite abbreviations: *PEP/PYR* phosphoenolpyruvate/pyruvate, *Ac* acetate, *CHO, ICIT* isocitrate/citrate, *MALOAA* L-malate-oxaloacetate, *OXG* 2-oxoglutarate, *SUC* succinate, *SUCCOA* succinyl-CoA, *GLX* glyoxylate, *OXG* 2-oxoglutarate

was indeed induced during slow growth rate but how and why the organism was utilising the glyoxylate shunt during slow growth on glycerol was a mystery. The mystery was compounded when we constructed an ICL mutant of *M. tuberculosis* and demonstrated that the mutant was unable to grow at slow growth rate in a glycerol limited chemostat [13]. To discover the metabolic pathways involved in slow (and fast) growth we recently, for the first time, performed ^{13}C-MFA on *M. bovis* BCG and *M. tuberculosis* at fast and slow growth rate. Tracer experiments were performed with steady-state chemostat cultures supplied with ^{13}C labelled glycerol. Through measurements of the ^{13}C isotopomer labelling patterns in protein-derived amino acids and enzymatic activity assays we identified the activity of a novel pathway (termed the GAS pathway) that is used for pyruvate dissimulation in *M. tuberculosis*. This pathway is characterised by significant flux through the glyoxylate shunt and also through the carbon fixing anaplerotic reactions at the PEP-pyruvate-oxaloacetate node combined with very low flux through the succinate–oxaloacetate segment of the TCA cycle (Fig. 4.4). The flux through the GAS pathway is increased at slow growth rate accounting for the essentiality of ICL at slow growth rate. An interesting feature of the GAS pathway is that it included a significant fraction of flux (far more than required for anaplerosis) going through one or more of the anaplerotic reactions between phosphoenolpyruvate/pyruvate and malate/oxaloacetate in the carbon-fixing direction. This prediction of ^{13}C-MFA was confirmed by feeding *M. tuberculosis* ^{13}C-labelled sodium bicarbonate and confirming that the pathogen is indeed able to incorporate this unusual carbon source into amino acids [13]. As the human host is abundant in CO_2 this finding and the operation of the GAS pathway requires further investigation in vivo as carbon dioxide fixation may provide a point of vulnerability that could be targeted with novel drugs.

7 Future Challenges

The application of systems biology to the study of TB is a science that is still in its infancy. Nevertheless, significant progress has already been made. Several in silico models of *M. tuberculosis* have been constructed and a reconstruction of the *M. tuberculosis*-macrophage system has been described. The model building process itself is a highly informative exercise that not only defines a minimal metabolic capacity necessary for making a *M. tuberculosis* cell but also provides clues to gene annotation and generates novel insights into the metabolic capability of this pathogen. The models have been shown to be useful tools for drug target prediction. One of the most powerful applications of these approaches has been to use the in silico models to interrogate experimental data to provide system-level insight into underlying metabolic responses associated with the response of *M. tuberculosis* to stress, drugs and growth in host cells. Each of these models is available online giving researchers across the world access to systems biology tools that can be used to investigate the biology of the tubercle bacillus but also to interrogate both published and new datasets.

There is no question that existing models are not yet realistic reconstructions of the *M. tuberculosis* cell. However, interrogating the models with experimental data in an iterative cycle of model refinements and experimentation will ensure that the current models will become more accurate descriptions of the metabolism of *M. tuberculosis*. A limitation of current FBA-type models is that they can only strictly be applied to steady-state systems. A future goal will be to extend these models to simulate dynamic states, such as during in vivo growth. This will require integration of metabolic networks with gene regulatory networks and kinetic models of enzyme action together with realistic models of host–pathogen interactions. Such multi-scale models may eventually be used to build an in silico *M. tuberculosis* cell. Such a model may be used for drug discovery and optimisation of treatment regimes but will also be an enormously powerful tool to investigate the fundamental biology of this important pathogen.

References

1. McKinney JD, Honer zu Bentrup K, Munoz-Elias EJ, Miczak ACB, Chan W-T, Senson D, Sacchettini JC, Jacobs-WR J, Russell DG (2000) Persistance of *Mycobacterium tuberculosis* in macrophages and mice requires the glyoxylate shunt enzyme isocitrate lyase. Nature 406:735–738

2. Munoz-Elias EJ, McKinney JD (2005) Mycobacterium tuberculosis isocitrate lyases 1 and 2 are jointly required for in vivo growth and virulence. Nat Med 11:638–644
3. Fritz C, Maass S, Kreft A, Bange FC (2002) Dependence of Mycobacterium bovis BCG on anaerobic nitrate reductase for persistence is tissue specific. Infect Immun 70:286–291
4. Miner MD, Chang JC, Pandey AK, Sassetti CM, Sherman DR (2009) Role of cholesterol in Mycobacterium tuberculosis infection. Indian J Exp Biol 47:407–411
5. Movahedzadeh F, Smith DA, Norman RA, Dinadayala P, Murray-Rust J, Russell DG, Kendall SL, Rison SC, McAlister MS, Bancroft GJ et al (2004) The Mycobacterium tuberculosis ino1 gene is essential for growth and virulence. Mol Microbiol 51:1003–1014
6. Singh A, Crossman DK, Mai D, Guidry L, Voskuil MI, Renfrow MB, Steyn AJ (2009) Mycobacterium tuberculosis WhiB3 maintains redox homeostasis by regulating virulence lipid anabolism to modulate macrophage response. PLoS Pathog 5:e1000545
7. Glickman MS, Cox JS, Jacobs WR Jr (2000) A novel mycolic acid cyclopropane synthetase is required for cording, persistence, and virulence of Mycobacterium tuberculosis. Mol Cell 5:717–727
8. Keating LA, Wheeler PR, Mansoor H, Inwald JK, Dale J, Hewinson RG, Gordon SV (2005) The pyruvate requirement of some members of the Mycobacterium tuberculosis complex is due to an inactive pyruvate kinase: implications for in vivo growth. Mol Microbiol 56:163–174
9. Munoz-Elias EJ, Upton AM, Cherian J, McKinney JD (2006) Role of the methylcitrate cycle in Mycobacterium tuberculosis metabolism, intracellular growth, and virulence. Mol Microbiol 60:1109–1122
10. Gengenbacher M, Rao SP, Pethe K, Dick T (2010) Nutrient-starved, non-replicating Mycobacterium tuberculosis requires respiration, ATP synthase and isocitrate lyase for maintenance of ATP homeostasis and viability. Microbiology 156:81–87
11. Baughn AD, Garforth SJ, Vilcheze C, Jacobs WR Jr (2009) An anaerobic-type alpha-ketoglutarate ferredoxin oxidoreductase completes the oxidative tricarboxylic acid cycle of Mycobacterium tuberculosis. PLoS Pathog 5:e1000662
12. Beste D, Hooper T, Stewart GS, Bonde B, Avignone-Rossa C, Bushell M, Wheeler PR, Klamt S, Kierzek AM, McFadden JJ (2007) GSMN-TB: a web-based genome scale network model of Mycobacterium tuberculosis metabolism. Genome Biol 8:R89
13. Beste DJ, Bonde B, Hawkins N, Ward JL, Beale MH, Noack S, Noh K, Kruger NJ, Ratcliffe RG, McFadden J (2011) ^{13}C metabolic flux analysis identifies an unusual route for pyruvate dissimilation in mycobacteria which requires isocitrate lyase and carbon dioxide fixation. PLoS Pathog 7:e1002091
14. Tian J, Bryk R, Shi S, Erdjument-Bromage H, Tempst P, Nathan C (2005) Mycobacterium tuberculosis appears to lack alpha-ketoglutarate dehydrogenase and encodes pyruvate dehydrogenase in widely separated genes. Mol Microbiol 57:859–868
15. Tian J, Bryk R, Itoh M, Suematsu M, Nathan C (2005) Variant tricarboxylic acid cycle in Mycobacterium tuberculosis: identification of alpha-ketoglutarate decarboxylase. Proc Natl Acad Sci USA 102:10670–10675
16. de Carvalho LP, Zhao H, Dickinson CE, Arango NM, Lima CD, Fischer SM, Ouerfelli O, Nathan C, Rhee KY (2010) Activity-based metabolomic profiling of enzymatic function: identification of Rv1248c as a mycobacterial 2-hydroxy-3-oxoadipate synthase. Chem Biol 17:323–332
17. Wagner T, Bellinzoni M, Wehenkel A, O'Hare HM, Alzari PM (2011) Functional plasticity and allosteric regulation of alpha-ketoglutarate decarboxylase in central mycobacterial metabolism. Chem Biol 18:1011–1020
18. Watanabe S, Zimmermann M, Goodwin MB, Sauer U, Barry CE III, Boshoff HI (2011) Fumarate reductase activity maintains an energized membrane in anaerobic Mycobacterium tuberculosis. PLoS Pathog 7:e1002287
19. Kitano H (2007) Towards a theory of biological robustness. Mol Syst Biol 3:137
20. Hoskisson PA, Hobbs G (2005) Continuous culture–making a comeback? Microbiology 151:3153–3159
21. Boer VM, de Winde JH, Pronk JT, Piper MD (2003) The genome-wide transcriptional responses of Saccharomyces cerevisiae grown on glucose in aerobic chemostat cultures limited for carbon, nitrogen, phosphorus, or sulfur. J Biol Chem 278:3265–3274

22. Boer VM, Tai SL, Vuralhan Z, Arifin Y, Walsh MC, Piper MD, de Winde JH, Pronk JT, Daran JM (2007) Transcriptional responses of Saccharomyces cerevisiae to preferred and nonpreferred nitrogen sources in glucose-limited chemostat cultures. FEMS Yeast Res 7:604–620
23. Hayes A, Zhang N, Wu J, Butler PR, Hauser NC, Hoheisel JD, Lim FL, Sharrocks AD, Oliver SG (2002) Hybridization array technology coupled with chemostat culture: tools to interrogate gene expression in Saccharomyces cerevisiae. Methods 26:281–290
24. James BW, Williams A, Marsh PD (2000) The physiology and pathogenicity of Mycobacterium tuberculosis grown under controlled conditions in a defined medium. J Appl Microbiol 88:669–677
25. Bacon J, James BW, Wernisch L, Williams A, Morley KA, Hatch GJ, Mangan JA, Hinds J, Stoker NG, Butcher PD et al (2004) The influence of reduced oxygen availability on pathogenicity and gene expression in Mycobacterium tuberculosis. Tuberculosis (Edinb) 84:205–217
26. Bacon J, Dover LG, Hatch KA, Zhang Y, Gomes JM, Kendall S, Wernisch L, Stoker NG, Butcher PD, Besra GS et al (2007) Lipid composition and transcriptional response of Mycobacterium tuberculosis grown under iron-limitation in continuous culture: identification of a novel wax ester. Microbiology 153:1435–1444
27. Jenkins C, Bacon J, Allnutt J, Hatch KA, Bose A, O'Sullivan DM, Arnold C, Gillespie SH, McHugh TD (2009) Enhanced heterogeneity of rpoB in Mycobacterium tuberculosis found at low pH. J Antimicrob Chemother 63:1118–1120
28. Daran-Lapujade P, Daran JM, Kotter P, Petit T, Piper MD, Pronk JT (2003) Comparative genotyping of the Saccharomyces cerevisiae laboratory strains S288C and CEN.PK113-7D using oligonucleotide microarrays. FEMS Yeast Res 4:259–269
29. Beste DJ, Peters J, Hooper T, Avignone-Rossa C, Bushell ME, McFadden J (2005) Compiling a molecular inventory for Mycobacterium bovis BCG at two growth rates: evidence for growth rate-mediated regulation of ribosome biosynthesis and lipid metabolism. J Bacteriol 187:1677–1684
30. Beste DJ, Espasa M, Bonde B, Kierzek AM, Stewart GR, McFadden J (2009) The genetic requirements for fast and slow growth in mycobacteria. PLoS One 4:e5349
31. Beste DJ, Laing E, Bonde B, Avignone-Rossa C, Bushell ME, McFadden JJ (2007) Transcriptomic analysis identifies growth rate modulation as a component of the adaptation of mycobacteria to survival inside the macrophage. J Bacteriol 189:3969–3976
32. Bumann D (2009) System-level analysis of Salmonella metabolism during infection. Curr Opin Microbiol 12:559–567
33. Feist AM, Herrgard MJ, Thiele I, Reed JL, Palsson BO (2009) Reconstruction of biochemical networks in microorganisms. Nat Rev Microbiol 7:129–143
34. Durot M, Bourguignon PY, Schachter V (2009) Genome-scale models of bacterial metabolism: reconstruction and applications. FEMS Microbiol Rev 33:164–190
35. Bettenbrock K, Fischer S, Kremling A, Jahreis K, Sauter T, Gilles ED (2006) A quantitative approach to catabolite repression in Escherichia coli. J Biol Chem 281:2578–2584
36. AbuOun M, Suthers PF, Jones GI, Carter BR, Saunders MP, Maranas CD, Woodward MJ, Anjum MF (2009) Genome scale reconstruction of a Salmonella metabolic model: comparison of similarity and differences with a commensal Escherichia coli strain. J Biol Chem 284:29480–29488
37. Duarte NC, Becker SA, Jamshidi N, Thiele I, Mo ML, Vo TD, Srivas R, Palsson BO (2007) Global reconstruction of the human metabolic network based on genomic and bibliomic data. Proc Natl Acad Sci USA 104:1777–1782
38. Feist AM, Henry CS, Reed JL, Krummenacker M, Joyce AR, Karp PD, Broadbelt LJ, Hatzimanikatis V, Palsson BO (2007) A genome-scale metabolic reconstruction for Escherichia coli K-12 MG1655 that accounts for 1260 ORFs and thermodynamic information. Mol Syst Biol 3:121
39. Mo ML, Jamshidi N, Palsson BO (2007) A genome-scale, constraint-based approach to systems biology of human metabolism. Mol Biosyst 3:598–603
40. Oberhardt MA, Puchalka J, Fryer KE, Martins dos Santos VA, Papin JA (2008) Genome-scale metabolic network analysis of the opportunistic pathogen Pseudomonas aeruginosa PAO1. J Bacteriol 190:2790–2803

41. Thiele I, Jamshidi N, Fleming RM, Palsson BO (2009) Genome-scale reconstruction of Escherichia coli's transcriptional and translational machinery: a knowledge base, its mathematical formulation, and its functional characterization. PLoS Comput Biol 5:e1000312
42. Kim HU, Kim TY, Lee SY (2008) Metabolic flux analysis and metabolic engineering of microorganisms. Mol Biosyst 4:113–120
43. Wiechert W (2001) 13C metabolic flux analysis. Metab Eng 3:195–206
44. Boyle NR, Morgan JA (2009) Flux balance analysis of primary metabolism in Chlamydomonas reinhardtii. BMC Syst Biol 3:4
45. Raman K, Chandra N (2009) Flux balance analysis of biological systems: applications and challenges. Brief Bioinform 10:435–449
46. Varma A, Palsson BO (1994) Stoichiometric flux balance models quantitatively predict growth and metabolic by-product secretion in wild-type Escherichia coli W3110. Appl Environ Microbiol 60:3724–3731
47. Antoniewicz MR, Kraynie DF, Laffend LA, Gonzalez-Lergier J, Kelleher JK, Stephanopoulos G (2007) Metabolic flux analysis in a nonstationary system: fed-batch fermentation of a high yielding strain of E. coli producing 1,3-propanediol. Metab Eng 9:277–292
48. Kayser A, Weber J, Hecht V, Rinas U (2005) Metabolic flux analysis of Escherichia coli in glucose-limited continuous culture. I. Growth-rate-dependent metabolic efficiency at steady state. Microbiology 151:693–706
49. Peng L, Arauzo-Bravo MJ, Shimizu K (2004) Metabolic flux analysis for a ppc mutant Escherichia coli based on 13C-labelling experiments together with enzyme activity assays and intracellular metabolite measurements. FEMS Microbiol Lett 235:17–23
50. Raman K, Rajagopalan P, Chandra N (2005) Flux balance analysis of mycolic acid pathway: targets for anti-tubercular drugs. PLoS Comput Biol 1:e46
51. Schuetz R, Kuepfer L, Sauer U (2007) Systematic evaluation of objective functions for predicting intracellular fluxes in Escherichia coli. Mol Syst Biol 3:119
52. Jamshidi N, Palsson BO (2007) Investigating the metabolic capabilities of Mycobacterium tuberculosis H37Rv using the in silico strain iNJ661 and proposing alternative drug targets. BMC Syst Biol 1:26
53. Sassetti CM, Boyd DH, Rubin EJ (2003) Genes required for mycobacterial growth defined by high density mutagenesis. Mol Microbiol 48:77–84
54. Feist AM, Palsson BO (2008) The growing scope of applications of genome-scale metabolic reconstructions using Escherichia coli. Nat Biotechnol 26:659–667
55. Sassetti CM, Rubin EJ (2003) Genetic requirements for mycobacterial survival during infection. Proc Natl Acad Sci USA 100:12989–12994
56. Sohaskey CD, Wayne LG (2003) Role of narK2X and narGHJI in hypoxic upregulation of nitrate reduction by Mycobacterium tuberculosis. J Bacteriol 185:7247–7256
57. Bordbar A, Lewis NE, Schellenberger J, Palsson BO, Jamshidi N (2010) Insight into human alveolar macrophage and M. tuberculosis interactions via metabolic reconstructions. Mol Syst Biol 6:422
58. Weber I, Fritz C, Ruttkowski S, Kreft A, Bange FC (2000) Anaerobic nitrate reductase (narGHJI) activity of Mycobacterium bovis BCG in vitro and its contribution to virulence in immunodeficient mice. Mol Microbiol 35:1017–1025
59. Oberhardt MA, Palsson BO, Papin JA (2009) Applications of genome-scale metabolic reconstructions. Mol Syst Biol 5:320
60. Covert MW, Knight EM, Reed JL, Herrgard MJ, Palsson BO (2004) Integrating high-throughput and computational data elucidates bacterial networks. Nature 429:92–96
61. Cappelli G, Volpe E, Grassi M, Liseo B, Colizzi V, Mariani F (2006) Profiling of Mycobacterium tuberculosis gene expression during human macrophage infection: upregulation of the alternative sigma factor G, a group of transcriptional regulators, and proteins with unknown function. Res Microbiol 157:445–455
62. Raju B, Hoshino Y, Belitskaya-Levy I, Dawson R, Ress S, Gold JA, Condos R, Pine R, Brown S, Nolan A et al (2008) Gene expression profiles of bronchoalveolar cells in pulmonary TB. Tuberculosis (Edinb) 88:39–51

63. Schnappinger D, Ehrt S, Voskuil MI, Liu Y, Mangan JA, Monahan IM, Dolganov G, Efron B, Butcher PD, Nathan C et al (2003) Transcriptional adaptation of Mycobacterium tuberculosis within macrophages: insights into the phagosomal environment. J Exp Med 198:693–704
64. Tailleux L, Waddell SJ, Pelizzola M, Mortellaro A, Withers M, Tanne A, Castagnoli PR, Gicquel B, Stoker NG, Butcher PD et al (2008) Probing host pathogen cross-talk by transcriptional profiling of both Mycobacterium tuberculosis and infected human dendritic cells and macrophages. PLoS One 3:e1403
65. Talaat AM, Lyons R, Howard ST, Johnston SA (2004) The temporal expression profile of Mycobacterium tuberculosis infection in mice. Proc Natl Acad Sci USA 101:4602–4607
66. Salmon K, Hung SP, Mekjian K, Baldi P, Hatfield GW, Gunsalus RP (2003) Global gene expression profiling in Escherichia coli K12. The effects of oxygen availability and FNR. J Biol Chem 278:29837–29855
67. Kacser H, Burns JA (1995) The control of flux. Biochem Soc Trans 23:341–366
68. Heinrich R, Rapoport TA (1974) A linear steady-state treatment of enzymatic chains. Critique of the crossover theorem and a general procedure to identify interaction sites with an effector. Eur J Biochem 42:97–105
69. Patil KR, Nielsen J (2005) Uncovering transcriptional regulation of metabolism by using metabolic network topology. Proc Natl Acad Sci USA 102:2685–2689
70. Shlomi T, Cabili MN, Herrgard MJ, Palsson BO, Ruppin E (2008) Network-based prediction of human tissue-specific metabolism. Nat Biotechnol 26:1003–1010
71. Colijn C, Brandes A, Zucker J, Lun DS, Weiner B, Farhat MR, Cheng TY, Moody DB, Murray M, Galagan JE (2009) Interpreting expression data with metabolic flux models: predicting Mycobacterium tuberculosis mycolic acid production. PLoS Comput Biol 5:e1000489
72. Boshoff HI, Myers TG, Copp BR, McNeil MR, Wilson MA, Barry CE III (2004) The transcriptional responses of Mycobacterium tuberculosis to inhibitors of metabolism: novel insights into drug mechanisms of action. J Biol Chem 279:40174–40184
73. Shi L, Sohaskey CD, Pfeiffer C, Datta P, Parks M, McFadden J, North RJ, Gennaro ML (2010) Carbon flux rerouting during Mycobacterium tuberculosis growth arrest. Mol Microbiol 78:1199–1215
74. Bonde BK, Beste D, Laing E, Kierzek A, McFadden J (2011) Differential Producibility Analysis (DPA) of transcriptomic data with metabolic networks: deconstructing the metabolic response of M. tuberculosis. PLoS Comput Biol 7:e1002060
75. Garton NJ, Waddell SJ, Sherratt AL, Lee SM, Smith RJ, Senner C, Hinds J, Rajakumar K, Adegbola RA, Besra GS et al (2008) Cytological and transcript analyses reveal fat and lazy persister-like bacilli in tuberculous sputum. PLoS Med 5:e75
76. de Carvalho LP, Fischer SM, Marrero J, Nathan C, Ehrt S, Rhee KY (2010) Metabolomics of Mycobacterium tuberculosis reveals compartmentalized co-catabolism of carbon substrates. Chem Biol 17:1122–1131
77. Thomas ST, Vanderven BC, Sherman DR, Russell DG, Sampson NS (2011) Pathway profiling in Mycobacterium tuberculosis: elucidation of cholesterol-derived catabolite and enzymes that catalyze its metabolism. J Biol Chem 286:43668–43678
78. Marrero J, Rhee KY, Schnappinger D, Pethe K, Ehrt S (2010) Gluconeogenic carbon flow of tricarboxylic acid cycle intermediates is critical for Mycobacterium tuberculosis to establish and maintain infection. Proc Natl Acad Sci USA 107:9819–9824
79. Zamboni N, Fendt SM, Ruhl M, Sauer U (2009) (13)C-based metabolic flux analysis. Nat Protoc 4:878–892
80. Munoz-Elias EJ, McKinney JD (2006) Carbon metabolism of intracellular bacteria. Cell Microbiol 8:10–22

Chapter 5
Protein–Protein Interaction in the -Omics Era: Understanding *Mycobacterium tuberculosis* Function

Adrie J.C. Steyn, D. Mai, V. Saini, and A. Farhana

Abstract An important challenge for TB investigators in the postgenomic era is to integrate distinct functional strategies to study the molecular mechanism of *Mycobacterium tuberculosis* (*Mtb*) virulence. However, the biological function of the majority of *Mtb* genes is unknown. This has revealed the need for an approach to convert raw genome sequence data into functional information. In the past decade, the yeast two-hybrid system (Y2H) has contributed significantly towards studying TB virulence and persistence, but has several drawbacks. Recently, several mycobacterial protein–protein interaction (PPI) technologies have been reported that helped propose functions for unknown proteins through "guilt by association" and will be discussed in this chapter. We will examine the advantages, disadvantages and limitations of these systems and how these technologies can be used to dissect signaling, drug resistance, and virulence pathways. We will also discuss how mycobacterial PPI technologies can be exploited to force proteins to interact and for the discovery of small-molecule inhibitors against protein complexes. In sum, by characterizing *Mtb* PPIs on a genomic scale, it will be possible to assemble physiologically relevant protein pathways in mycobacteria, the outcome of which will be invaluable for determining virulence mechanisms and the function of previously uncharacterized proteins.

A.J.C. Steyn (✉)
Department of Microbiology, University of Alabama at Birmingham, Birmingham, AL 35294, USA

Center for AIDS Research, University of Alabama at Birmingham, Birmingham, AL 35294, USA and

KwaZulu-Natal Research Institute for Tuberculosis and HIV, Congella, Durban, South Africa
e-mail: asteyn@uab.edu

D. Mai • V. Saini • A. Farhana
Department of Microbiology, University of Alabama at Birmingham, Birmingham, AL 35294, USA

J. McFadden et al. (eds.), *Systems Biology of Tuberculosis*,
DOI 10.1007/978-1-4614-4966-9_5,

Abbreviations

BACTH	Bacterial adenylate cyclase two-hybrid
BM	BacterioMatch
hDHFR	Human dihydrofolate reductase
Mbov	*Mycobacterium bovis*
M-PFC	Mycobacterial protein fragment complementation
Msm	*Mycobacterium smegmatis*
Mtb	*Mycobacterium tuberculosis*
NO	Nitric oxide
PFC	Protein fragment complementation
PPI	Protein–protein interaction
RNAi	RNA interference
RNAP	RNA polymerase
TRX	Thioredoxin
Y2H	Yeast two-hybrid system
Y3H	Yeast three-hybrid system

1 Introduction

Mycobacterium tuberculosis is an extremely successfully pathogen due to its ability to persist, and to latently infect more than one-third of the world's population [1, 2]. Annually, there are approximately eight million new cases of TB and two million deaths worldwide. The increase in multidrug-resistant (MDR), extensively drug-resistant (XDR) and super XDR *Mycobacterium tuberculosis* (*Mtb*) strains, together with the synergy with HIV infection is a frightening development [2, 3] and poses significant problems in the treatment and control of TB.

Genome-scale molecular networks such as protein interaction and gene regulatory pathways are taking a center stage in the emerging disciplines of systems biology and biocomplexity. As a result, an important challenge for TB investigators in the postgenomic era is to integrate functional strategies such as allelic replacement techniques [3–5], signature tagging mutagenesis [6, 7], in vivo expression technology [8, 9], proteomics, [10, 11], DNA microarrays [12–16], deep-genome sequence strategies [17, 18] and protein–protein interaction (PPI) approaches [19, 20], to study the molecular mechanism of *Mtb* virulence.

To fulfill their biological function in cells, most proteins function in association with protein partners or as large molecular assemblies. Not surprisingly, virulence pathways are also mediated by molecular connections that require PPIs. The rationale for studying PPI in bacterial pathogens such as *Mtb* is several fold. First, in dissecting these pathways, it has been established that physical association between a protein of unknown function and a known protein suggests that the former often has a function related to that of the latter. This "guilty by association" principle has led to the functional annotations of numerous proteins of unknown function.

Since over the past decade more than 1,000 microbial genomes have been sequenced it is anticipated that the focus on genes of unknown function will continue to increase, as it is these genes (of unknown function), which make the particular microbe unique. Second, an important feature of PPI networks is that most proteins associate with multiple interacting partners, suggesting that they fulfill multiple functions. Third, elucidation of PPI can rapidly provide detailed mechanistic information about a specific biological question. The above-mentioned approaches usually attempt to identify new drug targets, or to achieve a better understanding of the mechanistic basis of *Mtb* virulence.

While substantial efforts focused on prediction of protein–protein association by in silico analysis using phylogenetic profiles [21], domain fusion [22], and gene clustering methods [23, 24], these types of analyses must be supported by biological experimentation. Not surprisingly, due to the large number of PPIs studies over the past 20 years, a large number of protein interaction databases such as HPRD, DRP, MIPS, STRING, BIND [25] have been generated.

Mtb is a genetically intractable microbe and there is an urgent need to develop effective genome-wide tools to study protein–protein association in mycobacterial cells. By characterizing PPIs on a genomic scale it will be possible to assemble physiologically relevant protein pathways in mycobacteria, the outcome of which will be invaluable for determining the function of previously uncharacterized proteins and virulence mechanisms.

Thus far, despite the development of bacterial systems (BacterioMatch [BM] and bacterial adenylate cyclase two-hybrid [BACTH] and the mammalian two-hybrid system [M2H]), *Saccharomyces cerevisiae* is the most exploited surrogate host and represents the current standard. The first large-scale yeast two-hybrid (Y2H) interaction network was performed with the *Escherichia coli* bacteriophage T7 [26] and was rapidly followed by a whole-genome analysis of *S. cerevisiae* [27–30], *Drosophila* [31, 32], *Arabidopsis* [33], and *C. elegans* [34]. These studies predicted the function of a multitude of proteins and revealed numerous novel interactions, thereby allowing investigators to link biological functions together into larger cellular processes.

In this review, we will provide an overview about the different PPI techniques that have successfully been exploited to study *Mtb*. We will discuss different mycobacterial PPI technologies, how it could be exploited for the discovery of new antimycobacterial drugs, potential pitfalls of PPI technologies, and in silico methods for predicting PPI.

2 Microbial PPI Systems

2.1 *The Y2H System*

In the original Y2H assay, a bait protein is fused to the GAL4 DNA-binding domain (DNA-BD), and a library of prey proteins are expressed as fusions to the GAL4 activation domain (AD) [35] (Fig. 5.1a). When the "bait" protein interacts with a

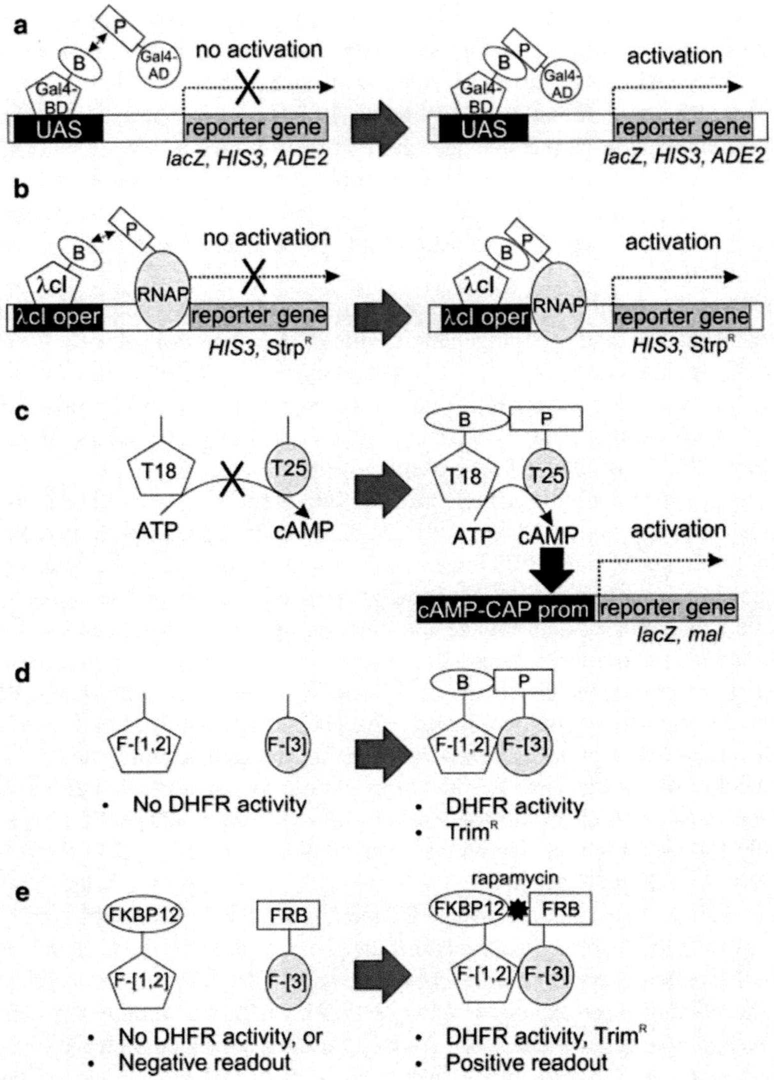

Fig. 5.1 Conceptual basis of PPI methods used to study protein function in mycobacteria and other pathogens. In the (**a**) Y2H, (**b**) BM, (**c**) BACTH, and (**d**) M-PFC systems the two interacting proteins (bait [B] and prey [P]) are independently fused to either a DNA-AD (e.g., Gal4-AD, α-subunit of RNAP) and DNA-BD (e.g., Gal4-BD, λcI) (**a**, **b**) or to two enzymatic subunits (**c**, **d**) that reconstitute enzymatic activity (e.g., AC or hDHFR). However, in the case of RAP-inducible M-PFC (**e**), rapamycin functions as a bridge that forces FKBP12 and FRB to "interact," thereby functionally reconstituting the reporter system consisting of F-[1,2] and F-[3] to generate Trim^R mycobacterial clones. Note that in case of **e**, F-[1,2] and F-[3] can be replaced by any two proteins that the investigator wishes to force to interact. Although the above systems examine bimolecular protein interactions, (**a**), (**b**), and (**d**) have been modified to examine tri-molecular protein interactions (see text for detail). *UAS* upstream activating sequence, *cAMP-CAP prom* cAMP-CAP promoter, *λcI oper* λcI operator, *T18 and T25* adenylate cyclase enzymatic domains

"prey" protein from the library, the DNA-BD and AD are brought into proximity to activate transcription of several reporter genes (e.g., *ADE2*, *HIS3*, *MEL1*, and *AUR1*). The Y2H system is an effective tool use to identify novel protein interactions, confirm putative interactions, and define interacting domains and residues. Subsequent to the development of the original Y2H system, the reverse Y2H [36] and yeast three-hybrid (Y3H) [37] systems were developed.

2.2 The E. coli BacterioMatch System

Similar to the Y2H system, the BM two-hybrid system is designed to examine PPIs between a pair of proteins cloned into separate "bait" and "prey" vectors. The bait protein is fused to the full-length bacteriophage λ repressor protein (λcI) containing the amino terminal DNA binding domain and the carboxyl terminal dimerization domain (Fig. 5.1b). When produced inside cell, the bait fusion is tethered to the operator sequence upstream of the reporter promoter through the DNA-BD of λcI [38]. The target or prey protein is fused to the N-terminal domain of the α-subunit of RNA polymerase (RNAP). When the bait and prey proteins associate, they recruit and stabilize binding of RNAP at the promoter and activate transcription of the *HIS3* and *aadA* (confers streptomycin resistance) reporter genes [39].

Very recently, the BM system was modified to study ternary mycobacterial protein complexes in *E. coli* [40]. Using this three-hybrid system, it was demonstrated that the interaction between CFP-10 and Rv3871 was strengthened in the presence of Esat-6. Lastly, the BM system was also used to examine PPIs between *Mtb* proteins and approximately 8,000 novel interactions were discovered [41]. Notably, validation of PPI using overexpression and surface plasmon resonance analyses demonstrated a success rate of approximately 60 %. Important findings include demonstrating a link between the *Mtb* ESX1 and ESX5 protein secretion systems, and that the Fe–S cluster proteins WhiB3 and WhiB7 are highly connected [41].

2.3 The Bacterial Adenylate Cyclase Two-Hybrid System

In the BACTH system, proteins of interest are fused with two fragments of the catalytic domain of the *Bordetella pertussis* adenylate cyclase (AC) and co-expressed in an *E. coli* Δ*cya* strain (Fig. 5.1c). Interaction of the two proteins results in the functional complementation between the two AC subunits, leading to cAMP synthesis and subsequent activation of catabolic operons [42] or the expression of the *lacZ* gene. Using BACTH, it was demonstrated that *Msm* PsPpm2 interacts with MsPpm1 to stabilize the synthase MsPpm1 in the bacterial membrane [43]. The BACTH system was also successfully used to examine the interactions between *Mtb* ClpX and FtsZ [44].

2.4 Protein Fragment Complementation

Recently, a different experimental system, coined protein fragment complementation (PFC), was shown to be highly effective in studying PPIs in a variety of organisms [45–48]. In PCF, a particular reporter enzyme is rationally dissected into two fragments and fused with two interacting proteins. Interaction among the two proteins results in active refolding and reconstitution of the enzyme activity of the two fragments. Since nuclear translocation of interacting proteins is not required, membrane proteins can also be analyzed.

For example, using human dihydrofolate reductase (hDHFR), any two proteins (X and Y) thought to interact are fused to two rationally dissected DHFR fragments called F-[1,2] and F-[3]. In vivo reassembly due to the interacting proteins X and Y, and subsequent reconstitution of hDHFR domains X-F-[1,2] and Y-F-[3] into active hDHFR can be monitored in vivo by cell survival under methotrexate selection, by fluorescence detection of fluorescein-conjugated methotrexate binding to reconstituted hDHFR, or by trimethoprim (Trim) resistance (Fig. 5.1d). hDHFR is a small 21-kDa monomeric protein that contains three structural fragments (F-1, F-2, and F-3) containing two domains; an adenine-binding domain (F-2) and a discontinuous domain (F-[1] and F-[3]). Previously, it has been shown that disruption of the disordered loop at the junction between F-[2] and F-[3] has no significant effect on activity [49]. This property was exploited to develop a eukaryotic DHFR PFC system to analyze reassembly of murine dihydrofolate reductase (hDHFR) fragments [47, 50, 51]. Using eukaryotic DHFR PFC, 148 combinations of 35 different PPIs in the RTK/FRAP signal transduction pathway were studied with no false-positive interactions observed among the pairs tested [47]. Importantly, the DHFR PFC system (albeit eukaryotic) is the only system that could validate the interactions through pharmacological perturbation of the interactions—even if the site of action of the perturbant is distant from the interaction studied [47].

The concept of PFC using hDHFR fragments was successfully exploited to develop a mycobacterial PFC system termed mycobacterial PFC (M-PFC) [52] (Fig. 5.1d). Using this system, the interactions between the two-component proteins DevS and DevR, and KdpD and KdpE were demonstrated. In addition, several previously undiscovered proteins were shown to interact with *Mtb* Cfp-10. Notably, proteins complexes were identified that form only in mycobacteria and not in the Y2H system [52]. It is likely that many interacting *Mtb* proteins will require the mycobacterial cytoplasmic environment to associate and is an important consideration in a PPI screen. In an independent study, M-PFC identified a strong interaction between Pup and the proteasome substrate FabD (malonyl coenzyme A acyl carrier protein), whereas this interaction was not detected using *E. coli* as surrogate host [53]. M-PFC was successful in demonstrating interaction between an essential DNA-binding protein (IdeR) and the enzymatic complexes (LeuC/LeuD) [54]. Lastly, M-PFC was also used to demonstrate interaction between *Mtb* ClpX and FtsZ [44].

The split-Trp system is another PFC assay that monitors the enzymatic reconstitution of tryptophan biosynthesis in a tryptophan autotrophic microbe. This system was originally developed in *S. cerevisiae* [55] and shown to be effective for examining the *Mtb* protein complexes Esat-6/CFP-10, RegX3 homodimerization, self-association of Rv3782 (galactosyl transferase), and the coiled-coil peptides C1 and C2 [56].

3 Shared Properties of Microbial Interaction Systems

The bacterial and Y2H systems (or variations thereof) have several properties in common that can profoundly affect postscreen analyses. For example, all systems have relatively strong promoters (Y2H: $ADH1_p$ or *GAL10*; M2H: CMV; BM: *lac-UV5*; BACT: *lac-UV5* and M-PFC: *hsp60*), and all PPI systems are based upon fusion technologies (Y2H; GAL4 AD and BD or LexA; M2H; JAK and GP130; BM; α-subunit of RNAP and λcI; BACT; AC; and M-PFC; DHFR). Lastly, all systems rely on unique peptide detection tags (HA, cMyC, FLAG, His, GP120, etc., or the reporter domains itself) to enable specific detection.

The two most widely used protein interaction validation approaches are to fuse a detection tag (e.g., GST) to the "bait" protein, in vitro transcribe/translate the "prey" protein followed by incubation of the mixtures and assessment of binding/elution of the labeled prey protein. A second widely-used validation experiment includes in vivo co-affinity purification in which one protein is tagged, overexpressed in *E. coli* (or native host) followed by a pull-down of the prey from the extract.

Important differences between the BM and PFC systems are that: (1) in PFC, protein interaction does not need to take place near the transcription machinery, (2) PFC is better suited for studying interactions among membrane proteins, (3) PFC requires no other host-specific processes or enzymes, (4) the structure of DHFR is known thereby allowing control over the way interactions can occur, and (5) it is advantageous to employ PFC in the native host rather than surrogate hosts such as yeast or *E. coli* wherein protein interactions are determined in the native host where they function in the context of other native proteins.

4 Is Yeast the Optimal Host for Studying Mycobacterial PPIs?

As is described in the section below, the Y2H system has been used successfully to study *Mtb* biology and pathogenesis. While in some cases it might be beneficial to use yeast as surrogate host, the Y2H system does have certain limitations. For example, (1) protein interactions occur in the nucleus, (2) membrane proteins are not fully compatible with the conventional Y2H system, (3) bacterial proteins do not undergo appropriate post-translational modification, (4) self-activation of bait proteins can occur, and finally, (5) high G+C DNA is sometimes not well tolerated in the Y2H system. A well-known class of Y2H false positives is "anti-sense" clones that contain anti-sense DNA fragments cloned in the library vector that when translated produce a nonphysiological peptide that associates with the bait protein. False positives are inherently present in all large-scale Y2H screens and are extensively documented in the literature.

5 Specificity and False Positives of PPI Technologies

In large-scale PPI studies, technical and biological false positives are typically being considered. Technical false positives resulting in experimental errors can be avoided. However, in order to eliminate false positives (e.g., those interacting clones that are genuinely observed in more than one assay, but do not occur in vivo) and to increase the verification rate, the following factors are taken in consideration during large-scale protein interaction screens: (1) overlapping (interacting) clones increases the confidence score [28, 31, 57], (2) literature curated interactions increases the confidence score [31, 57], (3) membrane proteins are underrepresented and negatively affect the confidence score [31], (4) post-translational modifications (e.g., phosphorylation) may be required for many interactions, (5) verification with other independent techniques increase the confidence score [27], (6) "masking" of bait or prey proteins and "self-activation" affect the screens [28, 31], (7) logistic regression models increase the probability of interactions [31], and (8) most studies validate interactions using detection tags or reporter fusions for "pull-down assays" [57].

6 How Can PPI Technologies Help Us Understand *Mtb* Virulence?

PPI technologies are flexible approaches that typically allow investigators to address previously unanswered questions. This is important to the mycobacteriology field as *Mtb* is a genetically intractable microbe for which few novel tools to determine virulence mechanisms are available. A widely cited rationale for exploiting PPI technologies in microbes is to ascribe function to genes of unknown function (e.g., those genes that are unique to the organism). It can be speculated that these genes distinguish the particular species from all other species and play a unique role in the biology of the microbe. Other areas in which PPI technologies can play an timportant role include the identification and dissection of virulence pathways, linking virulence pathways with each other, and examining the components of signaling cascades and drug resistance pathways. Particularly relevant to the study of *Mtb* is the effect of in vivo environmental conditions implicated in *Mtb* persistence (e.g., temperature, pH, NO, superoxide, etc.) on protein–protein association. Other important areas include the effect of post-translational modifications on *Mtb* PPI, screening for drugs that disrupt PPI, and construction of a complete protein linkage map of the *Mtb* proteome.

7 Impact of the Y2H System on *Mtb* Research

Over the past decade, PPI technologies have filled an important void in the mycobacterial field and opened up new avenues of TB research. The original discovery of *Mtb* WhiB3 in 2002 using the Y2H system [19] is a particularly good example for how PPI technologies can advance a particular research area.

7.1 Mtb WhiB3

It was previously established that a single-point mutation in the 4.2 region of the principal σ-factor *rpoV* causes loss of virulence in *Mycobacterium bovis* (*Mbov*), a member of the *Mtb* complex [58]. This mutation, known to result in an Arg_{515}-His change, was originally suggested to influence recognition of the –35 promoter region that abolished or altered expression of a gene or subsets of genes essential for virulence. However, it was hypothesized that this mutation might alter the interaction of RpoV with a transcription factor responsible for regulating the expression of one or more genes involved in virulence. An abundance of data have shown that mutations in, or close to the helix-turn-helix motif in region 4.2 of bacterial δ^{70}-type sigma factors results in either positive or negative effects on activation by transcription factors. Subsequently, it was hypothesized that the 4.2 domain of *Mtb* SigA interacts with a regulatory protein that controls a subset of genes involved in virulence. To screen for proteins that interact exclusively with the 4.2 region of SigA in which the Arg_{515}-His mutation is localized, a *sigA* DNA fragment (spanning region 4.2) was screened against a *Mtb* library using the Y2H system. Several clones contained in-frame fusions with the full-length open reading frame of *Mtb whiB3* (Rv3416) [19]. Since it was initially hypothesized that the Arg_{515}-His mutation abolished or reduced interaction of an unknown transcription factor with the 4.2 region of SigA, it was shown that $SigA_{R515-H}$ does not interact with WhiB3, suggesting that the single Arg_{515}-His mutation abolishes the interaction of WhiB3 with SigA. Knock-out studies have shown that the *Mtb whiB3* mutant behaved identically to the wild-type strain with respect to its ability to replicate in mice, but was attenuated in terms of host survival. In addition, the *whiB3* mutant strain showed much reduced lung pathology, compared to wild type infected mice [19]. Intriguingly, a *whiB3* mutant of virulent *Mbov* was completely impaired for growth in guinea pigs. These mutants define a new class ("path"; pathology) of virulence genes in *Mtb* and *Mbov*. It is notable that this virulence gene would not have been detected using conventional screens such as signature-tagged mutagenesis, which screen primarily for mutants defective in growth and not virulence. Notably, these findings led to the identification of WhiB3 as a 4Fe–4S cluster protein that reacts with NO and O_2 [59], and is implicated in the metabolic switchover from using glucose as carbon source to fatty acids. *Mtb* WhiB3 was also shown to regulate virulence lipid production, function as an intracellular redox sensor [60] and prevent the bacillus from experiencing reductive stress during infection of macrophages [61] (for a recent review on *Mtb* WhiB3, see [62]). The above findings illustrate the power of PPI technologies to study virulence mechanisms in a genetically intractable pathogen.

7.2 Secretion

Mtb Esat-6 and Cfp-10 are important secreted antigens that are part of the ESX-1 secretion system, which delivers virulence proteins during infection of host cells [63]. These small proteins interact strongly with each other as well as several other *Mtb* proteins. In recent years, the Y2H system has been particularly effective in mapping *Mtb* ESX PPIs [64], and identifying and characterizing the individual components of the ESX-1 secretion system [64–68], which has led to new testable hypotheses. A substantial advance in our understanding of Mtb protein secretion was the discovery of a C-terminal signal sequence in Cfp-10 using the Y2H system. This C-terminal signal sequence was shown to be necessary for targeting Cfp-10 and Esat-6 for secretion. Besides, the C-terminal seven amino acid signal sequence was sufficient for targeting unrelated proteins such as ubiquitin for secretion [65].

7.3 Mtb Two-Component Signaling Proteins and Sigma Factors

Two-component signal transduction pathways are typically comprised of a membrane bound histidine kinase and its cognate cytoplasmic response regulator. In response to a signal, auto-phosphorylation occurs at a conserved residue of the histidine kinase and subsequently the phosphate group is transferred to the conserved aspartate residue of the response regulator. Even though these interactions are likely transient the Y2H system was effective in examining these interactions. Interactions among different domains of *Mtb* HK1 (Rv0600c), HK2 (Rv0601c), and TcrA (Rv0602c) were examined using the Y2H system [69]. It was found that HK2, but not HK1 or TcrA self-interacted, and that HK2 interacted with HK1 and TcrA. Lastly, the conserved receiver domain of TcrA was shown to interact with HK2, but not HK1 [69]. In another study the Y2H system was used to identify proteins that interact with the sensing domain of the *Mtb* histidine kinase, KdpD. Two membrane lipoproteins, LprJ and LprF, were identified that specifically associated with KdpD [20].

Mtb contains 13 sigma factors that can associate with one or more components of RNAP (RpoB, RpoB′, α-subunit) under distinct environmental conditions. In addition, anti-anti-sigma factors can interact with anti-sigma factors (e.g., RsbW) or sigma factors. In extensive Y2H studies, it was shown that most anti-sigma factor antagonists interact with either RsbW or SigF or both [70]. In a separate study, it was shown that SigK positively regulates expression of the antigenic proteins MBP70 and MBP83 [71]. High-level expression of *sigK* was associated with a mutated Rv0444c, and Y2H analysis demonstrated that the N-terminal region of Rv0444c interacted with SigK. The authors concluded that Rv0444c functions as a regulator of SigK (RskA) that modulates MPT70/MPT83 expression [71]. As described earlier the principle sigma factor, SigA was used in Y2H screen to identify the virulence factor WhiB3 [19].

7.4 DNA Repair

The Y2H system has been effectively exploited to study DNA damage and repair in *Mtb* [72–75]. For example, in a genome wide screen UvrD1 was identified as a novel interacting partner of Ku, suggesting potential cross-talk between components of nonhomologous end-joining and nucleotide excision repair pathways [74]. In another study that examined the role of *Mtb* DinB homologs in DNA damage, Y2H analyses showed that DinB1, but not DinB2 interacts with the mycobacterial β clamp, which is consistent with its C-terminal DNA-binding motif [73]. In a related study Y2H analysis showed that ImuB interact with ImuA′ and DnaE2 as well as with the β clamp [75].

7.5 Other

The Y2H system has also been used to identify and characterize interacting partners of *Mtb* WhiB1, an iron–sulfur cluster protein [76], the *Mtb* SUF machinery [77, 78], components of FASII (KasA, KasB, mtFabH, InhA, and MabA) [79, 80], the ABC transporter Rv1747 [81], resuscitation promoting factors (Rpfs) [82], a GTP binding protein (Obg) [83], and VapBC toxin-antitoxin modules [84].

8 Protein–Protein Interaction in Other Pathogens

PPI networks of bacteria have not yet reached the same comprehensive level as their yeast counterpart. An exception is the protein network of the human gastric pathogen *Helicobacter pylori* [85]. A high-throughput Y2H systems was used to screen 261 *H. pylori* proteins against a highly complex library of genome-encoded polypeptides and yielded over 1,200 interactions; connecting 46.6 % of the proteome [85]. The success of this approach in detecting new protein interactions and assignment of previously un-annotated proteins to new pathways lead to many such studies using the Y2H system to develop PPI maps for *Plasmodium falciparum* [86], *Rickettsia sibirica* [87], *Bacillus anthracis*, *Francisella tularensis*, *Yersinia pestis* [88], *Campylobacter jejuni* [89], *Treponema pallidum* [90] and viruses including HIV and HCV [91]. Unfortunately, these high-throughput screens are plagued by many drawbacks including false positives and negatives, and the temporal or spatial requirement of expression and post-translational modifications. Consequently, high-throughput PPI approaches have been augmented by the addition of techniques such as protein arrays and mass spectrometry [92–94].

While evaluating intra-bacterial PPIs provides a unique resource to identify essential cell processes and protein targets for drug screens against pathogenic bacteria, assessing interactions between host and bacterial proteins are imperative for understanding the mechanism of disease pathogenesis. Using a high-throughput Y2H screen, extensive host–pathogen PPIs have been identified for the pathogens *B. anthracis*, *F. tularensis*, and *Y. pestis*. Though the three pathogens cause different diseases (anthrax, lethal acute pneumonic disease and bubonic plague, respectively), PPIs pointed to similar mechanisms of immune modulation. For example, both *B. anthracis* and *Y. pestis* proteins interact with host major histocompatibility complex proteins, whereas TGF-β1 was shown to interact with *Y. pestis* and *F. tularensis* proteins. In sum, a network of 3,073 human-*B. anthracis*, 1,383 human-*F. tularensis*, and 4,059 human-*Y. pestis* PPIs were identified. The networks included 304 uncharacterized proteins from *B. anthracis*, 52 from *F. tularensis*, and 330 from *Y. pestis* [88].

Using three datasets that include physical interaction assays, genome-wide RNA interference (RNAi) screens, and microarray assays, the first draft of the mosquito PPI network was developed for the Dengue virus (DENv) carrier. This PPI network included 4,214 *Aedes aegypti* proteins with 10,209 interactions [95]. The study identified 714 putative DENv-associated mosquito proteins, and RNAi-mediated gene silencing of some of the highly interconnected proteins reduced the dengue viral titer in mosquito midgets. This observation further underscores the importance of identifying critical host–pathogen PPIs, which can provide an immense resource for identifying prospective antimicrobial drug targets.

In an attempt to characterize essential cellular process in *Bacillus subtilis*, a PPI network was generated that comprised 793 interactions that connected 287 proteins. Further evaluation of these hubs provided insights into distinct subgroups of PPI corresponding to protein networks or regulatory pathways differentially expressed under diverse conditions [96]. These PPI network data are a valuable resource for the functional annotation of genes of unknown function and integration of cellular pathways.

In addition to the Y2H system, high-throughput pull-down strategies combined with quantitative proteomics have also been used to decipher interacting circuits in methicillin-resistant *Staphylococcus aureus* [97]. Several highly connected hub proteins were identified. Notably, examination of the PPI network of *S. aureus* drug targets indicated that most of the clinical or experimental drugs targets lie at the periphery of the interacting circuit with few interacting partners. In contrast, the proteins that lie at the network hub, which could logically serve as a better target, were overlooked as drug targets [97].

9 Considerations for Mycobacterial PPIs

9.1 Some Mycobacterial Proteins Interact Exclusively in Their Native Environment

Although many bacterial PPIs have been identified in the Y2H system, it is logical to expect that some bacterial protein interactions may require the native cytoplasmic or membrane environment. For example, using M-PFC some *Mtb* proteins were shown to only interact in mycobacterial cytoplasmic environment, but not in yeast [52]. In an independent study, detection of interactions between Pup and other *Mtb* proteasome components in *E. coli* was unsuccessful. However, using M-PFC and therefore *Msm* as host, a strong interaction was observed between Pup and the proteasome substrate FabD [53].

9.2 Some Mycobacterial Proteins Require More Than One Protein for Interaction

Since all in vivo PPI methods (with the exception of the Y3H system) are binary systems that can detect interaction between only two proteins, interacting partners that require the presence of two or more proteins might be missed. In a recent study that made use of a modified M-PFC system, the *Mtb* ESX secretory system was examined by using a single fusion protein comprised of EsxB and EsxA as bait [98]. Three novel prey proteins, Rv3869, Rv3884 and Rv3885 were identified, whereas the single bait protein EsxB was unable to interact with any of these three proteins [98]. Exploiting fusion proteins that naturally associated in mycobacteria as bait has broad implications for the characterization of *Mtb* protein complexes, and may open new avenues of research.

The Y3H system was also exploited to delineate the molecular interactions between two membrane proteins and the *Mtb* two-component sensor kinases KdpD [20]. In this system, a third protein acts as a bridge between two proteins and can stabilize, enhance, or prevent interaction between proteins. The third protein is under the control of the inducible methionine promoter that is positively regulated in media lacking methionine. In this study LprJ and LprF were shown to modulate the interaction between N-KdpD and C-KdpD, and it was speculated that it is this ternary protein complex that modulates the

KdpE-specific phosphatase activity of KdpD to regulate the expression of the KdpFABC system [20].

9.3 Post-translational Modification Can Affect PPI in the Y2H System

In 2003, a Y2H assay was developed to examine nitric oxide (NO)-dependent PPI [99]. Deleting yeast hemoglobin, which consumes NO very efficiently, was essential to the success of this approach. In this study, the authors screened a library of proteins that interact with procaspase-3 only in the presence of NO and identified four clones, iNOS, ASM, IRG and PGM [99]. These findings suggest that S-nitrosylation regulates PPI and may profoundly influence cellular signaling.

In another study, in vitro proteomic analysis identified numerous thioredoxin (TRX) targets. However, in vivo approaches failed to identify the expected number of TRX targets [100]. This problem was solved by constructing a specific yeast strain that contains deletions of genes encoding cytosolic TRX1 and TRX2. Subsequently, numerous TRX interacting partners were identified, whereas the same interactions could not be detected in the classic Y2H strain [100]. The above findings are highly relevant for studying mycobacterial PPIs, and illustrate the fundamental concepts that (1) proteins only interact when functionally required, (2) essential genes can be studied since genetic knockouts are not required, and (3) genes that are transcriptionally switched off can be studied since constitutive promoters are being used in the PPI systems.

10 Molecules That Dissociate or Force Protein–Protein Interaction

Despite recent successes [101–103], no new effective anti-tuberculosis drugs have been developed in the past 40 years. As a result, a high priority of the Global Alliance for TB Drug Development is the generation of new drugs with activity against dormant bacilli as well as the discovery of agents which could shorten or simplify the treatment of active TB. TB can be cured with existing drugs; however, the 6–9 months of treatment lead to patient noncompliance, which enhances drug resistance. Approximately 50 million people are already infected with MDR-TB [2]. While drug-sensitive TB can be cured with isoniazid, rifampin, ethambutol and pyrazinamide following a 6-month regimen, treatment of MDR-TB can exceed 2 years, thus dramatically increasing costs.

How can PPI in pathogenic microbes contribute to the discovery of new drugs? Tightly regulated PPIs are required for cellular functions in all living systems. The necessity for proper protein placement within enzymatic and receptor–ligand complexes, cell signaling pathways, and PPIs lend to the appeal of disruption of critical PPIs as therapeutic intervention. However, it should be noted that PPIs that participate in virulence or persistence pathways may only be induced in vivo and therefore, not be susceptible to drugs in in vitro screens and must be identified through other means. PPIs in particular share complimentary interfaces and "hot spots" with one another [104, 105], in which the primary forces that drive two proteins to interact are: van der Waal's forces, electrostatic interactions, hydrogen bonds, and hydrophobic interactions [106–110]. Successful "disruption" of these interactions by an inhibitor, while not necessarily always in the context of protein–protein separation would be considered as any compound which modulates a protein interacting complex to achieve a desired therapeutic outcome and/or downstream effect.

Among several well-known inhibitors that modulate PPI, common mechanisms of action have emerged: prevention of PPI via protein binding, allosteric inhibition and, forced dissociation and association. More importantly, their method of action differs from drugs that prevent substrates from binding to active sites on enzyme complexes, as these sites are often marked with clear, defined pockets [111]. PPI inhibitors can include peptides, drugs, and small molecule compounds. These PPI inhibitors exert their functions over a range of target protein complexes in different cell types and have been reviewed over recent years [112–115].

Many inhibitors that prevent protein interaction have been shown to bind with amino acid residues that comprise "hot spots" at the protein–protein interface. Inhibitors form a complex with a protein at the binding site to structurally alter or prevent natural association of the cognate partner protein. For example, structural biology studies revealed that nutlins, a series of *cis*-imidazoline analogs identified via high-throughput screening, act by binding to three dominant residues of the p53 binding site on MDM2 and display in vitro and in vivo antitumor activity [116, 117]. Virstatin is an example of a compound that targets the dimerization domain of ToxT, a homodimer that regulates the production of cholera toxin and toxin co-regulated pilus in *Vibrio cholera*. Bacterial two-hybrid assays with ToxT truncation mutants demonstrated that virstatin specifically targets the N-terminal dimerization domain of ToxT [118, 119].

PPI can also be modulated when compounds bind distally to the protein interaction interface, that cause structural changes that prevent PPI without competition for protein binding sites. Such allosteric modifications have been documented for compounds that inhibit iNOS dimerization [120, 121]. PPA250, BBS-2, and clotrimazole are compounds that bind to the heme cofactor in the protein active site, which subsequently distort the α-helices [120] or the 8b and 9b β-strands [121] to prevent iNOS dimerization. In addition, other examples of allosteric inhibitors have been demonstrated for CBFβ-RUNX1, LFA-1-ICAM-1, and β-lactamase [122–124].

Inhibitors that dissociate preexisting protein complexes are functionally different from those that prevent protein dimerization. The most notable example of this method is TNF-α, whose active complex is maintained as a homotrimer when bound to its receptor. He et al. [125] demonstrated that at low TNF-α concentrations, the compound SPD00000034 bound to the pre-associated TNF-α trimer and promoted the dissociation of the active complex into dimer and monomer subunits. Similarly, previous studies have shown that the GroEL multimeric complex can exist in an "open" state, which allows 4,4′-dithiodipyridene to bind to an otherwise inaccessible Cys^{458}, leading to GroEL subunit disassembly [126]. More recently, a proof-of-concept quantitative HTS screen was developed to screen for small-molecule inhibitors of *Mtb* PPI [54], which demonstrated the versatility of M-PFC.

Finally, several compounds modulate PPI by inducing the formation of previously unassociated complexes or by stabilizing protein complexes. Chemical inducers exist for the p66 and p51 subunits of HIV-1 reverse transcriptase [127]. However, the most well-known example of forced protein association comes from studies involving a physical relationship between immunophilins, their ligands, and their target [128]. FK506, rapamycin, and cyclosporine, are examples of hydrophobic, immunosuppressive ligands that contain two protein binding surfaces which mediate interactions between FKBP12 or cyclophilin [128, 129] and their corresponding target protein. In mammalian cells, the FKBP12–FK506 complex binds to and inhibits calcineurin phosphatase activity [130]. The FKBP12–rapamycin complex binds to the rapamycin binding domain (FRB) of FRAP [131]. The resulting complexes affect different immune responses and can lead to programmable physiological responses. Furthermore, these binding partners have led many researchers to exploit forcible ligand binding of effector molecules for the development of inducible PFC assays (PCA) [132–135]. In mycobacteria, a rapamycin inducible mycobacterial-PFC (RAP-inducible M-PFC) assay was developed as proof-of-concept to show forced interaction in bacterial cells, where FKBP12 and FRB were independently fused to the DHFR reporter fragments F-[1,2] and F-[3], respectively. Association of FKBP12 and FRB could only be detected in the presence of the selective drug trimethoprim and nanomolar concentrations of the rapamycin ligand [54]. Taken together, the M-PFC and the RAP-inducible M-PFC systems are powerful methods used to identify interacting proteins in protein networks, where in future studies, vehicles like FKBP12-ligand binding can be designed and utilized to manipulate PPIs in mycobacteria (Fig. 5.1e). In short, the ability of these effector molecules to bridge or induce dimeric and multimeric complexes paves the way for potential applications in controlling protein pathways for therapeutic and experimental studies [129, 136, 137].

11 In Silico Methods for Predicting PPI

Over the past few decades, knowledge of PPI has been generated primarily from biochemical and genetic experimentation approaches such as Y2H systems, pull-down assays, mass spectrometry, co-related mRNA expression, and protein arrays. However, despite the best attempts to collect experimental data on different organisms, the rate of discovery remains slow (e.g., approximately <10 % of interactions in humans have been experimentally characterized). With the advent of the genomic era, several computational and bioinformatics-based approaches have been developed to infer PPI. These in silico approaches exploit annotated information from established observations and use the structural, genomic, and biological context of proteins and genes from completely sequenced genomes to predict protein interaction networks [138, 139]. These in silico methods may rely on information gleaned from protein structure, gene sequence and the presence or absence of genes across numerous genomes, conserved gene neighborhoods across different species, co-expression of genes in transcriptome studies, involvement of proteins in a common metabolic pathway, curation of published literature or a combination of these datasets [140–145]. High-resolution three-dimensional (3D) structures of interacting proteins provide the best source of information with atomic description of the binding interfaces based on hydrophobicity, charge, and thermodynamic constraints of the interaction [140, 146]. Several approaches have been developed that include computational modeling of homologous proteins based on previously known structures, or domain or sequence signature analysis if the complete structure of a homologous protein is not available [147, 148]. For example, Inter PreTS (EMBL Heidelberg) is a popular resource, which for any pair of query sequences first searches for homologues in a database of interacting domains of known 3D complex structures [149]. Pairs of sequences homologous to a known interacting pair are scored for how well they preserve the atomic contacts at the interaction interface and a priority ranking is used to score for possible interacting partners.

A number of structure-based computational methods have been developed for the prediction of PPIs, which utilize advances in the field of genomics. One such popular approach, known as phylogenetic profiling [150], is based on the pattern of the presence or absence of a given gene in a given set of genomes. This method could, for example, ascertain the distribution of a specific gene in different species [151]. Any similarity of phylogenetic profiles might then be interpreted as being indicative of the functional need for corresponding proteins to be present simultaneously to perform a given function together. This approach stems from the idea that functionally linked proteins would co-occur in genomes and that the phylogenetic trees for known interacting protein families tend to show a higher degree of similarity than trees for noninteracting proteins. In several cases, the similarity in topology of phylogenetic trees has been considered as a positive indication towards establishing the likelihood of interacting proteins pairs, especially in the case of protein partners that may have co-evolved (mirror tree approach) [150]. Likewise, co-localization-based approaches are based on the notion that physically interacting (or functionally associated) proteins must co-evolve to preserve their ability to interact with one another [152]. This is especially relevant in the case of prokaryotes, which have operonic transcription units.

Genomic context-based approaches also exploit gene fusion events, which can be considered as the ultimate form of co-localization as the fusion of two independent genes to encode a single unrelated polypeptide (called a Rosetta stone protein) retains the physical proximity of the two peptides, but also makes them a single entity [142]. Publically available databases that provide support for gene context and co-localization analyses include FUSION DB, STRING and PHYDBAC.

Another robust tool implementing genome context-based analysis is based on the Integrated Microbial Genomes (IMG) database. The IMG provides one of the largest genome integrations, containing ~7,000 complete and draft genomes across all three domains of life [153]. Similarly, an in silico two-hybrid method has been proposed, based on the study of correlated mutations in multiple sequence alignments. In this method, pairs of multiple sequence alignments with a distinctive co-variation signal are analyzed based on the hypothesis that co-adaptation of interacting proteins can be detected by the presence of a distinctive number of compensatory mutations in the corresponding proteins of different species [154].

Similar to wet lab-based approaches, most computational approaches have intrinsic limitations [155]. For example, the success of most of sequence and genomic context-based approaches requires extensive analysis of completely sequenced genomes, whereas the success of phylogenetic tree-based methods depends on the number and distribution of genomes used to build the tree [156]. Similarly, gene fusion-based methods may be confounded by errors caused by the occurrence of lateral gene transfer events in prokaryotes and the longer multi-gene architecture of eukaryotes [157]. Likewise, despite providing the highest quality information on PPI, protein structure-based approaches are restricted in their scope because of the limited availability of high quality protein structures in the databases and the high cost associated with determination of protein structures.

There is a clear need to unify genome sequencing and functional genomics data using computational tools to minimize the discrepancies associated with the use of a single approach. Several worthy attempts have been made in this direction. In addition, there is an encouraging community-driven initiative in the form of guidelines such as "MIMIx" and "MIAPE, which are the minimum information required for reporting a molecular interaction experiment" or a proteomics experiment, respectively [158, 159]. Under this initiative, a checklist of information has been provided, which every scientist must furnish when describing experimental molecular interaction data in an article, displaying data on a website or depositing data directly into a public database.

12 Integrative Physiology: The Emergence of Systems Biology?

Proteins are the catalytic effectors that carry out the intent of the microbial cell, but protein levels do not necessarily correlate with gene expression. For example, a lack of correlation was found between mRNA level and the corresponding protein level in *Haemophilus influenza* exposed to antibiotics [160], increased cell density in *E. coli* cultures [161], *Bacillus subtilis* exposed to peroxide stress [162], exponentially growing *S. cerevisiae* cells [163], and *S. cerevisiae* exposure to lithium [164]. This demonstrates the challenges of correlating mRNA expression levels with protein levels, and highlights the role of post-transcriptional regulatory control. Furthermore, some studies have observed a disparity between gene expression profiles and metabolic flux. This was elegantly demonstrated by analysis of the transcriptome, metabolome, and fluxome of *Corynebacterium glutamicum* [165]. Integrating PPI data with complementary high-throughput techniques such as transcriptomes,

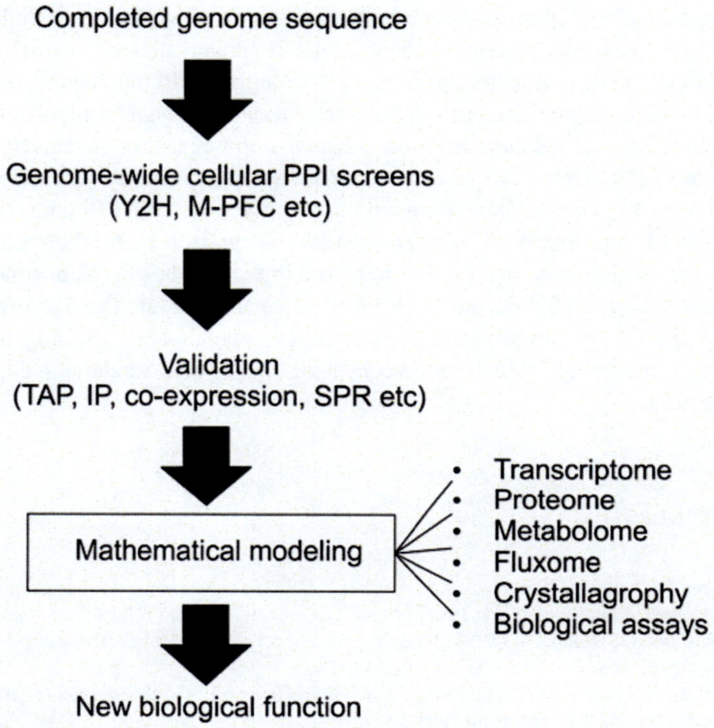

Fig. 5.2 Integrated analyses methodology depicting the role of PPI in TB systems biology. Towards this end, regulatory networks (gene expression arrays), proteomics, fluxomics and PPI networks have already begun to be established, but are commonly represented as static set of nodes to represent the components of the network (mRNA, proteins, metabolites, etc.). The ultimate goal will be to develop, test, and validate mathematical models that represent cellular components and their interactions to eventually predict cellular function. *TAP* tandem affinity purification, *IP* immunoprecipitation, *SPR* surface plasmon resonance

proteomics, metabolomics, and fluxomics represent unique opportunities to study and predict *Mtb* protein function through systems biology (Fig. 5.2).

13 Conclusions

Tuberculosis research is primarily driven by the quest for a better understanding of how *Mtb* causes disease. The past decade, the Y2H system and mycobacterial PPI technologies gave mechanistic insight into distinct aspects of *Mtb* virulence, pathogenesis and have stimulated antimycobacterial drug discovery efforts. PPI studies are particularly powerful to provide information about the function of genes with unknown function through "guilt by association." Not surprisingly, it is anticipated that the integration of functional data from PPI networks with the emerging discipline of systems biology could prove particularly useful to provide a better understanding of *Mtb* persistence. Although mycobacterial PPI networks have already begun to be established, the current focus is still on high-throughput PPI tool development, which is still lacking for mycobacteria. In addition, despite the generation of a single *Mtb* PPI map using *E. coli* as surrogate host, the more important stage of data interpretation, validation and integration with mycobacterial physiology is lacking. A future challenge would be to interconnect increasing amounts of mycobacterial PPI data with the PPI networks of other bacterial pathogens and its integration with other genome-wide databases, which should lead to new testable hypotheses. The generation of high-throughput global datasets will be an expensive venture that requires detailed knowledge about mycobacterial physiology, metabolism, pathogenesis, and computer modeling, which will contribute to a goal understanding of *Mtb* pathogenesis.

References

1. Arno PS, Murray CJ, Bonuck KA, Alcabes P (1993) The economic impact of tuberculosis in hospitals in New York City: a preliminary analysis. J Law Med Ethics 21(3–4):317–323
2. Murray CJ, Salomon JA (1998) Modeling the impact of global tuberculosis control strategies. Proc Natl Acad Sci USA 95(23):13881–13886
3. Pelicic V, Jackson M, Reyrat JM, Jacobs WR Jr, Gicquel B, Guilhot C (1997) Efficient allelic exchange and transposon mutagenesis in *Mycobacterium tuberculosis*. Proc Natl Acad Sci USA 94(20):10955–10960
4. Bardarov S, Kriakov J, Carriere C, Yu S, Vaamonde C, McAdam RA, Bloom BR, Hatfull GF, Jacobs WR Jr (1997) Conditionally replicating mycobacteriophages: a system for transposon delivery to *Mycobacterium tuberculosis*. Proc Natl Acad Sci USA 94(20):10961–10966
5. Bardarov S, Bardarov S Jr, Pavelka MS Jr, Sambandamurthy V, Larsen M, Tufariello J, Chan J, Hatfull G, Jacobs WR Jr (2002) Specialized transduction: an efficient method for generating marked and unmarked targeted gene disruptions in *Mycobacterium tuberculosis*, *M. bovis* BCG and *M. smegmatis*. Microbiology 148(Pt 10):3007–3017
6. Cox JS, Chen B, McNeil M, Jacobs WR Jr (1999) Complex lipid determines tissue-specific replication of *Mycobacterium tuberculosis* in mice. Nature 402(6757):79–83
7. Camacho LR, Ensergueix D, Perez E, Gicquel B, Guilhot C (1999) Identification of a virulence gene cluster of *Mycobacterium tuberculosis* by signature-tagged transposon mutagenesis. Mol Microbiol 34(2):257–267
8. Dubnau E, Fontan P, Manganelli R, Soares-Appel S, Smith I (2002) *Mycobacterium tuberculosis* genes induced during infection of human macrophages. Infect Immun 70(6):2787–2795
9. Dubnau E, Chan J, Mohan VP, Smith I (2005) Responses of *Mycobacterium tuberculosis* to growth in the mouse lung. Infect Immun 73(6):3754–3757
10. Mollenkopf HJ, Grode L, Mattow J, Stein M, Mann P, Knapp B, Ulmer J, Kaufmann SH (2004) Application of mycobacterial proteomics to vaccine design: improved protection by Mycobacterium bovis BCG prime-Rv3407 DNA boost vaccination against tuberculosis. Infect Immun 72(11):6471–6479
11. Mattow J, Schaible UE, Schmidt F, Hagens K, Siejak F, Brestrich G, Haeselbarth G, Muller EC, Jungblut PR, Kaufmann SH (2003) Comparative proteome analysis of culture supernatant proteins from virulent *Mycobacterium tuberculosis* H37Rv and attenuated *M. bovis* BCG Copenhagen. Electrophoresis 24(19–20):3405–3420
12. Schnappinger D, Ehrt S, Voskuil MI, Liu Y, Mangan JA, Monahan IM, Dolganov G, Efron B, Butcher PD, Nathan C, Schoolnik GK (2003) Transcriptional adaptation of *Mycobacterium*

tuberculosis within macrophages: insights into the phagosomal environment. J Exp Med 198(5):693–704

13. Sherman DR, Voskuil M, Schnappinger D, Liao R, Harrell MI, Schoolnik GK (2001) Regulation of the *Mycobacterium tuberculosis* hypoxic response gene encoding alpha-crystallin. Proc Natl Acad Sci USA 98(13):7534–7539
14. Voskuil MI, Schnappinger D, Visconti KC, Harrell MI, Dolganov GM, Sherman DR, Schoolnik GK (2003) Inhibition of respiration by nitric oxide induces a *Mycobacterium tuberculosis* dormancy program. J Exp Med 198(5):705–713
15. Voskuil MI, Visconti KC, Schoolnik GK (2004) *Mycobacterium tuberculosis* gene expression during adaptation to stationary phase and low-oxygen dormancy. Tuberculosis (Edinb) 84(3–4):218–227
16. Sassetti CM, Boyd DH, Rubin EJ (2003) Genes required for mycobacterial growth defined by high density mutagenesis. Mol Microbiol 48(1):77–84
17. Ford CB, Lin PL, Chase MR, Shah RR, Iartchouk O, Galagan J, Mohaideen N, Ioerger TR, Sacchettini JC, Lipsitch M (2011) Use of whole genome sequencing to estimate the mutation rate of *Mycobacterium tuberculosis* during latent infection. Nat Genet 43(5):482–486
18. Ioerger TR, Koo S, No EG, Chen X, Larsen MH, Jacobs WR Jr, Pillay M, Sturm AW, Sacchettini JC (2009) Genome analysis of multi-and extensively-drug-resistant tuberculosis from KwaZulu-Natal, South Africa. PLoS One 4(11):e7778
19. Steyn AJ, Collins DM, Hondalus MK, Jacobs WR Jr, Kawakami RP, Bloom BR (2002) *Mycobacterium tuberculosis* WhiB3 interacts with RpoV to affect host survival but is dispensable for in vivo growth. Proc Natl Acad Sci USA 99(5):3147–3152
20. Steyn AJ, Joseph J, Bloom BR (2003) Interaction of the sensor module of *Mycobacterium tuberculosis* H37Rv KdpD with members of the Lpr family. Mol Microbiol 47(4):1075–1089
21. Pellegrini M, Marcotte EM, Thompson MJ, Eisenberg D, Yeates TO (1999) Assigning protein functions by comparative genome analysis: protein phylogenetic profiles. Proc Natl Acad Sci USA 96(8):4285–4288
22. Marcotte EM, Pellegrini M, Thompson MJ, Yeates TO, Eisenberg D (1999) A combined algorithm for genome-wide prediction of protein function. Nature 402(6757):83–86
23. Overbeek R, Fonstein M, D'Souza M, Pusch GD, Maltsev N (1999) The use of gene clusters to infer functional coupling. Proc Natl Acad Sci USA 96(6):2896–2901
24. Dandekar T, Snel B, Huynen M, Bork P (1998) Conservation of gene order: a fingerprint of proteins that physically interact. Trends Biochem Sci 23(9):324–328
25. De Las RJ, De Luis A (2004) Interactome data and databases: different types of protein interaction. Comp Funct Genomics 5(2):173–178
26. Bartel PL, Roecklein JA, SenGupta D, Fields S (1996) A protein linkage map of Escherichia coli bacteriophage T7. Nat Genet 12(1):72–77
27. Ito T, Chiba T, Ozawa R, Yoshida M, Hattori M, Sakaki Y (2001) A comprehensive two-hybrid analysis to explore the yeast protein interactome. Proc Natl Acad Sci USA 98(8):4569–4574
28. Uetz P, Giot L, Cagney G, Mansfield TA, Judson RS, Knight JR, Lockshon D, Narayan V, Srinivasan M, Pochart P, Qureshi-Emili A, Li Y, Godwin B, Conover D, Kalbfleisch T, Vijayadamodar G, Yang M, Johnston M, Fields S, Rothberg JM (2000) A comprehensive analysis of protein-protein interactions in Saccharomyces cerevisiae. Nature 403(6770):623–627
29. Schwikowski B, Uetz P, Fields S (2000) A network of protein-protein interactions in yeast. Nat Biotechnol 18(12):1257–1261
30. Ito T, Tashiro K, Muta S, Ozawa R, Chiba T, Nishizawa M, Yamamoto K, Kuhara S, Sakaki Y (2000) Toward a protein-protein interaction map of the budding yeast: a comprehensive system to examine two-hybrid interactions in all possible combinations between the yeast proteins. Proc Natl Acad Sci USA 97(3):1143–1147
31. Formstecher E, Aresta S, Collura V, Hamburger A, Meil A, Trehin A, Reverdy C, Betin V, Maire S, Brun C, Jacq B, Arpin M, Bellaiche Y, Bellusci S, Benaroch P, Bornens M, Chanet R, Chavrier P, Delattre O, Doye V, Fehon R, Faye G, Galli T, Girault JA, Goud B, de Gunzburg J, Johannes L, Junier MP, Mirouse V, Mukherjee A, Papadopoulo D, Perez F, Plessis A, Rosse

C, Saule S, Stoppa-Lyonnet D, Vincent A, White M, Legrain P, Wojcik J, Camonis J, Daviet L (2005) Protein interaction mapping: a Drosophila case study. Genome Res 15(3):376–384
32. Giot L, Bader JS, Brouwer C, Chaudhuri A, Kuang B, Li Y, Hao YL, Ooi CE, Godwin B, Vitols E, Vijayadamodar G, Pochart P, Machineni H, Welsh M, Kong Y, Zerhusen B, Malcolm R, Varrone Z, Collis A, Minto M, Burgess S, McDaniel L, Stimpson E, Spriggs F, Williams J, Neurath K, Ioime N, Agee M, Voss E, Furtak K, Renzulli R, Aanensen N, Carrolla S, Bickelhaupt E, Lazovatsky Y, DaSilva A, Zhong J, Stanyon CA, Finley RL Jr, White KP, Braverman M, Jarvie T, Gold S, Leach M, Knight J, Shimkets RA, McKenna MP, Chant J, Rothberg JM (2003) A protein interaction map of Drosophila melanogaster. Science 302(5651):1727–1736
33. de Folter S, Immink RG, Kieffer M, Parenicova L, Henz SR, Weigel D, Busscher M, Kooiker M, Colombo L, Kater MM, Davies B, Angenent GC (2005) Comprehensive interaction map of the Arabidopsis MADS box transcription factors. Plant Cell 17(5):1424–1433
34. Li S, Armstrong CM, Bertin N, Ge H, Milstein S, Boxem M, Vidalain PO, Han JD, Chesneau A, Hao T, Goldberg DS, Li N, Martinez M, Rual JF, Lamesch P, Xu L, Tewari M, Wong SL, Zhang LV, Berriz GF, Jacotot L, Vaglio P, Reboul J, Hirozane-Kishikawa T, Li Q, Gabel HW, Elewa A, Baumgartner B, Rose DJ, Yu H, Bosak S, Sequerra R, Fraser A, Mango SE, Saxton WM, Strome S, Van Den Heuvel S, Piano F, Vandenhaute J, Sardet C, Gerstein M, Doucette-Stamm L, Gunsalus KC, Harper JW, Cusick ME, Roth FP, Hill DE, Vidal M (2004) A map of the interactome network of the metazoan *C. elegans*. Science 303(5657):540–543
35. Fields S, Song O (1989) A novel genetic system to detect protein protein interactions. Nature 340(6230):245–246
36. Vidal M, Brachmann RK, Fattaey A, Harlow E, Boeke JD (1996) Reverse two-hybrid and one-hybrid systems to detect dissociation of protein-protein and DNA-protein interactions. Proc Natl Acad Sci 93(19):10315
37. SenGupta DJ, Zhang B, Kraemer B, Pochart P, Fields S, Wickens M (1996) A three-hybrid system to detect RNA-protein interactions *in vivo*. Proc Natl Acad Sci 93(16):8496
38. Dove SL, Joung JK, Hochschild A (1997) Activation of prokaryotic transcription through arbitrary protein-protein contacts. Nature 386(6625):627–630
39. Dove SL, Hochschild A (2004) A bacterial two-hybrid system based on transcription activation. Methods Mol Biol 261:231–246
40. Tharad M, Samuchiwal SK, Bhalla K, Ghosh A, Kumar K, Kumar S, Ranganathan A (2011) A three-hybrid system to probe *in vivo* protein-protein interactions: application to the essential proteins of the RD1 complex of M. tuberculosis. PLoS One 6(11):e27503
41. Wang Y, Cui T, Zhang C, Yang M, Huang Y, Li W, Zhang L, Gao C, He Y, Li Y (2010) A global protein-protein interaction network in the human pathogen *Mycobacterium tuberculosis* H37Rv. J Proteome Res 9(12):6665–6677
42. Karimova G, Dautin N, Ladant D (2005) Interaction network among Escherichia coli membrane proteins involved in cell division as revealed by bacterial two-hybrid analysis. J Bacteriol 187(7):2233–2243
43. Baulard AR, Gurcha SS, Engohang-Ndong J, Gouffi K, Locht C, Besra GS (2003) In vivo interaction between the polyprenol phosphate mannose synthase Ppm1 and the integral membrane protein Ppm2 from Mycobacterium smegmatis revealed by a bacterial two-hybrid system. J Biol Chem 278(4):2242
44. Dziedzic R, Kiran M, Plocinski P, Ziolkiewicz M, Brzostek A, Moomey M, Vadrevu IS, Dziadek J, Madiraju M, Rajagopalan M (2010) *Mycobacterium tuberculosis* ClpX interacts with FtsZ and interferes with FtsZ assembly. PLoS One 5(7):e11058
45. Galarneau A, Primeau M, Trudeau LE, Michnick SW (2002) Beta-lactamase protein fragment complementation assays as in vivo and in vitro sensors of protein protein interactions. Nat Biotechnol 20(6):619–622
46. Nyfeler B, Michnick SW, Hauri HP (2005) Capturing protein interactions in the secretory pathway of living cells. Proc Natl Acad Sci USA 102(18):6350–6355
47. Remy I, Michnick SW (2001) Visualization of biochemical networks in living cells. Proc Natl Acad Sci USA 98(14):7678–7683

48. Subramaniam R, Desveaux D, Spickler C, Michnick SW, Brisson N (2001) Direct visualization of protein interactions in plant cells. Nat Biotechnol 19(8):769–772
49. Gegg CV, Bowers KE, Matthews CR (1997) Probing minimal independent folding units in dihydrofolate reductase by molecular dissection. Protein Sci 6(9):1885–1892
50. Remy I, Michnick SW (1999) Clonal selection and in vivo quantitation of protein interactions with protein-fragment complementation assays. Proc Natl Acad Sci USA 96(10):5394–5399
51. Remy I, Galarneau A, Michnick SW (2002) Detection and visualization of protein interactions with protein fragment complementation assays. Methods Mol Biol 185:447–459
52. Singh A, Mai D, Kumar A, Steyn AJ (2006) Dissecting virulence pathways of *Mycobacterium tuberculosis* through protein-protein association. Proc Natl Acad Sci USA 103(30): 11346–11351
53. Pearce MJ, Mintseris J, Ferreyra J, Gygi SP, Darwin KH (2008) Ubiquitin-like protein involved in the proteasome pathway of *Mycobacterium tuberculosis*. Science 322(5904):1104
54. Mai D, Jones J, Rodgers JW, Hartman JL, Kutsch O, Steyn AJC (2011) A screen to identify small molecule inhibitors of protein-protein interactions in Mycobacteria. Assay Drug Dev Technol 9(3):299–310
55. Tafelmeyer P, Johnsson N, Johnsson K (2004) Transforming a ([beta]/[alpha]) 8-barrel enzyme into a split-protein sensor through directed evolution. Chem Biol 11(5):681–689
56. O'Hare H, Juillerat A, Dianiskova P, Johnsson K (2008) A split-protein sensor for studying protein-protein interaction in mycobacteria. J Microbiol Methods 73(2):79–84
57. Rual JF, Venkatesan K, Hao T, Hirozane-Kishikawa T, Dricot A, Li N, Berriz GF, Gibbons FD, Dreze M, Ayivi-Guedehoussou N, Klitgord N, Simon C, Boxem M, Milstein S, Rosenberg J, Goldberg DS, Zhang LV, Wong SL, Franklin G, Li S, Albala JS, Lim J, Fraughton C, Llamosas E, Cevik S, Bex C, Lamesch P, Sikorski RS, Vandenhaute J, Zoghbi HY, Smolyar A, Bosak S, Sequerra R, Doucette-Stamm L, Cusick ME, Hill DE, Roth FP, Vidal M (2005) Towards a proteome-scale map of the human protein-protein interaction network. Nature 437(7062):1173–1178
58. Collins DM, Kawakami RP, de Lisle GW, Pascopella L, Bloom BR, Jacobs WR Jr (1995) Mutation of the principal sigma factor causes loss of virulence in a strain of the *Mycobacterium tuberculosis* complex. Proc Natl Acad Sci USA 92(17):8036–8040
59. Singh A, Guidry L, Narasimhulu KV, Mai D, Trombley J, Redding KE, Giles GI, Lancaster JR, Steyn AJC (2007) *Mycobacterium tuberculosis* WhiB3 responds to O2 and nitric oxide via its [4Fe-4S] cluster and is essential for nutrient starvation survival. Proc Natl Acad Sci 104(28):11562
60. Singh A, Crossman DK, Mai D, Guidry L, Voskuil MI, Renfrow MB, Steyn AJC (2009) *Mycobacterium tuberculosis* WhiB3 maintains redox homeostasis by regulating virulence lipid anabolism to modulate macrophage response. PLoS Pathog 5(8):e1000545
61. Farhana A, Guidry L, Srivastava A, Singh A, Hondalus MK, Steyn AJC (2010) Reductive stress in microbes: implications for understanding *Mycobacterium tuberculosis* disease and persistence. Adv Microb Physiol 57:43–117
62. Saini V, Farhana A, Steyn AJ (2012) *Mycobacterium tuberculosis* WhiB3: a novel iron-sulfur cluster protein that regulates redox homeostasis and virulence. Antioxid Redox Signal 16(7):687–697
63. Abdallah AM, van Pittius NCG, Champion PADG, Cox J, Luirink J, Vandenbroucke-Grauls CMJE, Appelmelk BJ, Bitter W (2007) Type VII secretion: mycobacteria show the way. Nat Rev Microbiol 5(11):883–891
64. Teutschbein J, Schumann G, Möllmann U, Grabley S, Cole ST, Munder T (2009) A protein linkage map of the ESAT-6 secretion system 1 (ESX-1) of *Mycobacterium tuberculosis*. Microbiol Res 164(3):253–259
65. Champion PADG, Stanley SA, Champion MM, Brown EJ, Cox JS (2006) C-terminal signal sequence promotes virulence factor secretion in *Mycobacterium tuberculosis*. Science 313(5793):1632

66. DiGiuseppe Champion PA, Champion MM, Manzanillo P, Cox JS (2009) ESX-1 secreted virulence factors are recognized by multiple cytosolic AAA ATPases in pathogenic mycobacteria. Mol Microbiol 73(5):950–962
67. Lightbody KL, Renshaw PS, Collins ML, Wright RL, Hunt DM, Gordon SV, Hewinson RG, Buxton RS, Williamson RA, Carr MD (2004) Characterisation of complex formation between members of the *Mycobacterium tuberculosis* complex CFP-10/ESAT-6 protein family: towards an understanding of the rules governing complex formation and thereby functional flexibility. FEMS Microbiol Lett 238(1):255–262
68. MacGurn JA, Raghavan S, Stanley SA, Cox JS (2005) A non-RD1 gene cluster is required for Snm secretion in *Mycobacterium tuberculosis*. Mol Microbiol 57(6):1653–1663
69. Shrivastava R, Ghosh AK, Das AK (2009) Intra-and intermolecular domain interactions among novel two-component system proteins coded by Rv0600c, Rv0601c and Rv0602c of *Mycobacterium tuberculosis*. Microbiology 155(3):772
70. Parida BK, Douglas T, Nino C, Dhandayuthapani S (2005) Interactions of anti-sigma factor antagonists of *Mycobacterium tuberculosis* in the yeast two-hybrid system. Tuberculosis 85(5):347–355
71. Saïd-Salim B, Mostowy S, Kristof AS, Behr MA (2006) Mutations in *Mycobacterium tuberculosis* Rv0444c, the gene encoding anti-SigK, explain high level expression of MPB70 and MPB83 in Mycobacterium bovis. Mol Microbiol 62(5):1251–1263
72. Handa P, Acharya N, Thanedar S, Purnapatre K, Varshney U (2000) Distinct properties of *Mycobacterium tuberculosis* single-stranded DNA binding protein and its functional characterization in Escherichia coli. Nucleic Acids Res 28(19):3823
73. Kana BD, Abrahams GL, Sung N, Warner DF, Gordhan BG, Machowski EE, Tsenova L, Sacchettini JC, Stoker NG, Kaplan G (2010) Role of the DinB homologs Rv1537 and Rv3056 in *Mycobacterium tuberculosis*. J Bacteriol 192(8):2220
74. Sinha KM, Stephanou NC, Gao F, Glickman MS, Shuman S (2007) Mycobacterial UvrD1 is a Ku-dependent DNA helicase that plays a role in multiple DNA repair events, including double-strand break repair. J Biol Chem 282(20):15114
75. Warner DF, Ndwandwe DE, Abrahams GL, Kana BD, Machowski EE, Venclovas Ä, Mizrahi V (2010) Essential roles for imuA′- and imuB-encoded accessory factors in DnaE2-dependent mutagenesis in *Mycobacterium tuberculosis*. Proc Natl Acad Sci 107(29):13093–13098
76. Garg S, Alam MS, Bajpai R, Kishan KVR, Agrawal P (2009) Redox biology of *Mycobacterium tuberculosis* H37Rv: protein-protein interaction between GlgB and WhiB1 involves exchange of thiol-disulfide. BMC Biochem 10(1):1
77. Huet G, Castaing JP, Fournier D, Daffé M, Saves I (2006) Protein splicing of SufB is crucial for the functionality of the *Mycobacterium tuberculosis* SUF machinery. J Bacteriol 188(9):3412–3414
78. Huet G, Daffa M, Saves I (2005) Identification of the *Mycobacterium tuberculosis* SUF machinery as the exclusive mycobacterial system of [Fe-S] cluster assembly: evidence for its implication in the pathogen's survival. J Bacteriol 187(17):6137
79. Veyron-Churlet R, Guerrini O, Mourey L, Daffa M, Zerbib D (2004) Protein-protein interactions within the Fatty Acid Synthase-II system of *Mycobacterium tuberculosis* are essential for mycobacterial viability. Mol Microbiol 54(5):1161–1172
80. Veyron-Churlet R, Bigot S, Guerrini O, Verdoux S, Malaga W, Daffe M, Zerbib D (2005) The biosynthesis of mycolic acids in *Mycobacterium tuberculosis* relies on multiple specialized elongation complexes interconnected by specific protein-protein interactions. J Mol Biol 353(4):847–858
81. Curry JM, Whalan R, Hunt DM, Gohil K, Strom M, Rickman L, Colston MJ, Smerdon SJ, Buxton RS (2005) An ABC transporter containing a forkhead-associated domain interacts with a serine-threonine protein kinase and is required for growth of *Mycobacterium tuberculosis* in mice. Infect Immun 73(8):4471–4477
82. Hett EC, Chao MC, Steyn AJ, Fortune SM, Deng LL, Rubin EJ (2007) A partner for the resuscitation-promoting factors of *Mycobacterium tuberculosis*. Mol Microbiol 66(3): 658–668

83. Sasindran S, Saikolappan S, Scofield V, Dhandayuthapani S (2011) Biochemical and physiological characterization of the GTP-binding protein Obg of *Mycobacterium tuberculosis*. BMC Microbiol 11(1):43
84. Ahidjo BA, Kuhnert D, McKenzie JL, Machowski EE, Gordhan BG, Arcus V, Abrahams GL, Mizrahi V (2011) VapC toxins from *Mycobacterium tuberculosis* are ribonucleases that differentially inhibit growth and are neutralized by cognate VapB antitoxins. PLoS One 6(6):e21738
85. Rain JC, Selig L, De Reuse H, Battaglia V, Reverdy C, Simon S, Lenzen G, Petel F, Wojcik J, Schachter V, Chemama Y, Labigne A, Legrain P (2001) The protein-protein interaction map of *Helicobacter pylori*. Nature 409(6817):211–215
86. LaCount DJ, Vignali M, Chettier R, Phansalkar A, Bell R, Hesselberth JR, Schoenfeld LW, Ota I, Sahasrabudhe S, Kurschner C (2005) A protein interaction network of the malaria parasite *Plasmodium falciparum*. Nature 438(7064):103–107
87. Malek JA, Wierzbowski JM, Tao W, Bosak SA, Saranga DJ, Doucette-Stamm L, Smith DR, McEwan PJ, McKernan KJ (2004) Protein interaction mapping on a functional shotgun sequence of *Rickettsia sibirica*. Nucleic Acids Res 32(3):1059–1064
88. Dyer MD, Neff C, Dufford M, Rivera CG, Shattuck D, Bassaganya-Riera J, Murali TM, Sobral BW (2010) The human-bacterial pathogen protein interaction networks of *Bacillus anthracis, Francisella tularensis*, and *Yersinia pestis*. PLoS One 5(8):e12089
89. Parrish JR, Yu J, Liu G, Hines JA, Chan JE, Mangiola BA, Zhang H, Pacifico S, Fotouhi F, DiRita VJ (2007) A proteome-wide protein interaction map for *Campylobacter jejuni*. Genome Biol 8(7):R130
90. Titz B, Rajagopala SV, Goll J, Häuser R, McKevitt MT, Palzkill T, Uetz P (2008) The binary protein interactome of *Treponema pallidum*: the syphilis spirochete. PLoS One 3(5):e2292
91. Zhang L, Villa NY, Rahman MM, Smallwood S, Shattuck D, Neff C, Dufford M, Lanchbury JS, LaBaer J, McFadden G (2009) Analysis of vaccinia virus-host protein-protein interactions: validations of yeast two-hybrid screenings. J Proteome Res 8(9):4311–4318
92. Gavin AC, Basche M, Krause R, Grandi P, Marzioch M, Bauer A, Schultz J, Rick JM, Michon AM, Cruciat CM (2002) Functional organization of the yeast proteome by systematic analysis of protein complexes. Nature 415(6868):141–147
93. Gerber D, Maerkl SJ, Quake SR (2008) An *in vitro* microfluidic approach to generating protein-interaction networks. Nat Methods 6(1):71–74
94. Ho Y, Gruhler A, Heilbut A, Bader GD, Moore L, Adams SL, Millar A, Taylor P, Bennett K, Boutilier K (2002) Systematic identification of protein complexes in *Saccharomyces cerevisiae* by mass spectrometry. Nature 415(6868):180–183
95. Xiang G, Yao X, Guowu B, Yan X, Zhiyong X (2010) Response of the mosquito protein interaction network to dengue infection. BMC Genomics 11:380
96. Marchadier E, Carballidoa-Lapez R, Brinster S, Fabret C, Mervelet P, Bessiares P, Noirota-Gros MF, Fromion V, Noirot P (2011) An expanded protein-protein interaction network in *Bacillus subtilis* reveals a group of hubs: exploration by an integrative approach. Proteomics 11(15):2981–2991
97. Cherkasov A, Hsing M, Zoraghi R, Foster LJ, See RH, Stoynov N, Jiang J, Kaur S, Lian T, Jackson L (2011) Mapping the protein interaction network in methicillin-resistant *Staphylococcus aureus*. J Proteome Res 10(3):1139–1150
98. Callahan B, Nguyen K, Collins A, Valdes K, Caplow M, Crossman DK, Steyn AJC, Eisele L, Derbyshire KM (2010) Conservation of structure and protein-protein interactions mediated by the secreted mycobacterial proteins EsxA, EsxB, and EspA. J Bacteriol 192(1):326
99. Matsumoto A, Comatas KE, Liu L, Stamler JS (2003) Screening for nitric oxide-dependent protein-protein interactions. Science 301(5633):657
100. Vignols F, Brahalin C, Surdin-Kerjan Y, Thomas D, Meyer Y (2005) A yeast two-hybrid knockout strain to explore thioredoxin-interacting proteins *in vivo*. Proc Natl Acad Sci USA 102(46):16729
101. Andries K, Verhasselt P, Guillemont J, Gohlmann HW, Neefs JM, Winkler H, Van Gestel J, Timmerman P, Zhu M, Lee E, Williams P, de Chaffoy D, Huitric E, Hoffner S, Cambau E, Truffot-Pernot C, Lounis N, Jarlier V (2005) A diarylquinoline drug active on the ATP synthase of *Mycobacterium tuberculosis*. Science 307(5707):223–227

102. Manjunatha UH, Boshoff H, Dowd CS, Zhang L, Albert TJ, Norton JE, Daniels L, Dick T, Pang SS, Barry CE 3rd (2006) Identification of a nitroimidazo-oxazine-specific protein involved in PA-824 resistance in *Mycobacterium tuberculosis*. Proc Natl Acad Sci USA 103(2):431–436
103. Stover CK, Warrener P, VanDevanter DR, Sherman DR, Arain TM, Langhorne MH, Anderson SW, Towell JA, Yuan Y, McMurray DN, Kreiswirth BN, Barry CE, Baker WR (2000) A small-molecule nitroimidazopyran drug candidate for the treatment of tuberculosis. Nature 405(6789):962–966
104. Bogan AA, Thorn KS (1998) Anatomy of hot spots in protein interfaces. J Mol Biol 280(1):1–9
105. Clackson T, Wells JA (1995) A hot spot of binding energy in a hormone-receptor interface. Science 267(5196):383
106. Tsai CJ, Lin SL, Wolfson HJ, Nussinov R (1997) Studies of protein-protein interfaces: a statistical analysis of the hydrophobic effect. Protein Sci 6(1):53–64
107. Xu D, Tsai CJ, Nussinov R (1997) Hydrogen bonds and salt bridges across protein-protein interfaces. Protein Eng 10(9):999–1012
108. McCoy AJ, Chandana Epa V, Colman PM (1997) Electrostatic complementarity at protein/protein interfaces. J Mol Biol 268(2):570–584
109. Jones S, Thornton JM (1996) Principles of protein-protein interactions. Proc Natl Acad Sci USA 93(1):13–20
110. Gadek TR, Nicholas JB (2003) Small molecule antagonists of proteins. Biochem Pharmacol 65(1):1–8
111. Laskowski RA, Luscombe NM, Swindells MB, Thornton JM (1996) Protein clefts in molecular recognition and function. Protein Sci 5(12):2438–2452
112. Loregian A, Palu G (2005) Disruption of protein-protein interactions: towards new targets for chemotherapy. J Cell Physiol 204(3):750–762
113. Wells JA, McClendon CL (2007) Reaching for high-hanging fruit in drug discovery at protein-protein interfaces. Nature 450(7172):1001–1009
114. Arkin MR, Wells JA (2004) Small-molecule inhibitors of protein-protein interactions: progressing towards the dream. Nat Rev Drug Discov 3(4):301–317
115. Strosberg AD (2007) Protein-protein interactions as targets for novel therapeutics. Drug Discov
116. Vassilev LT, Vu BT, Graves B, Carvajal D, Podlaski F, Filipovic Z, Kong N, Kammlott U, Lukacs C, Klein C, Fotouhi N, Liu EA (2004) In vivo activation of the p53 pathway by small-molecule antagonists of MDM2. Science 303(5659):844–848
117. Kussie PH, Gorina S, Marechal V, Elenbaas B, Moreau J, Levine AJ, Pavletich NP (1996) Structure of the MDM2 oncoprotein bound to the p53 tumor suppressor transactivation domain. Science 274(5289):948–953
118. Hung DT, Shakhnovich EA, Pierson E, Mekalanos JJ (2005) Small-molecule inhibitor of *Vibrio cholerae* virulence and intestinal colonization. Science 310(5748):670–674
119. Shakhnovich EA, Hung DT, Pierson E, Lee K, Mekalanos JJ (2007) Virstatin inhibits dimerization of the transcriptional activator ToxT. Proc Natl Acad Sci USA 104(7):2372–2377
120. McMillan K, Adler M, Auld DS, Baldwin JJ, Blasko E, Browne LJ, Chelsky D, Davey D, Dolle RE, Eagen KA, Erickson S, Feldman RI, Glaser CB, Mallari C, Morrissey MM, Ohlmeyer MH, Pan G, Parkinson JF, Phillips GB, Polokoff MA, Sigal NH, Vergona R, Whitlow M, Young TA, Devlin JJ (2000) Allosteric inhibitors of inducible nitric oxide synthase dimerization discovered via combinatorial chemistry. Proc Natl Acad Sci USA 97(4):1506–1511
121. Sennequier N, Wolan D, Stuehr DJ (1999) Antifungal imidazoles block assembly of inducible NO synthase into an active dimer. J Biol Chem 274(2):930–938
122. Gorczynski MJ, Grembecka J, Zhou Y, Kong Y, Roudaia L, Douvas MG, Newman M, Bielnicka I, Baber G, Corpora T, Shi J, Sridharan M, Lilien R, Donald BR, Speck NA, Brown ML, Bushweller JH (2007) Allosteric inhibition of the protein-protein interaction between the leukemia-associated proteins Runx1 and CBFbeta. Chem Biol 14(10):1186–1197

123. Last-Barney K, Davidson W, Cardozo M, Frye LL, Grygon CA, Hopkins JL, Jeanfavre DD, Pav S, Qian C, Stevenson JM, Tong L, Zindell R, Kelly TA (2001) Binding site elucidation of hydantoin-based antagonists of LFA-1 using multidisciplinary technologies: evidence for the allosteric inhibition of a protein-protein interaction. J Am Chem Soc 123(24):5643–5650
124. Horn JR, Shoichet BK (2004) Allosteric inhibition through core disruption. J Mol Biol 336(5):1283–1291
125. He MM, Smith AS, Oslob JD, Flanagan WM, Braisted AC, Whitty A, Cancilla MT, Wang J, Lugovskoy AA, Yoburn JC, Fung AD, Farrington G, Eldredge JK, Day ES, Cruz LA, Cachero TG, Miller SK, Friedman JE, Choong IC, Cunningham BC (2005) Small-molecule inhibition of TNF-alpha. Science 310(5750):1022–1025. doi:10.1126/science.1116304
126. Bochkareva E, Safro M, Girshovich A (1999) Interaction of 4,4′-dithiodipyridine with Cys(458) triggers disassembly of GroEL. J Biol Chem 274(30):20756–20758
127. Tachedjian G, Orlova M, Sarafianos SG, Arnold E, Goff SP (2001) Nonnucleoside reverse transcriptase inhibitors are chemical enhancers of dimerization of the HIV type 1 reverse transcriptase. Proc Natl Acad Sci USA 98(13):7188–7193
128. Crabtree GR, Schreiber SL (1996) Three-part inventions: intracellular signaling and induced proximity. Trends Biochem Sci 21(11):418–422
129. Michnick SW (2000) Chemical biology beyond binary codes. Chem Biol 7(12):R217–221, pii: S1074-5521(00)00040-5
130. Liu J, Farmer JD Jr, Lane WS, Friedman J, Weissman I, Schreiber SL (1991) Calcineurin is a common target of cyclophilin-cyclosporin A and FKBP-FK506 complexes. Cell 66(4):807–815, pii: 0092-8674(91)90124-H
131. Brown EJ, Albers MW, Shin TB, Ichikawa K, Keith CT, Lane WS, Schreiber SL (1994) A mammalian protein targeted by G1-arresting rapamycin-receptor complex. Nature 369(6483):756–758. doi:10.1038/369756a0
132. Williams DJ, Puhl HL, Ikeda SR, Brezina V (2009) Rapid modification of proteins using a rapamycin-inducible tobacco etch virus protease system. PLoS One 4(10):e7474
133. Remy I, Michnick SW (2006) A highly sensitive protein-protein interaction assay based on *Gaussia* luciferase. Nat Methods 3(12):977–979
134. Janse DM, Crosas B, Finley D, Church GM (2004) Localization to the proteasome is sufficient for degradation. J Biol Chem 279(20):21415–21420
135. Joshi PB, Hirst M, Malcolm T, Parent J, Mitchell D, Lund K, Sadowski I (2007) Identification of protein interaction antagonists using the repressed transactivator two-hybrid system. Biotechniques 42(5):635–644
136. Kley N (2004) Chemical dimerizers and three-hybrid systems: scanning the proteome for targets of organic small molecules. Chem Biol 11(5):599–608
137. Veselovsky AV, Ivanov YD, Ivanov AS, Archakov AI, Lewi P, Janssen P (2002) Protein-protein interactions: mechanisms and modification by drugs. J Mol Recognit 15(6):405–422
138. Valencia A, Pazos F (2002) Computational methods for the prediction of protein interactions. Curr Opin Struct Biol 12(3):368–373
139. Marcotte EM, Pellegrini M, Ng HL, Rice DW, Yeates TO, Eisenberg D (1999) Detecting protein function and protein-protein interactions from genome sequences. Science 285(5428):751
140. Bock JR, Gough DA (2001) Predicting protein-protein interactions from primary structure. Bioinformatics 17(5):455–460
141. Eisenberg D, Marcotte EM, Xenarios I, Yeates TO (2000) Protein function in the post-genomic era. Nature 405(6788):823–826
142. Enright AJ, Iliopoulos I, Kyrpides NC, Ouzounis CA (1999) Protein interaction maps for complete genomes based on gene fusion events. Nature 402(6757):86–90
143. Fraser HB, Hirsh AE, Wall DP, Eisen MB (2004) Coevolution of gene expression among interacting proteins. Proc Natl Acad Sci USA 101(24):9033
144. Blaschke C, Andrade MA, Ouzounis C, Valencia A (1999) Automatic extraction of biological information from scientific text: protein-protein interactions. Proc Int Conf Intell Syst Mol Biol 1999:60–67

145. Cusick ME, Yu H, Smolyar A, Venkatesan K, Carvunis AR, Simonis N, Rual JF, Borick H, Braun P, Dreze M (2008) Literature-curated protein interaction datasets. Nat Methods 6(1):39–46
146. Russell RB, Alber F, Aloy P, Davis FP, Korkin D, Pichaud M, Topf M, Sali A (2004) A structural perspective on protein-protein interactions. Curr Opin Struct Biol 14(3):313–324
147. Neuvirth H, Raz R, Schreiber G (2004) ProMate: a structure based prediction program to identify the location of protein-protein binding sites. J Mol Biol 338(1):181–199
148. Stein A, Caol A, Aloy P (2011) 3did: identification and classification of domain-based interactions of known three-dimensional structure. Nucleic Acids Res 39(suppl 1):D718
149. Aloy P, Russell RB (2003) InterPreTS: protein interaction prediction through tertiary structure. Bioinformatics 19(1):161
150. Pazos F, Valencia A (2001) Similarity of phylogenetic trees as indicator of protein-protein interaction. Protein Eng 14(9):609
151. Pagel P, Wong P, Frishman D (2004) A domain interaction map based on phylogenetic profiling. J Mol Biol 344(5):1331–1346
152. Shoemaker BA, Panchenko AR (2007) Deciphering protein-protein interactions. Part II. Computational methods to predict protein and domain interaction partners. PLoS Comput Biol 3(4):e43
153. Mavromatis K, Chu K, Ivanova N, Hooper SD, Markowitz VM, Kyrpides NC (2009) Gene context analysis in the Integrated Microbial Genomes (IMG) data management system. PLoS One 4(11):e7979
154. Pazos F, Valencia A (2002) *In silico* two-hybrid system for the selection of physically interacting protein pairs. Proteins 47(2):219–227
155. Pazos F, Valencia A (2008) Protein co-evolution, co-adaptation and interactions. EMBO J 27(20):2648–2655
156. Friedberg I (2006) Automated protein function prediction: the genomic challenge. Brief Bioinform 7(3):225–242
157. Snitkin E, Gustafson A, Mellor J, Wu J, DeLisi C (2006) Comparative assessment of performance and genome dependence among phylogenetic profiling methods. BMC Bioinformatics 7(1):420
158. Orchard S, Salwinski L, Kerrien S, Montecchi-Palazzi L, Oesterheld M, Stümpflen V, Ceol A, Chatr-aryamontri A, Armstrong J, Woollard P (2007) The minimum information required for reporting a molecular interaction experiment (MIMIx). Nat Biotechnol 25(8):894–898
159. Taylor CF, Paton NW, Lilley KS, Binz PA, Julian RK, Jones AR, Zhu W, Apweiler R, Aebersold R, Deutsch EW (2007) The minimum information about a proteomics experiment (MIAPE). Nat Biotechnol 25(8):887–893
160. Gmuender H, Kuratli K, Di Padova K, Gray CP, Keck W, Evers S (2001) Gene expression changes triggered by exposure of Haemophilus influenzae to novobiocin or ciprofloxacin: combined transcription and translation analysis. Genome Res 11(1):28–42
161. Yoon SH, Han MJ, Lee SY, Jeong KJ, Yoo JS (2003) Combined transcriptome and proteome analysis of Escherichia coli during high cell density culture. Biotechnol Bioeng 81(7):753–767
162. Mostertz J, Scharf C, Hecker M, Homuth G (2004) Transcriptome and proteome analysis of Bacillus subtilis gene expression in response to superoxide and peroxide stress. Microbiology 150(2):497
163. Gygi SP, Rochon Y, Franza BR, Aebersold R (1999) Correlation between protein and mRNA abundance in yeast. Mol Cell Biol 19(3):1720–1730
164. Bro C, Regenberg B, Lagniel G, Labarre J, Montero-Lomelí M, Nielsen J (2003) Transcriptional, proteomic, and metabolic responses to lithium in galactose-grown yeast cells. J Biol Chem 278(34):32141–32149
165. Kramer JO, Sorgenfrei O, Klopprogge K, Heinzle E, Wittmann C (2004) In-depth profiling of lysine-producing *Corynebacterium glutamicum* by combined analysis of the transcriptome, metabolome, and fluxome. J Bacteriol 186(6):1769–1784

Chapter 6
Host–Pathogen Interactions

Simon J. Waddell, Axel von Kamp, Steffen Klamt, and Olivier Neyrolles

Abstract The ability to genome sequence mycobacteria and host organisms has enabled a range of system-wide approaches to be developed to explore the interplay between host and pathogen. These global analyses offer an unbiased means of generating new hypotheses to further understand bacterial pathogenesis and immune activation states. *Mycobacterium tuberculosis* high-throughput mutant screening has identified key genes and pathways involved in mycobacterial physiology or pathogenicity that are required *in vivo* or during macrophage infection. Reciprocal genome-wide RNAi-based screening approaches have highlighted host genes that play crucial roles in the immune and metabolic crosstalk with infecting bacilli. In addition to these loss-of-function screens, transcriptional profiling of the pathogen, of the host, or of both together has provided clues into the divergent metabolic states and key signalling events that characterise *M. tuberculosis* infection. Such global analyses, linked in a systems approach through interaction databases and network mapping, allow descriptive and predictive models of infection and disease to be constructed. In this chapter we review the recent developments and applications of these system-wide approaches to better understand the interactions of *M. tuberculosis* with its host.

S.J. Waddell (✉)
Brighton and Sussex Medical School, University of Sussex,
Brighton, UK
e-mail: s.waddell@bsms.ac.uk

A. von Kamp • S. Klamt
Max Planck Institute for Dynamics of Complex Technical Systems, Magdeburg, Germany

O. Neyrolles (✉)
Centre National de la Recherche Scientifique, Institut de Pharmacologie et de Biologie Structurale, Toulouse, France

Université de Toulouse, Université Paul Sabatier, Institut de Pharmacologie et de Biologie Structurale, Toulouse, France
e-mail: olivier.neyrolles@ipbs.fr

J. McFadden et al. (eds.), *Systems Biology of Tuberculosis*,
DOI 10.1007/978-1-4614-4966-9_6,

1 Introduction

In the post-genomic era, the development of a variety of system-wide approaches has allowed host–pathogen interactions to be examined on a global level. Genomic analyses offer an unbiased means of generating new hypotheses to further understand bacterial pathogenesis. In the case of the tuberculosis bacillus, several high-throughput *Mycobacterium tuberculosis* mutant screening studies performed during macrophage infection or in vivo have identified key genes and pathways involved in mycobacterial physiology and required for virulence. More recently, genome-wide RNAi-based screening approaches have highlighted host genes that play crucial roles in the immune and metabolic crosstalk with infecting bacilli. In addition, global gene expression profiling of the pathogen, of the host, or of both together has provided clues into the divergent metabolic states and key signalling events that characterise *M. tuberculosis* pathogenesis. As such, temporal analyses describing the changing interplay between bacilli and macrophage as infection progresses are particularly useful, allowing descriptive and predictive models to be constructed. In this chapter we review the recent developments and applications of these system-wide approaches to better understand the interactions of *M. tuberculosis* with its host. We illustrate how transcriptome analysis coupled to models of signalling and transcription networks can help to suggest novel interactions of potential importance during infection. This systems approach to interpreting host–mycobacterial interplay is summarised in Fig. 6.1.

2 Functional Genomics to Identify Mycobacterial Virulence Genes

Understanding how a pathogen and its host adapt to each other during the course of infection is key to developing new tools and better strategies to combat infectious disease. Over 10 years ago sequencing the *M. tuberculosis* genome [1], together with the development of genetic tools to inactivate genes in random or targeted approaches [2], allowed novel virulence genes and loci involved in pathogenesis and host parasitism to be discovered on a genome-wide level. Two studies published in 1999 made use of a functional genomics approach developed earlier in *Salmonella* [3], signature transposon-tagged mutagenesis (STM), using medium-size pools of *M. tuberculosis* mutants to identify *M. tuberculosis* genes important during infection in the mouse model [4, 5]. Both studies highlighted phthiocerol dimycocerosates, complex lipids of the mycobacterial cell wall, as key components of mycobacterial pathogenicity. PDIMs now constitute a prototypic example of a complex molecule of the mycobacterial cell envelope involved in pathogenesis; yet their exact function and mode of action still remain to be fully understood [6, 7]. A few years later, the generation of novel tools for transposition and tracking of transposon mutants using a microarray-based strategy, termed transposon site hybridization, allowed gene insertion events to reach saturation levels. This enabled the authors to classify virtually all mycobacterial genes required for

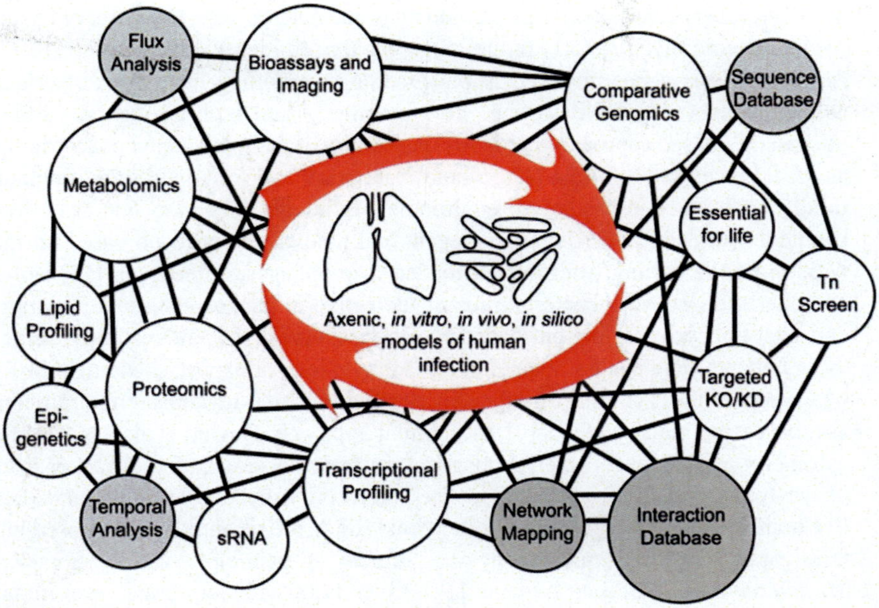

Fig. 6.1 An interaction network of techniques and approaches used to study host–pathogen interplay. The complementary methodologies are linked together by bioinformatics tools and databases (*shaded grey*) in a systems approach to understanding infection and disease

successful infection in the mouse model in a high-throughput and system-wide manner [8–10]. In these studies, Sassetti and Rubin used a library of 100,000 transposon mutants, in which almost all non-essential genes were inactivated and looked for mutants impaired in their ability to grow in various *in vitro* and *in vivo* conditions, including murine lungs. Although the mouse model is not ideal for studying mycobacterial virulence, it nevertheless provides an indication of the relative importance of mycobacterial pathways for in vivo survival. In this way, a number of genes predicted to be involved in secretion, lipid metabolism, carbohydrate transport and metabolism, inorganic ion transport and metabolism, cell envelope biogenesis, and amino acid transport and metabolism were recognised. Many of these genes had not previously been shown to play a part in mycobacterial virulence in vivo. A number of genes with unknown function specific to mycobacteria were also discovered; this raises the intriguing question of the role of ancient horizontal gene transfer events in mycobacterial physiology and pathogenicity (see below) [11]. Similar approaches have been used in other animal models that more closely resemble human disease, such as non-human primates (NHPs) [12]. Again in this study, a number of previously underappreciated genes, for example, involved in lipid metabolism and transport, biosynthesis of the cell wall, and sterol metabolism were classified to be functionally significant for mycobacterial pathogenicity in vivo.

More recently, screening approaches have been developed to detect mycobacterial genes involved in pathogenic processes at the host cell or sub-cellular levels.

M. tuberculosis genes mediating parasitism of the macrophage, the primary host cell for pathogenic mycobacteria in the lungs, have been identified through STM [13]. *M. tuberculosis* genes involved in phagosome remodelling have been determined using magnetic organelle sorting, flow cytometry and high-throughput confocal microscopy-based approaches [14–16]. The ability of pathogenic mycobacteria, such as *M. tuberculosis*, to arrest phagosome maturation and to remain in an immature, mildly acidic and non-proteolytic environment is thought to be a key feature of mycobacterial virulence [17]. Using an elegant and innovative approach based on magnetic organelle sorting from ferritin-loaded macrophages, Pethe et al. [15] isolated and characterised mycobacterial mutants defective in phagosome maturation arrest and thus trafficked to ferritin-loaded phago-lysosomes [15]. Interestingly, some of these mutants were again impaired in PDIMs synthesis or export, shedding new light on the role of these complex lipid moieties in intracellular mycobacterial trafficking, as recently reported by others [6]. In a similar approach, Stewart et al. [16] used flow cytometry to sort mycobacterial phagosomes from fluorescent LysoTracker-loaded phago-lysosomes, and was able to isolate and identify mycobacterial mutants defective in inhibition of phagosome acidification [16]. More recently, we have used high-throughput confocal microscopy to screen a genome-covering library of *M. tuberculosis* W-Beijing mutants [14]. Over 11,000 random transposon mutants were used to infect human macrophages in high-density 384-well plates in a one-well one-mutant manner. Infected cells were stained with the acid-specific dye LysoTracker. Mutants that colocalised with the dye were selected and their transposon insertion sites were sequenced. Two independent mutants in *Rv1503c* and *Rv1506c*, which belong to the same genetic locus in the mycobacterial chromosome, and two other mutants in *moaC1* and *moaD1*, which belong to another locus likely involved in synthesis of the molybdopterin cofactor, were isolated. Furthermore, we showed that the *Rv1503c/6c* locus is involved in the synthesis of trehalose-containing glycolipids, thus establishing a link between these lipids and the ability of pathogenic mycobacteria to prevent phagosome acidification. These studies illustrate how system-wide functional genomics approaches help to identify mycobacterial virulence genes and gene clusters in an unbiased manner. Strikingly, all these studies reported mutants in intergenic regions of the mycobacterial chromosome. This raises the intriguing question of the functional significance of non-coding small RNAs (sRNA) in *M. tuberculosis* pathogenicity [18]. As in other bacterial species, it is likely that sRNA play key roles in *M. tuberculosis* virulence by regulating the expression of other genes. Such findings lay the foundations for functional epigenomics in mycobacteria which will benefit from the development of new genomics tools in the future.

3 *In Silico* Mycobacterial Genomics

As more mycobacterial genomes have been sequenced over the years, genome comparison and in silico genomics have provided clues to mycobacterial pathogenicity. Comparative genomics identified the attenuation of the vaccine strain, *M. bovis*

BCG, to be a major deletion in its chromosome (the RD1 region of difference) as compared to the *M. tuberculosis* or the *M. bovis* chromosomes [19]. More recently, we and others have used *in silico* and comparative genomics to identify several chromosomal regions in *M. tuberculosis* that were most likely acquired by the ancestor of the *M. tuberculosis* complex through horizontal gene exchanges [20–23]. Strikingly, most of these regions are predicted to be acquired from environmental bacterial species, thus strengthening the long-thought hypothesis that the ancestor of *M. tuberculosis* was an environmental species that has gradually been "educated" to become pathogenic, and in particular to parasitise phagocytes [11]. Studying the role of these ancient horizontally acquired genes in mycobacterial physiology and virulence is now ongoing in several laboratories.

4 Functional Genomics to Recognise Host Genes Mediating the Response to Mycobacteria

A system-wide exploration of the role of host immuno-regulatory pathways in mycobacteria interactions is now possible because of the development of genetic tools to silence gene expression in eukaryotes using RNA interference (RNAi). Such approaches have been successfully used to identify host genes contributing to mycobacterial growth restriction in drosophila cells [24, 25] and more recently in mammalian cells [26, 27]. The future use of RNAi-based genetic screening techniques in multicellular organisms, such as the zebra fish, that can be infected by *Mycobacterium marinum*, a close relative of *M. tuberculosis*, will undoubtedly allow further understanding of the importance of specific host genes in immunity to mycobacteria in vivo. Thus, the application of whole genome approaches screening for mycobacterial survival or observable changes in macrophage–mycobacteria interactions, such as differential phagosome trafficking, has identified both host and pathogen genes that influence the outcome of infection. Comparative genomics have provided historical and geographic context to these genes and enabled mycobacterial pathogenicity to be directly associated with particular gene clusters. The transcriptional regulation of host and *M. tuberculosis* genes during infection provides yet another key perspective into these multi-factorial interactions.

5 Transcriptional Profiling Mycobacteria Interactions with Phagocytes

Techniques that exploit the differential regulation of genes during infection have been employed for many years to define the dialogue between *M. tuberculosis* bacilli and host immune cells. Selective approaches such as subtractive hybridisation [28, 29], promoter trap library screening [30], *in situ* hybridisation [31], and quantitative RT-PCR [32] have identified key genes highlighting pathways involved in the phagocytosis and survival of *M. tuberculosis* in host cells. Sequencing of the *M. tuberculosis* H37Rv genome [1], and subsequent mouse [33] and human genomes [34], heralded the age of genome-wide expression profiling using microarrays, qRT-PCR panels or more recently RNAseq [35]. These whole genome approaches together with the continued development of mRNA extraction, stabilisation, and amplification methodologies [36–40] have enabled previously intractable scenarios to be investigated, generating rich datasets describing host and pathogen responses to infection.

The first studies measuring transcriptional changes in host cells contrasted the gene expression patterns of macrophages after infection with different pathogens. For example, by comparing the macrophage responses to *M. tuberculosis* with six Gram-positive or Gram-negative bacteria, Nau et al. [41] defined a common macrophage activation signature and observed that interleukin (IL)-12 and IL-15 were not induced by *M. tuberculosis* infection. This distinguished the macrophage response to *M. tuberculosis* from other bacterial pathogens and suggested that *M. tuberculosis* may actively suppress macrophage pro-inflammatory processes. A similar approach has been employed to understand how events diverge between phagocytes and virulent or attenuated *M. tuberculosis* laboratory strains (H37Rv or H37Ra, respectively). Spira et al. [42] recognised a pro-apoptotic signature in alveolar macrophages after infection with H37Ra versus H37Rv, which was abrogated upon neutralisation of tumour necrosis factor (TNF). Thus, contributing to the hypothesis that virulent *M. tuberculosis* bacilli prevent macrophage programmed cell death mediated by TNF. In the converse experimental approach, Chaussabel et al. [43] contrasted the responses of different immune cell subtypes (monocyte-derived macrophages and dendritic cells (DCs)) to infection with the same pathogens. Such analyses have identified microbe-specific and cell-specific activation programmes that reflect the multi-factorial interplay of immune cell colonisation, providing insight into novel pathways influencing bacterial control and evasion of these processes by pathogens. We used the disparate ability of human monocyte-derived macrophages and DCs to control *M. tuberculosis* infection to compare the transcriptional responses of both host cell and infecting bacilli to the development of permissive and non-permissive intracellular microenvironments (in macrophages and DCs, respectively) [44]. This study revealed that a number of zinc-responsive genes were up-regulated in macrophages after *M. tuberculosis* infection and that correspondingly *M. tuberculosis* genes encoding heavy metal transporters were also induced after phagocytosis. Extension of this work demonstrated that zinc accumulation in phagosomes was toxic to engulfed non-tuberculous bacteria, uncovering a new macrophage anti-microbial strategy, and that *M. tuberculosis* bacilli are able to avoid zinc poisoning by inducing metal cation efflux pumps during macrophage infection [45].

Exploring macrophage transcriptional adaptations to *M. tuberculosis* infection may also contribute to understanding how genetic background influences susceptibility to tuberculosis. Keller et al. [46] compared the responses of murine bone marrow-derived macrophages extracted from C57BL/6 and

BALB/c (representing *M. tuberculosis*-resistant) with DBA/2 and CBA/J (*M. tuberculosis*-susceptible) mouse strains. The authors highlighted over 100 genes whose expression during the early phases of infection may, in part, be responsible for the contrary progression of tuberculosis disease in these genetically distinct mice lineages. Thuong et al. [47] extended this concept to investigate human genetic susceptibility to tuberculosis, examining the transcriptional responses of monocyte-derived macrophages from patients with latent versus pulmonary tuberculosis to *M. tuberculosis* antigen stimulation. By combining gene expression profiling with single nucleotide polymorphism mapping, the authors showed that the function of chemokine (C–C motif) ligand 1, CCL1, may be associated with pulmonary tuberculosis in man. The combination of mRNA profiling and targeted gene inactivation is a powerful tool for recognising key host immune-mediators. Ehrt et al. [48] mapped the transcriptional signatures of bone marrow-derived murine macrophages from WT, iNOS-deficient, or phox91-deficient mice to *M. tuberculosis* infection, framing roles for nitric oxide synthase 2 (iNOS) and phagocyte oxidase (phox) in the control of *M. tuberculosis*. Furthermore, this strategy has been extended to characterise MyD88 (myeloid differentiation primary response gene 88)-dependent and MyD88-independent pathways of macrophage activation following *M. tuberculosis* infection and continues to delineate signal transduction pathways that mediate macrophage activity by comparing the signatures of *M. tuberculosis*-infected macrophages derived from panels of knockout mice [49]. In this way, unbiased gene expression analyses, providing a snapshot of changing host cell status, have enabled novel mechanisms affecting mycobacterial control to be elucidated.

On the other side of this hostile engagement, transcriptional profiling of *M. tuberculosis* during macrophage infection has revealed how mycobacterial metabolism adapts after phagocytosis and has acted as a bioprobe to survey the internal microenvironments that bacilli encounter (recently reviewed in [50, 51]). Schnappinger et al. [52] demonstrated that multiple gene families involved in the beta-oxidation of fatty acids were induced after murine macrophage infection, hypothesising that intracellular *M. tuberculosis* adopt a lipolytic lifestyle. This key feature of *M. tuberculosis* infection has been observed after infection of human macrophage-like THP-1 cells [53] and human monocyte-derived macrophages [44] and, together with the identification of a cluster of genes that likely metabolise cholesterol [54], highlights these metabolic changes as a common strategy for *M. tuberculosis* intracellular survival. In addition, the respiratory state of *M. tuberculosis* also changes during infection shifting from aerobic to microaerophilic to anaerobic respiration dependent on changes to the dynamic immune-environment [49]. The differential regulation of these metabolic and respiratory pathways together with stress-responsive regulons (such as *dosR* and *phoP*) is most clearly highlighted by comparing *M. tuberculosis* signatures from different intracellular environments. For example, the impact of interferon (IFN)γ-mediated murine macrophage activation [52] or the development of a non-permissive environment after DC infection [44, 55] results in similar *M. tuberculosis* adaptive responses. Rhode et al. [56] used concanamycin A to limit the acidification of murine macrophages to pH 7.0 versus pH 6.4, enabling the authors to differentiate acid-inducible *M. tuberculosis* genes. This study led to the characterisation of an *M. tuberculosis* acid and phagosome-regulated locus (*aprA/B/C*) that is required for successful macrophage infection and whose expression is likely regulated by the PhoP/R two-component system [57]. Further evidence that the PhoP/R system is important for sensing and controlling *M. tuberculosis* responses to the intracellular environment comes from a study contrasting the transcriptional signatures of H37Rv with H37Ra (which contains a point mutation in *phoP*). Li et al. [58] observed that PhoP-regulated genes were differentially expressed between H37Rv and H37Ra during murine macrophage infection and concluded that the limited ability of *M. tuberculosis* H37Ra to react to the intracellular environment may account for its attenuated phenotype. Continuing this theme, the impact of genetic variation amongst *M. tuberculosis* clinical isolates on interactions with murine macrophages was explored further by Homolka et al. [59], who compared the intracellular gene expression profiles of 15 phylo-geographically distinct *M. tuberculosis* complex strains. The authors mapped genome-wide *M. tuberculosis* responses that reflected the diverse intracellular fates of these clinical strains and detailed a common programme of bacterial adaptation encompassing oxidative and/or nitrosative stresses and metabolic and physiological alterations. This analysis also detected clade-specific and strain-specific intracellular transcriptional patterns, providing a basis for further investigation into the consequences that geographical and genetic *M. tuberculosis* diversity may have on tuberculosis disease worldwide [59].

6 Transcriptional Profiling the Interplay Between Host and Pathogen

Global mRNA profiling of host and *M. tuberculosis* bacilli from invasive or non-invasive sampling of tissues offers an overview of the impact of mycobacterial infection as disease progresses. These studies explore the complex organ environments made up of diverse cell types and distinct populations of bacteria and survey the interactions between multiple cells. As such, these models are able to examine host–pathogen interplay in a heterogeneous environment capturing changes in cell populations as well as divergent gene regulation. Many of these studies are aimed at identifying biomarkers of tuberculosis disease states (reviewed recently by Walzl et al. [60]). For example, the mRNA abundance profiles of murine lungs and spleens after infection or vaccination have been used to follow changes in immune-mediators over time and to determine indicators of a protective response [61]. *M. tuberculosis* gene expression profiling from murine lungs has defined in vivo signatures and revealed divergent responses to infection contrasting immune-compromised with immune-competent murine hosts [62, 63]. Mehra et al. [64] described the temporal mRNA abundance profiles of NHP granulomas during early and late disease, observing that the expression of inflammatory markers significantly decreased in NHP granulomas through the course of disease. This approach has been translated to human tuberculosis disease by Kim et al. [65] who mapped the mRNA signature of human lung caseous granulomas using a combination of laser capture dissection microscopy and microarray analysis. The authors distinguished a gene expression pattern reflective of a change in lipid metabolism in caseous granulomas that likely results in the accumulation of host lipids. Correspondingly, *M. tuberculosis* genes involved in fatty acid metabolism were induced in human lung sections (extracted during surgery for untreatable tuberculosis) compared to axenic culture [66]. Moreover, a transcriptional signature of enhanced cholesterol metabolism was observed in *M. tuberculosis* bacilli extracted from human sputum, where slow or non-replicating lipid body-positive "fat and lazy" bacilli were characterised [67]. The activation state of human immune cells at the site of tuberculosis infection has been sampled by harvesting cells from bronchoalveolar lavage fluid, providing a readout of immune cell migration and shifting immuno-regulatory processes during active disease [68, 69]. Systemic host responses to *M. tuberculosis* infection have been measured from whole blood to define factors that influence relapse of disease [70] or active versus latent infection [71]. Thus, whole genome approaches to understanding mycobacterial disease continue to generate novel hypotheses, recently illustrated by the unexpected discovery of a neutrophil-mediated type I-interferon signature in the peripheral blood of patients with active tuberculosis [72].

Transcriptional profiling the crosstalk between host immune cells and *M. tuberculosis* bacilli in vitro and in vivo has identified common and specific responses to phagocyte or *M. tuberculosis* genotype, revealing novel mechanisms of bacterial control and immune-modulation and providing an interpretive framework for future studies. The techniques to generate genome-wide datasets at DNA, mRNA, protein, and whole cell levels are now established; the challenge, and the focus of the remainder of this chapter, is to integrate these layers of information to build predictive models describing host–pathogen interactions. For example, a greater understanding of the order of events during infection, mapping how interactions change over time, combined with targeted gene knockout/knockdown approaches promises to further unravel this destructive host–*M. tuberculosis* relationship. Such approaches may expose the functional significance of genes whose roles are currently unknown and which make up around 40 % of *M. tuberculosis* genes differentially regulated intracellularly [59]. To do this effectively we need mathematical models that are capable of mapping and forecasting these dynamic interactions between host immune cells and infecting pathogen.

7 Systems Biology and Modelling the Dialogue Between Host and Pathogen

The modelling of host–pathogen interactions is being actively pursued [73]; however, this approach is still in its infancy. Although mathematical models have a long history in biological science [74], their widespread application is a more recent phenomenon, linked to the field of systems biology, that has emerged over the last 15 years. Modelling can be performed on many scales (from molecular dynamics to whole organisms) and the entities that are modelled can be discrete (number of molecules) or continuous (concentrations, probabilities). Similarly, time (discrete time points, continuous time) and space (well-stirred solution, continuous concentration gradients, discrete neighbourhoods/microenvironments) can be considered in various ways. The choice of modelling method depends on the available knowledge and the phenomena that are to be investigated. Generally speaking, when the processes involved are known in sufficient detail, differential equations are often applied as they have been used extensively in the natural sciences, in particular physics, and are amenable to the mathematical analysis of large datasets. In a typical scenario, when the available knowledge is incomplete, discrete (variables and time) models are a good starting point. Modelling strategies used in host–pathogen systems biology have been reviewed by Forst [75], as has the use of models to complement experimentation by Kirschner and Lindermann [76]. The application of systems biology to tuberculosis research was reviewed by Young et al. [77]. Modelling host–pathogen interactions represents a particular challenge due to the multitude of different cell types that participate in the immune response to infection. Even if only direct connections between pathogens and host cells are considered the situation remains complex as infection can proceed in various ways. Since any modelling effort seeks to start with simple models, construction of models describing complex host–pathogen interactions has only begun in recent years.

8 Interaction Databases and Network Maps

Many models operate at the molecular level; therefore, it is a necessary first step to generate an overview of the possible interactions in the system. These may be taken from the relevant literature as well as interaction and pathway databases. A number of such databases exist and are detailed at http://www.pathguide.org [78]. Of particular interest is InnateDB [79], a database of interactions relevant to innate immunity in human and murine cells. Besides integrating data from external sources, InnateDB employs a curation team that uses the literature to specifically collect experimentally validated interactions in innate immunity. These interactions may be viewed in a pathway context mapping gene expression data onto them. This makes it possible to find pathways in which modulated genes are overrepresented. For this analysis, the pathways can be considered as models, because they represent the context in which interactions are thought to have a functional relevance. As an additional feature, InnateDB can use gene expression data to look for enrichment of transcription factor binding sites in up- and down-regulated genes. The putative TF-binding sites are mined from the cisRED database [80] which specialises in the prediction of these sites.

In addition to information about these molecular interactions, every modelling effort also requires data for model evaluation. This can be found in the literature or deposited in databases. Databases with particular relevance to host–pathogen interplay include http://www.macrophages.com, http://www.signaling-gateway.org [81], http://www.tbdb.org [82], or BugsBase (http://www.bugs.sgul.ac.uk/bugsbase).

These sites provide many types of datasets, in particular microarray, protein expression and protein regulation studies. Simple models built from interactions without specifying type or function may be generated and interrogated. For instance, Brodsky and Medzhitov [83] investigated targets of bacterial pathogens in protein–protein interaction networks of immune signalling. Their analysis suggests that pathogens which cause acute infection tend to target highly interconnected nodes of the network, while in chronic infections nodes with only a few connections are primarily targeted. Dyer et al. [84] surveyed the landscape of human proteins that interact with pathogens. Interestingly, the vast majority of interactions that they observed were from viral systems. They found that pathogens often target proteins that act as hubs, directly participating in a large number of interactions or involved in many different signalling pathways. At the next level of complexity, simple interaction networks may be annotated more richly to distinguish between the many different types of processes and components involved. This can be achieved in a standardised manner using existing ontologies and description standards (for example, gene ontology [85] and Systems Biology Graphical Notation [86]). In recent years, several descriptive models (pathway maps) relevant for host–pathogen interaction have been published [87–90]. These maps can be viewed as a kind of systematic knowledge representation which is complementary to classical review articles. In addition, it is often possible to overlay genome-wide data onto these maps for visualisation and analysis purposes. This provides a quick overview of the key features of the dataset and allows users to recognise interactions that may potentially form functional units in the experimental conditions tested. Although network maps cannot at present be used to calculate signalling outcomes or to make predictions about interference with the network, they serve as an excellent basis for new computational modelling efforts.

9 Models of Host–Pathogen Interactions

A basic model to predict cell-mediated immune-regulatory mechanisms during *M. tuberculosis* infection was proposed by Wigginton and Kirschner [91]. Ordinary differential equations were used to model the interplay between macrophages (resting *vs.* activated), *M. tuberculosis* (extra- and intracellular) and $Th_{0/1/2}$-cells as mediated by four cytokines (IL-4, IL-10, IL-12, and IFNγ). Most parameters were derived from published experimental data and if that was not possible their order of magnitude was estimated by sensitivity analysis. The main goal of this study was to explore which elements of the host–pathogen dialogue led to active disease or latency (and possible reactivation). Extensive model analysis concluded that if the initial immune response was dominated by Th_2-type cells, then the infection resulted in active tuberculosis. The prediction was not definitive when the initial immune response was predominantly mediated by Th_1-type cells. This model was extended by Sud et al. [92] to investigate the effects of CD8+ T-cells on disease outcome. The authors found that the cytotoxic and IFNγ-producing subpopulations of CD8+

cells contribute differently to the outcome of disease and that disease may still be controlled if either subpopulation is removed. However, if all CD8+ T-cells are deleted then the result was always active disease. As a further extension of these two models, Marino et al. [93] investigated the reactivation of tuberculosis following anti-TNF treatment and suggested several strategies for minimising the reactivation risk during anti-TNF treatment. In a closely related model, partly constructed from those previously mentioned, Day et al. [94] explored the effect of early appearance of classically activated macrophages in the lung upon *M. tuberculosis* infection. Under normal conditions, alveolar macrophages were considered to be alternatively activated and hence have reduced pro-inflammatory potential. The simulations showed that a reduced time delay for classical activation led to lower bacterial loads; this model was used to investigate the effectiveness of IFNγ therapy aimed at reducing this delay.

Raman et al. [95] developed a qualitative model of host–pathogen crosstalk in tuberculosis geared towards the prediction of disease outcome which can either be active disease, persistent infection or bacterial clearance. The interactions, between *M. tuberculosis* and different types of immune cells (innate and adaptive), were primarily mediated by cytokines and *M. tuberculosis* virulence factors; however, the molecular basis of these effects was included only in limited detail. Most interactions were modelled as Boolean functions, but there were additional parameters of time (*e.g.* onset of adaptive immunity) and the growth/clearance rates for *M. tuberculosis* affecting bacterial load which were modelled as continuous variables. For model simulation, an asynchronous update rule was used with each time interval corresponding to roughly 1 day. The primary result, the statistical evaluation of disease outcome, was determined after multiple (*e.g.* 100) model runs. This scheme made it possible to systematically study how changes in parameters or node deletions modified disease outcome. For instance, disabling phagocytosis always resulted in active disease which would only occur in 13 % of simulation runs with default parameters. Although the latter result was expected, the model may also be utilised to build more intricate predictions. For example, the knockout of TGFβ or IL-10 increases bacterial clearance, although these cytokines are typically classified as anti-inflammatory. This highlights that such simple classifications may not always be helpful because the effects of many signalling molecules are strongly dependent on the context. Similar to the previous study, Thakar et al. [96, 97] have developed models for infection of the lung by two *Bordetella* species. In the first version of the model [96], the authors concentrated on discrete dynamics to investigate basic effects like persistence and clearance of the bacteria. As the approach used by Thakar et al. [96] is analogous to that used by Raman et al. [95] described above, we concentrate here on the second version of this model published in 2009 [97], which uses a hybrid dynamic approach to better describe available quantitative data. In the hybrid dynamic model, each node is described using both a discrete (Boolean) and a continuous variable. The value of a discrete variable depends on whether the continuous variable exceeds a certain threshold, with the threshold being a parameter of the model. To describe the time evolution of the continuous variables, the Boolean rules from the first model are used for the activation of the nodes. The deactivation is modelled with separate linear decay terms which together yields a system of piecewise linear differential equations. In this hybrid model the parameters do not directly correspond to kinetic or binding parameters that are usually considered in differential equation models. In order to identify actual parameter values, a large range was sampled and only such parameter combinations were selected that reproduced certain well-known qualitative features of the infection dynamics. The parameters found in this manner were analysed further, searching for correlations to develop novel hypotheses for future experimental testing.

While the previous models consider the interactions of pathogens with different immune cell populations in an abstract manner, the model developed by Franke et al. [98] describes the crosstalk between *H. pylori* and epithelial cells in molecular detail. *H. pylori* is able, in a CagA-mediated process, to translocate into the host cell, triggering several events. In particular, the receptor tyrosine kinase c-Met, which normally plays a role in the context of human growth factor (HGF) signalling, may be recruited by CagA. The main target of CagA-induced immune-modulation is considered to be the MAP kinase ERK1/2, which is activated by stimulation with HGF or CagA. The interactions in this model are represented by Boolean functions and as a first step the interaction graph underlying the logical network was analysed. In this graph, only the information concerning positive and negative regulatory events was retained. The dependency matrix, which collects network-wide interdependencies, was calculated on the basis of the interaction graph. This revealed that HGF can exert both activating and inhibiting influences on ERK1/2, while CagA acts solely as an activator. Following on from this, the logical states in the network after stimulation with either HGF or CagA were determined, which showed that the signal was propagated through partially distinct pathways. This resulted in the systematic search for interventions that would prevent ERK1/2 activation upon CagA stimulation without affecting HGF signalling. Several of the predictions generated in this manner were then tested and confirmed. This indicates that the model captured important features of a real signalling network and could thus be used to generate new hypotheses for experimental testing. Additional Boolean models of within-host immune interactions are reviewed by Thakar and Albert [99]. To summarise, modelling complex host–pathogen interactions is well under way; however, one particular challenge remains the detailed modelling of the gene expression layer. Although many models contain transcription factors and interactions with their binding sites, these are currently far from comprehensive for both host and pathogen.

10 Future Perspective

The crosstalk between *M. tuberculosis* and its human host is both complex and dynamic, as such genome-wide approaches are invaluable tools for the unbiased discovery of novel interactions which serve to inspire testable hypotheses. Computational models are becoming increasingly useful for mapping and interrogating these multi-layered datasets, as evidenced by the chapters in this book. Advances in single cell manipulations together with the development of relevant infection models will enable single cell interactions between host and pathogen to be characterised, revealing the population dynamics of *M. tuberculosis* infection. Such analyses will aid the development of new drugs and vaccines which are desperately needed to reduce the burden of tuberculosis disease worldwide. Recent exciting progress classifying disease states and exploring the impact of genetic variation in both *M. tuberculosis* and human populations strengthens the prospect of elucidating valuable biomarkers of disease and determining the genetic basis of disease susceptibility. Finally, the emerging significance of small regulatory RNAs and epigenetics in the field of infectious disease promises to uncover novel mechanisms affecting immune-modulation, offering multiple opportunities for future intervention.

References

1. Cole ST, Brosch R, Parkhill J, Garnier T, Churcher C, Harris D, Gordon SV, Eiglmeier K, Gas S, Barry CE 3rd, Tekaia F, Badcock K, Basham D, Brown D, Chillingworth T, Connor R, Davies R, Devlin K, Feltwell T, Gentles S, Hamlin N, Holroyd S, Hornsby T, Jagels K, Krogh A, McLean J, Moule S, Murphy L, Oliver K, Osborne J, Quail MA, Rajandream MA, Rogers J, Rutter S, Seeger K, Skelton J, Squares R, Squares S, Sulston JE, Taylor K, Whitehead S, Barrell BG (1998) Deciphering the biology of *Mycobacterium tuberculosis* from the complete genome sequence. Nature 393(6685):537–544. doi:10.1038/31159
2. Pelicic V, Jackson M, Reyrat JM, Jacobs WR Jr, Gicquel B, Guilhot C (1997) Efficient allelic exchange and transposon mutagenesis in *Mycobacterium tuberculosis*. Proc Natl Acad Sci USA 94(20):10955–10960
3. Hensel M, Shea JE, Gleeson C, Jones MD, Dalton E, Holden DW (1995) Simultaneous identification of bacterial virulence genes by negative selection. Science 269(5222):400–403
4. Camacho LR, Ensergueix D, Perez E, Gicquel B, Guilhot C (1999) Identification of a virulence gene cluster of *Mycobacterium tuberculosis* by signature-tagged transposon mutagenesis. Mol Microbiol 34(2):257–267
5. Cox JS, Chen B, McNeil M, Jacobs WR Jr (1999) Complex lipid determines tissue-specific replication of *Mycobacterium tuberculosis* in mice. Nature 402(6757):79–83
6. Astarie-Dequeker C, Le Guyader L, Malaga W, Seaphanh FK, Chalut C, Lopez A, Guilhot C (2009) Phthiocerol dimycocerosates of *M. tuberculosis* participate in macrophage invasion by inducing changes in the organization of plasma membrane lipids. PLoS Pathog 5(2):e1000289. doi:10.1371/journal.ppat.1000289
7. Caminero JA, Pena MJ, Campos-Herrero MI, Rodriguez JC, Garcia I, Cabrera P, Lafoz C, Samper S, Takiff H, Afonso O, Pavon JM, Torres MJ, van Soolingen D, Enarson DA, Martin C (2001) Epidemiological evidence of the spread of a *Mycobacterium tuberculosis* strain of the Beijing genotype on Gran Canaria Island. Am J Respir Crit Care Med 164(7):1165–1170
8. Sassetti CM, Boyd DH, Rubin EJ (2001) Comprehensive identification of conditionally essential genes in mycobacteria. Proc Natl Acad Sci USA 98(22):12712–12717. doi:10.1073/pnas.231275498
9. Sassetti CM, Boyd DH, Rubin EJ (2003) Genes required for mycobacterial growth defined by high density mutagenesis. Mol Microbiol 48(1):77–84
10. Sassetti CM, Rubin EJ (2003) Genetic requirements for mycobacterial survival during infection. Proc Natl Acad Sci USA 100(22):12989–12994
11. Jang J, Becq J, Gicquel B, Deschavanne P, Neyrolles O (2008) Horizontally acquired genomic islands in the tubercle bacilli. Trends Microbiol 16(7):303–308. doi:10.1016/j.tim.2008.04.005

12. Dutta NK, Mehra S, Didier PJ, Roy CJ, Doyle LA, Alvarez X, Ratterree M, Be NA, Lamichhane G, Jain SK, Lacey MR, Lackner AA, Kaushal D (2010) Genetic requirements for the survival of tubercle bacilli in primates. J Infect Dis 201(11):1743–1752. doi:10.1086/652497
13. Rosas-Magallanes V, Stadthagen-Gomez G, Rauzier J, Barreiro LB, Tailleux L, Boudou F, Griffin R, Nigou J, Jackson M, Gicquel B, Neyrolles O (2007) Signature-tagged transposon mutagenesis identifies novel *Mycobacterium tuberculosis* genes involved in the parasitism of human macrophages. Infect Immun 75(1):504–507. doi:10.1128/IAI.00058-06
14. Brodin P, Peguillet I, Christophe T, Fenistein D, Rauzier J, Levillain F, Poquet Y, Jang J, Carralot J, Schrimpton R, Genovesio A, Gonzalo Asensio J, Martin C, Stewart G, Gicquel B, Neyrolles O (2010) High content phenotypic cell-based visual screen identifies *Mycobacterium tuberculosis* genes involved in early phagosome maturation arrest. PLoS Pathog 6(9)
15. Pethe K, Swenson DL, Alonso S, Anderson J, Wang C, Russell DG (2004) Isolation of *Mycobacterium tuberculosis* mutants defective in the arrest of phagosome maturation. Proc Natl Acad Sci USA 101(37):13642–13647
16. Stewart GR, Patel J, Robertson BD, Rae A, Young DB (2005) Mycobacterial mutants with defective control of phagosomal acidification. PLoS Pathog 1(3):269–278
17. Rohde K, Yates RM, Purdy GE, Russell DG (2007) *Mycobacterium tuberculosis* and the environment within the phagosome. Immunol Rev 219:37–54. doi:10.1111/j.1600-065X.2007.00547.x
18. Arnvig KB, Young DB (2009) Identification of small RNAs in *Mycobacterium tuberculosis*. Mol Microbiol 73(3):397–408. doi:10.1111/j.1365-2958.2009.06777.x
19. Pym AS, Brodin P, Brosch R, Huerre M, Cole ST (2002) Loss of RD1 contributed to the attenuation of the live tuberculosis vaccines *Mycobacterium bovis* BCG and Mycobacterium microti. Mol Microbiol 46(3):709–717. doi:3237
20. Becq J, Gutierrez MC, Rosas-Magallanes V, Rauzier J, Gicquel B, Neyrolles O, Deschavanne P (2007) Contribution of horizontally acquired genomic islands to the evolution of the tubercle bacilli. Mol Biol Evol 24:1861–1871
21. Rosas-Magallanes V, Deschavanne P, Quintana-Murci L, Brosch R, Gicquel B, Neyrolles O (2006) Horizontal transfer of a virulence operon to the ancestor of *Mycobacterium tuberculosis*. Mol Biol Evol 23(6):1129–1135
22. Stinear TP, Seemann T, Harrison PF, Jenkin GA, Davies JK, Johnson PD, Abdellah Z, Arrowsmith C, Chillingworth T, Churcher C, Clarke K, Cronin A, Davis P, Goodhead I, Holroyd N, Jagels K, Lord A, Moule S, Mungall K, Norbertczak H, Quail MA, Rabbinowitsch E, Walker D, White B, Whitehead S, Small PL, Brosch R, Ramakrishnan L, Fischbach MA, Parkhill J, Cole ST (2008) Insights from the complete genome sequence of *Mycobacterium marinum* on the evolution of *Mycobacterium tuberculosis*. Genome Res 18(5):729–741. doi:10.1101/gr.075069.107
23. Veyrier F, Pletzer D, Turenne C, Behr MA (2009) Phylogenetic detection of horizontal gene transfer during the step-wise genesis of *Mycobacterium tuberculosis*. BMC Evol Biol 9:196. doi:10.1186/1471-2148-9-196
24. Philips JA, Rubin EJ, Perrimon N (2005) Drosophila RNAi screen reveals CD36 family member required for mycobacterial infection. Science 309(5738):1251–1253. doi:10.1126/science.1116006
25. Koo IC, Ohol YM, Wu P, Morisaki JH, Cox JS, Brown EJ (2008) Role for lysosomal enzyme beta-hexosaminidase in the control of mycobacteria infection. Proc Natl Acad Sci USA 105(2):710–715. doi:10.1073/pnas.0708110105
26. Kuijl C, Savage ND, Marsman M, Tuin AW, Janssen L, Egan DA, Ketema M, van den Nieuwendijk R, van den Eeden SJ, Geluk A, Poot A, van der Marel G, Beijersbergen RL, Overkleeft H, Ottenhoff TH, Neefjes J (2007) Intracellular bacterial growth is controlled by a kinase network around PKB/AKT1. Nature 450(7170):725–730. doi:10.1038/nature06345
27. Kumar D, Nath L, Kamal MA, Varshney A, Jain A, Singh S, Rao KV (2010) Genome-wide analysis of the host intracellular network that regulates survival of *Mycobacterium tuberculosis*. Cell 140(5):731–743. doi:10.1016/j.cell.2010.02.012

28. Graham JE, Clark-Curtiss JE (1999) Identification of *Mycobacterium tuberculosis* RNAs synthesized in response to phagocytosis by human macrophages by selective capture of transcribed sequences (SCOTS). Proc Natl Acad Sci USA 96(20):11554–11559
29. Li MS, Monahan IM, Waddell SJ, Mangan JA, Martin SL, Everett MJ, Butcher PD (2001) cDNA-RNA subtractive hybridization reveals increased expression of mycocerosic acid synthase in intracellular *Mycobacterium bovis* BCG. Microbiology 147(Pt 8):2293–2305
30. Dubnau E, Fontan P, Manganelli R, Soares-Appel S, Smith I (2002) *Mycobacterium tuberculosis* genes induced during infection of human macrophages. Infect Immun 70(6):2787–2795
31. Fenhalls G, Stevens L, Moses L, Bezuidenhout J, Betts JC, Helden Pv P, Lukey PT, Duncan K (2002) In situ detection of *Mycobacterium tuberculosis* transcripts in human lung granulomas reveals differential gene expression in necrotic lesions. Infect Immun 70(11):6330–6338
32. Volpe E, Cappelli G, Grassi M, Martino A, Serafino A, Colizzi V, Sanarico N, Mariani F (2006) Gene expression profiling of human macrophages at late time of infection with *Mycobacterium tuberculosis*. Immunology 118(4):449–460. doi:10.1111/j.1365-2567.2006.02378.x
33. Gregory SG, Sekhon M, Schein J, Zhao S, Osoegawa K, Scott CE, Evans RS, Burridge PW, Cox TV, Fox CA, Hutton RD, Mullenger IR, Phillips KJ, Smith J, Stalker J, Threadgold GJ, Birney E, Wylie K, Chinwalla A, Wallis J, Hillier L, Carter J, Gaige T, Jaeger S, Kremitzki C, Layman D, Maas J, McGrane R, Mead K, Walker R, Jones S, Smith M, Asano J, Bosdet I, Chan S, Chittaranjan S, Chiu R, Fjell C, Fuhrmann D, Girn N, Gray C, Guin R, Hsiao L, Krzywinski M, Kutsche R, Lee SS, Mathewson C, McLeavy C, Messervier S, Ness S, Pandoh P, Prabhu AL, Saeedi P, Smailus D, Spence L, Stott J, Taylor S, Terpstra W, Tsai M, Vardy J, Wye N, Yang G, Shatsman S, Ayodeji B, Geer K, Tsegaye G, Shvartsbeyn A, Gebregeorgis E, Krol M, Russell D, Overton L, Malek JA, Holmes M, Heaney M, Shetty J, Feldblyum T, Nierman WC, Catanese JJ, Hubbard T, Waterston RH, Rogers J, de Jong PJ, Fraser CM, Marra M, McPherson JD, Bentley DR (2002) A physical map of the mouse genome. Nature 418(6899):743–750. doi:10.1038/nature00957
34. International Human Genome Sequencing Consortium (2004) Finishing the euchromatic sequence of the human genome. Nature 431(7011):931–945. doi:10.1038/nature03001
35. Hegedus Z, Zakrzewska A, Agoston VC, Ordas A, Racz P, Mink M, Spaink HP, Meijer AH (2009) Deep sequencing of the zebrafish transcriptome response to mycobacterium infection. Mol Immunol 46(15):2918–2930. doi:10.1016/j.molimm.2009.07.002
36. Rainen L, Oelmueller U, Jurgensen S, Wyrich R, Ballas C, Schram J, Herdman C, Bankaitis-Davis D, Nicholls N, Trollinger D, Tryon V (2002) Stabilization of mRNA expression in whole blood samples. Clin Chem 48(11):1883–1890
37. Mangan JA, Sole KM, Mitchison DA, Butcher PD (1997) An effective method of RNA extraction from bacteria refractory to disruption, including mycobacteria. Nucleic Acids Res 25(3):675–676
38. Rachman H, Lee JS, Angermann J, Kowall J, Kaufmann SH (2006) Reliable amplification method for bacterial RNA. J Biotechnol 126(1):61–68. doi:10.1016/j.jbiotec.2006.02.020
39. Van Gelder RN, von Zastrow ME, Yool A, Dement WC, Barchas JD, Eberwine JH (1990) Amplified RNA synthesized from limited quantities of heterogeneous cDNA. Proc Natl Acad Sci USA 87(5):1663–1667
40. Waddell SJ, Laing K, Senner C, Butcher PD (2008) Microarray analysis of defined *Mycobacterium tuberculosis* populations using RNA amplification strategies. BMC Genomics 9:94. doi:10.1186/1471-2164-9-94
41. Nau GJ, Richmond JF, Schlesinger A, Jennings EG, Lander ES, Young RA (2002) Human macrophage activation programs induced by bacterial pathogens. Proc Natl Acad Sci USA 99(3):1503–1508. doi:10.1073/pnas.022649799
42. Spira A, Carroll JD, Liu G, Aziz Z, Shah V, Kornfeld H, Keane J (2003) Apoptosis genes in human alveolar macrophages infected with virulent or attenuated *Mycobacterium tuberculosis*: a pivotal role for tumor necrosis factor. Am J Respir Cell Mol Biol 29(5):545–551. doi:10.1165/rcmb.2002-0310OC
43. Chaussabel D, Semnani RT, McDowell MA, Sacks D, Sher A, Nutman TB (2003) Unique gene expression profiles of human macrophages and dendritic cells to phylogenetically distinct parasites. Blood 102(2):672–681. doi:10.1182/blood-2002-10-3232

44. Tailleux L, Waddell SJ, Pelizzola M, Mortellaro A, Withers M, Tanne A, Castagnoli PR, Gicquel B, Stoker NG, Butcher PD, Foti M, Neyrolles O (2008) Probing host pathogen cross-talk by transcriptional profiling of both *Mycobacterium tuberculosis* and infected human dendritic cells and macrophages. PLoS One 3(1):e1403
45. Botella H, Peyron P, Levillain F, Poincloux R, Poquet Y, Brandli I, Wang C, Tailleux L, Tilleul S, Charrière G, Waddell S, Foti M, Lugo-Villarino G, de Chastellier C, Neyrolles O (2011) Mycobacterial P1-type ATPases mediate resistance to zinc poisoning in human macrophages. Cell Host Microbe 10(3):248–259
46. Keller C, Lauber J, Blumenthal A, Buer J, Ehlers S (2004) Resistance and susceptibility to tuberculosis analysed at the transcriptome level: lessons from mouse macrophages. Tuberculosis (Edinb) 84(3–4):144–158. doi:10.1016/j.tube.2003.12.003
47. Thuong NT, Dunstan SJ, Chau TT, Thorsson V, Simmons CP, Quyen NT, Thwaites GE, Thi Ngoc Lan N, Hibberd M, Teo YY, Seielstad M, Aderem A, Farrar JJ, Hawn TR (2008) Identification of tuberculosis susceptibility genes with human macrophage gene expression profiles. PLoS Pathog 4(12):e1000229. doi:10.1371/journal.ppat.1000229
48. Ehrt S, Schnappinger D, Bekiranov S, Drenkow J, Shi S, Gingeras TR, Gaasterland T, Schoolnik G, Nathan C (2001) Reprogramming of the macrophage transcriptome in response to interferon-gamma and *Mycobacterium tuberculosis*: signaling roles of nitric oxide synthase-2 and phagocyte oxidase. J Exp Med 194(8):1123–1140
49. Shi S, Blumenthal A, Hickey CM, Gandotra S, Levy D, Ehrt S (2005) Expression of many immunologically important genes in *Mycobacterium tuberculosis*-infected macrophages is independent of both TLR2 and TLR4 but dependent on IFN-alphabeta receptor and STAT1. J Immunol 175(5):3318–3328
50. Russell DG, VanderVen BC, Lee W, Abramovitch RB, Kim MJ, Homolka S, Niemann S, Rohde KH (2010) *Mycobacterium tuberculosis* wears what it eats. Cell Host Microbe 8(1):68–76. doi:10.1016/j.chom.2010.06.002
51. Waddell SJ (2010) Reprogramming the *Mycobacterium tuberculosis* transcriptome during host pathogenesis. Drug Discov Today 7(1):e67–e73
52. Schnappinger D, Ehrt S, Voskuil MI, Liu Y, Mangan JA, Monahan IM, Dolganov G, Efron B, Butcher PD, Nathan C, Schoolnik GK (2003) Transcriptional Adaptation of *Mycobacterium tuberculosis* within Macrophages: insights into the phagosomal environment. J Exp Med 198(5):693–704. doi:10.1084/jem.20030846
53. Fontan P, Aris V, Ghanny S, Soteropoulos P, Smith I (2008) Global transcriptional profile of *Mycobacterium tuberculosis* during THP-1 human macrophage infection. Infect Immun 76(2):717–725. doi:10.1128/IAI.00974-07
54. Van der Geize R, Yam K, Heuser T, Wilbrink MH, Hara H, Anderton MC, Sim E, Dijkhuizen L, Davies JE, Mohn WW, Eltis LD (2007) A gene cluster encoding cholesterol catabolism in a soil actinomycete provides insight into *Mycobacterium tuberculosis* survival in macrophages. Proc Natl Acad Sci USA 104(6):1947–1952. doi:10.1073/pnas.0605728104
55. Tailleux L, Neyrolles O, Honore-Bouakline S, Perret E, Sanchez F, Abastado JP, Lagrange PH, Gluckman JC, Rosenzwajg M, Herrmann JL (2003) Constrained intracellular survival of *Mycobacterium tuberculosis* in human dendritic cells. J Immunol 170(4):1939–1948
56. Rohde KH, Abramovitch RB, Russell DG (2007) *Mycobacterium tuberculosis* invasion of macrophages: linking bacterial gene expression to environmental cues. Cell Host Microbe 2(5):352–364. doi:10.1016/j.chom.2007.09.006
57. Abramovitch RB, Rohde KH, Hsu FF, Russell DG (2011) aprABC: a *Mycobacterium tuberculosis* complex-specific locus that modulates pH-driven adaptation to the macrophage phagosome. Mol Microbiol 80(3):678–694. doi:10.1111/j.1365-2958.2011.07601.x
58. Li AH, Waddell SJ, Hinds J, Malloff CA, Bains M, Hancock RE, Lam WL, Butcher PD, Stokes RW (2010) Contrasting transcriptional responses of a virulent and an attenuated strain of *Mycobacterium tuberculosis* infecting macrophages. PLoS One 5(6):e11066. doi:10.1371/journal.pone.0011066
59. Homolka S, Niemann S, Russell DG, Rohde KH (2010) Functional genetic diversity among *Mycobacterium tuberculosis* complex clinical isolates: delineation of conserved core and

lineage-specific transcriptomes during intracellular survival. PLoS Pathog 6(7):e1000988. doi:10.1371/journal.ppat.1000988
60. Walzl G, Ronacher K, Hanekom W, Scriba TJ, Zumla A (2011) Immunological biomarkers of tuberculosis. Nat Rev Immunol 11(5):343–354. doi:10.1038/nri2960
61. Mollenkopf HJ, Hahnke K, Kaufmann SH (2006) Transcriptional responses in mouse lungs induced by vaccination with *Mycobacterium bovis* BCG and infection with *Mycobacterium tuberculosis*. Microbes Infect 8(1):136–144. doi:10.1016/j.micinf.2005.06.015
62. Talaat AM, Lyons R, Howard ST, Johnston SA (2004) The temporal expression profile of *Mycobacterium tuberculosis* infection in mice. Proc Natl Acad Sci USA 101(13):4602–4607. doi:10.1073/pnas.0306023101
63. Talaat AM, Ward SK, Wu CW, Rondon E, Tavano C, Bannantine JP, Lyons R, Johnston SA (2007) Mycobacterial bacilli are metabolically active during chronic tuberculosis in murine lungs: insights from genome-wide transcriptional profiling. J Bacteriol 189(11):4265–4274. doi:10.1128/JB.00011-07
64. Mehra S, Pahar B, Dutta NK, Conerly CN, Philippi-Falkenstein K, Alvarez X, Kaushal D (2010) Transcriptional reprogramming in nonhuman primate (rhesus macaque) tuberculosis granulomas. PLoS One 5(8):e12266. doi:10.1371/journal.pone.0012266
65. Kim MJ, Wainwright HC, Locketz M, Bekker LG, Walther GB, Dittrich C, Visser A, Wang W, Hsu FF, Wiehart U, Tsenova L, Kaplan G, Russell DG (2010) Caseation of human tuberculosis granulomas correlates with elevated host lipid metabolism. EMBO Mol Med 2(7):258–274. doi:10.1002/emmm.201000079
66. Rachman H, Strong M, Ulrichs T, Grode L, Schuchhardt J, Mollenkopf H, Kosmiadi GA, Eisenberg D, Kaufmann SH (2006) Unique transcriptome signature of *Mycobacterium tuberculosis* in pulmonary tuberculosis. Infect Immun 74(2):1233–1242. doi:10.1128/IAI.74.2.1233-1242.2006
67. Garton NJ, Waddell SJ, Sherratt AL, Lee SM, Smith RJ, Senner C, Hinds J, Rajakumar K, Adegbola RA, Besra GS, Butcher PD, Barer MR (2008) Cytological and transcript analyses reveal fat and lazy persister-like bacilli in tuberculous sputum. PLoS Med 5(4):e75. doi:10.1371/journal.pmed.0050075
68. Grassi M, Bocchino M, Marruchella A, Volpe E, Saltini C, Colizzi V, Mariani F (2006) Transcriptional profile of the immune response in the lungs of patients with active tuberculosis. Clin Immunol 121(1):100–107. doi:10.1016/j.clim.2006.06.008
69. Raju B, Hoshino Y, Belitskaya-Levy I, Dawson R, Ress S, Gold JA, Condos R, Pine R, Brown S, Nolan A, Rom WN, Weiden MD (2008) Gene expression profiles of bronchoalveolar cells in pulmonary TB. Tuberculosis (Edinb) 88(1):39–51. doi:10.1016/j.tube.2007.07.003
70. Mistry R, Cliff JM, Clayton CL, Beyers N, Mohamed YS, Wilson PA, Dockrell HM, Wallace DM, van Helden PD, Duncan K, Lukey PT (2007) Gene-expression patterns in whole blood identify subjects at risk for recurrent tuberculosis. J Infect Dis 195(3):357–365. doi:10.1086/510397
71. Maertzdorf J, Repsilber D, Parida SK, Stanley K, Roberts T, Black G, Walzl G, Kaufmann SH (2011) Human gene expression profiles of susceptibility and resistance in tuberculosis. Genes Immun 12(1):15–22. doi:10.1038/gene.2010.51
72. Berry MP, Graham CM, McNab FW, Xu Z, Bloch SA, Oni T, Wilkinson KA, Banchereau R, Skinner J, Wilkinson RJ, Quinn C, Blankenship D, Dhawan R, Cush JJ, Mejias A, Ramilo O, Kon OM, Pascual V, Banchereau J, Chaussabel D, O'Garra A (2010) An interferon-inducible neutrophil-driven blood transcriptional signature in human tuberculosis. Nature 466(7309):973–977. doi:10.1038/nature09247
73. Aderem A, Adkins JN, Ansong C, Galagan J, Kaiser S, Korth MJ, Law GL, McDermott JG, Proll SC, Rosenberger C, Schoolnik G, Katze MG (2011) A systems biology approach to infectious disease research: innovating the pathogen-host research paradigm. MBio 2(1):e00325–e00310. doi:10.1128/mBio.00325-10
74. Lotka AJ (1920) Analytical note on certain rhythmic relations in organic systems. Proc Natl Acad Sci USA 6(7):410–415
75. Forst CV (2006) Host–pathogen systems biology. Drug Discov Today 11(5–6):220–227. doi:10.1016/S1359-6446(05)03735-9

76. Kirschner DE, Linderman JJ (2009) Mathematical and computational approaches can complement experimental studies of host–pathogen interactions. Cell Microbiol 11(4):531–539. doi:10.1111/j.1462-5822.2008.01281.x
77. Young D, Stark J, Kirschner D (2008) Systems biology of persistent infection: tuberculosis as a case study. Nat Rev Microbiol 6(7):520–528. doi:10.1038/nrmicro1919
78. Bader GD, Cary MP, Sander C (2006) Pathguide: a pathway resource list. Nucleic Acids Res 34(Database issue):D504–D506. doi:10.1093/nar/gkj126
79. Lynn DJ, Winsor GL, Chan C, Richard N, Laird MR, Barsky A, Gardy JL, Roche FM, Chan TH, Shah N, Lo R, Naseer M, Que J, Yau M, Acab M, Tulpan D, Whiteside MD, Chikatamarla A, Mah B, Munzner T, Hokamp K, Hancock RE, Brinkman FS (2008) InnateDB: facilitating systems-level analyses of the mammalian innate immune response. Mol Syst Biol 4:218. doi:10.1038/msb.2008.55
80. Robertson G, Bilenky M, Lin K, He A, Yuen W, Dagpinar M, Varhol R, Teague K, Griffith OL, Zhang X, Pan Y, Hassel M, Sleumer MC, Pan W, Pleasance ED, Chuang M, Hao H, Li YY, Robertson N, Fjell C, Li B, Montgomery SB, Astakhova T, Zhou J, Sander J, Siddiqui AS, Jones SJ (2006) cisRED: a database system for genome-scale computational discovery of regulatory elements. Nucleic Acids Res 34(Database issue):D68–D73. doi:10.1093/nar/gkj075
81. Natarajan M, Lin KM, Hsueh RC, Sternweis PC, Ranganathan R (2006) A global analysis of cross-talk in a mammalian cellular signalling network. Nat Cell Biol 8(6):571–580. doi:10.1038/ncb1418
82. Reddy TB, Riley R, Wymore F, Montgomery P, DeCaprio D, Engels R, Gellesch M, Hubble J, Jen D, Jin H, Koehrsen M, Larson L, Mao M, Nitzberg M, Sisk P, Stolte C, Weiner B, White J, Zachariah ZK, Sherlock G, Galagan JE, Ball CA, Schoolnik GK (2009) TB database: an integrated platform for tuberculosis research. Nucleic Acids Res 37(Database issue):D499–D508. doi:10.1093/nar/gkn652
83. Brodsky IE, Medzhitov R (2009) Targeting of immune signalling networks by bacterial pathogens. Nat Cell Biol 11(5):521–526. doi:10.1038/ncb0509-521
84. Dyer MD, Murali TM, Sobral BW (2008) The landscape of human proteins interacting with viruses and other pathogens. PLoS Pathog 4(2):e32. doi:10.1371/journal.ppat.0040032
85. Ashburner M, Ball CA, Blake JA, Botstein D, Butler H, Cherry JM, Davis AP, Dolinski K, Dwight SS, Eppig JT, Harris MA, Hill DP, Issel-Tarver L, Kasarskis A, Lewis S, Matese JC, Richardson JE, Ringwald M, Rubin GM, Sherlock G (2000) Gene ontology: tool for the unification of biology. The Gene Ontology Consortium. Nat Genet 25(1):25–29. doi:10.1038/75556
86. Le Novere N, Hucka M, Mi H, Moodie S, Schreiber F, Sorokin A, Demir E, Wegner K, Aladjem MI, Wimalaratne SM, Bergman FT, Gauges R, Ghazal P, Kawaji H, Li L, Matsuoka Y, Villeger A, Boyd SE, Calzone L, Courtot M, Dogrusoz U, Freeman TC, Funahashi A, Ghosh S, Jouraku A, Kim S, Kolpakov F, Luna A, Sahle S, Schmidt E, Watterson S, Wu G, Goryanin I, Kell DB, Sander C, Sauro H, Snoep JL, Kohn K, Kitano H (2009) The systems biology graphical notation. Nat Biotechnol 27(8):735–741. doi:10.1038/nbt.1558
87. Cavalieri D, Rivero D, Beltrame L, Buschow SI, Calura E, Rizzetto L, Gessani S, Gauzzi MC, Reith W, Baur A, Bonaiuti R, Brandizi M, De Filippo C, D'Oro U, Draghici S, Dunand-Sauthier I, Gatti E, Granucci F, Gundel M, Kramer M, Kuka M, Lanyi A, Melief CJ, van Montfoort N, Ostuni R, Pierre P, Popovici R, Rajnavolgyi E, Schierer S, Schuler G, Soumelis V, Splendiani A, Stefanini I, Torcia MG, Zanoni I, Zollinger R, Figdor CG, Austyn JM (2010) DC-ATLAS: a systems biology resource to dissect receptor specific signal transduction in dendritic cells. Immunome Res 6:10. doi:10.1186/1745-7580-6-10
88. Oda K, Kitano H (2006) A comprehensive map of the toll-like receptor signaling network. Mol Syst Biol 2:2006.0015. doi:10.1038/msb4100057
89. Patil S, Pincas H, Seto J, Nudelman G, Nudelman I, Sealfon SC (2010) Signaling network of dendritic cells in response to pathogens: a community-input supported knowledgebase. BMC Syst Biol 4:137. doi:10.1186/1752-0509-4-137
90. Raza S, McDerment N, Lacaze PA, Robertson K, Watterson S, Chen Y, Chisholm M, Eleftheriadis G, Monk S, O'Sullivan M, Turnbull A, Roy D, Theocharidis A, Ghazal P,

Freeman TC (2010) Construction of a large scale integrated map of macrophage pathogen recognition and effector systems. BMC Syst Biol 4:63. doi:10.1186/1752-0509-4-63
91. Wigginton JE, Kirschner D (2001) A model to predict cell-mediated immune regulatory mechanisms during human infection with *Mycobacterium tuberculosis*. J Immunol 166(3):1951–1967
92. Sud D, Bigbee C, Flynn JL, Kirschner DE (2006) Contribution of CD8+ T cells to control of *Mycobacterium tuberculosis* infection. J Immunol 176(7):4296–4314
93. Marino S, Sud D, Plessner H, Lin PL, Chan J, Flynn JL, Kirschner DE (2007) Differences in reactivation of tuberculosis induced from anti-TNF treatments are based on bioavailability in granulomatous tissue. PLoS Comput Biol 3(10):1909–1924. doi:10.1371/journal.pcbi.0030194
94. Day J, Friedman A, Schlesinger LS (2009) Modeling the immune rheostat of macrophages in the lung in response to infection. Proc Natl Acad Sci USA 106(27):11246–11251. doi:10.1073/pnas.0904846106
95. Raman K, Bhat AG, Chandra N (2010) A systems perspective of host–pathogen interactions: predicting disease outcome in tuberculosis. Mol Biosyst 6(3):516–530. doi:10.1039/b912129c
96. Thakar J, Pilione M, Kirimanjeswara G, Harvill ET, Albert R (2007) Modeling systems-level regulation of host immune responses. PLoS Comput Biol 3(6):e109. doi:10.1371/journal.pcbi.0030109
97. Thakar J, Saadatpour-Moghaddam A, Harvill ET, Albert R (2009) Constraint-based network model of pathogen-immune system interactions. J R Soc Interface 6(36):599–612. doi:10.1098/rsif.2008.0363
98. Franke R, Muller M, Wundrack N, Gilles ED, Klamt S, Kahne T, Naumann M (2008) Host–pathogen systems biology: logical modelling of hepatocyte growth factor and Helicobacter pylori induced c-Met signal transduction. BMC Syst Biol 2:4. doi:10.1186/1752-0509-2-4
99. Thakar J, Albert R (2010) Boolean models of within-host immune interactions. Curr Opin Microbiol 13(3):377–381. doi:10.1016/j.mib.2010.04.003

Chapter 7
A Systems Biology Approach for Understanding Granuloma Formation and Function in Tuberculosis

Mohammad Fallahi-Sichani, Simeone Marino, JoAnne L. Flynn, Jennifer J. Linderman, and Denise E. Kirschner

Abstract The pathologic hallmark of tuberculosis is the granuloma. A granuloma is a multifaceted cellular structure that serves to focus the host immune response, contain infection and pathology, and provide a niche for the bacillus to persist within the host. Granulomas form in response to *Mycobacterium tuberculosis* infection, and if a granuloma is capable of inhibiting or killing most of the *M. tuberculosis* present, humans develop a clinically latent infection. However, if a granuloma is impaired in function, infection progresses, granulomas enlarge, and bacteria seed new granulomas; this results in progressive pathology and disease, i.e., active tuberculosis. In clinical latency, immunologic perturbation at the level of the granuloma can result in reactivation of infection. In humans, there are a variety of granuloma types, even within the lungs of a single host.

The roles and interactions of various cells (macrophages, T cells, B cells, and neutrophils) and molecules (cytokines, chemokines, and effector molecules) within a granuloma are complex and challenging to address by experimental methods alone. Computational approaches, in particular agent-based modeling, can be used

M. Fallahi-Sichani, Ph.D.
Department of Systems Biology, Harvard Medical School, Boston, MA 02115, USA
e-mail: fallahi@umich.edu

S. Marino, Ph.D. • D.E. Kirschner, Ph.D. (✉)
Department of Microbiology and Immunology, University of Michigan Medical School, Ann Arbor, MI 48109, USA
e-mail: simeonem@umich.edu; kirschne@umich.edu

J.L. Flynn, Ph.D.
Department of Microbiology and Molecular Genetics, University of Pittsburgh School of Medicine, Pittsburgh, PA 15261, USA
e-mail: joanne@pitt.edu

J.J. Linderman, Ph.D. (✉)
Department of Chemical Engineering, University of Michigan, Ann Arbor, MI 48109, USA
e-mail: linderma@umich.edu

J. McFadden et al. (eds.), *Systems Biology of Tuberculosis*,
DOI 10.1007/978-1-4614-4966-9_7, © Springer Science+Business Media, LLC 2013

to dissect the temporal and spatial aspects of granuloma formation and function. Here we explain how a systems biology approach can integrate experimental and computational work to address critical questions necessary to understanding granulomas and contribute to the development and testing of strategies for prevention and treatment.

1 Introduction

1.1 Granuloma Formation and Function

Mycobacterium tuberculosis is primarily a respiratory pathogen, transmitted via aerosol from a person with active tuberculosis to another person. Once in the airways, the bacillus encounters various cells, primarily alveolar macrophages and dendritic cells (DCs). Although the early events in transmission are poorly understood, it is believed that many bacilli may be destroyed by alveolar macrophages or other airway cells. Surviving bacilli transit to the parenchyma, begin to replicate, and initiate an inflammatory response. Dendritic cells engulf bacilli and migrate to the thoracic lymph nodes (LNs) to prime a T cell response [1]. At least one thoracic LN becomes infected, and often stays infected (the classic Ghon complex is the initial lung lesion and the associated infected LN).

The inflammatory response in the lung likely results in recruitment of cells from the blood, including macrophages and neutrophils, which attempt to contain the infection, but also contributes to additional inflammation. In the LN, the T cell response takes at least 2 weeks to be primed [2–4], at which point the T cells migrate to the lungs and participate in granuloma formation. A granuloma forms in response to a chronic antigenic stimulus in the context of macrophage inflammation and is the pathologic hallmark of mycobacterial infections, including tuberculosis. A tuberculous granuloma can take many different forms (see Fig. 7.1), but is generally composed primarily of macrophages and lymphocytes, organized into a spherical structure (for review, see [5]). The classic caseous granuloma consists of a centrally necrotic area (grossly resembling cheese and hence the term "caseous"), surrounded by layers of macrophages, which are in turn surrounded by a smaller cuff of lymphocytes. The lymphocytic cuff can contain both CD4+ and CD8+ T cells, but B cells, including plasma cells, are also prominent. Multiple other cell types, including neutrophils, DCs, and fibroblasts, can also be found in granulomas.

There are several "types" of macrophages within granulomas, including epithelioid macrophages, foamy macrophages, Langhans' giant cells, classically activated macrophages, and alternatively activated macrophages (see [6] for review). The roles and functions of each macrophage subset in the granuloma in terms of control of bacterial proliferation and pathology are not known, but one can speculate that classically activated macrophages participate in bacterial killing while alternatively activated macrophages may be important in preventing excessive pathology or promoting fibrosis.

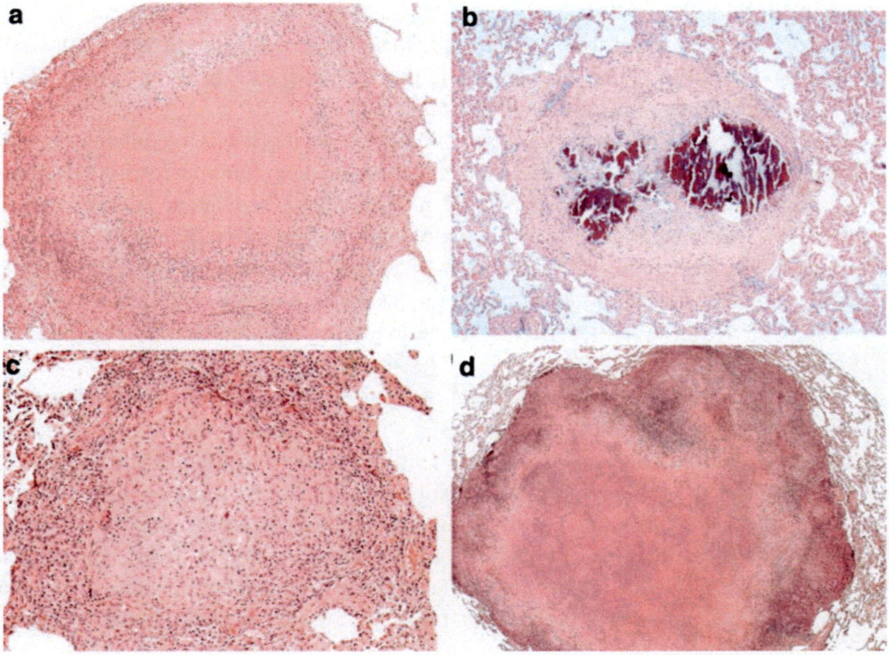

Fig. 7.1 Microscopic histopathology images of different types of tuberculosis granulomas from lungs of cynomolgus macaque model. (**a**) A caseous granuloma consisting of a central area of caseum surrounded by a mantle of epithelioid macrophages and peripherally located cuff of lymphocytes from a monkey with active disease, ×5 H&E. (**b**) A well-circumscribed non-necrotizing (solid cellular) granuloma consisting of a core of epithelioid macrophages and peripheral lymphocytes from a monkey with active disease ×5 H&E. (**c**) A fibrocalcific granuloma consisting of centrally located mineral (darkly staining) and fibroblasts ×5 H&E. (**d**) A large caseous granuloma from an anti-TNF antibody treated monkey at 8 weeks post-infection (×2, H&E). (**a**) and (**b**) were reprinted with permission from American Society for Microbiology and originally appeared in Lin et al., Infection and Immunity, 2009 Vol 77, p 4637 (Fig. 5) DOI 10.1128/iai.00592-09. (**d**) was reprinted with permission from John Wiley & Sons Publishing, and originally appeared in Lin et al., 2010, Arthritis and Rheumatism, Vol 62, p 344 (Fig. 3) DOI 10.1002/art.27271

In addition to the classic caseous granuloma, there are other types of granulomas seen in humans, including non-necrotic (composed primarily of macrophages with sparse lymphocytes), suppurative (neutrophil-rich), fibrotic (primarily fibroblasts), and mineralized (calcified often in the setting of caseous necrosis). All granuloma types, even mineralized granulomas, can be seen in active TB, although a person may have a predominance of one type or another. In latent TB, the granuloma types appear to be primarily caseous, mineralized, or fibrotic, rather than non-necrotizing [7]. In active TB, granulomas can be multifocal and coalescing, and large consolidations or TB pneumonia can also occur. Granulomas can also break through to an airway, resulting in cavity formation, which appears to be a major factor in transmission [8], as the cavity allows escape of large numbers of bacilli from granuloma into the airways. Granulomas are also present in the thoracic LN that was initially infected, and more than one LN can be involved in a host with active or latent infection. Granulomas can also be seen in other organs, such as liver, spleen, brain, or bone in the case of extra-pulmonary disease.

The role of a granuloma from the host-centric point of view is to contain infection, destroy bacilli, and limit the pathology of the infection. A granuloma forms in response to infection, and granulomas are found in both active and latent *M. tuberculosis* infections, indicating that mere formation of a granuloma is insufficient to control the infection. The granuloma must function adequately to be a useful barrier to disease, i.e., to allow the host to either eliminate the bacilli completely or maintain sufficient control over bacterial replication that the infection remains clinically asymptomatic (i.e., latent infection).

From the mycobacteria-centric point of view, the granuloma serves a purpose as well. *M. tuberculosis* has evolved to persist within the granuloma for years, and can cause reactivation TB decades after the initial infection. Thus, the granuloma is a survival niche for the bacillus. In addition, the pathology caused by a granuloma, particularly one that results in cavity formation, is essential for efficient transmission of infection to a new host. It has been demonstrated in the zebrafish system that mycobacterial virulence factors actively participate in granuloma formation and cell recruitment [9], supporting the view that this structure is important in the pathogenesis of the infection.

Thus a granuloma, the hallmark of tuberculosis, is a structure that benefits both the host and the microbe: the central battle between host and microbe likely occurs at the level of the granuloma. Understanding and dissecting mechanisms, both host and bacterial, which occur during granuloma formation and function within each type of granuloma, will lead to a better understanding of this complex disease. This in turn will direct development of new therapeutic and preventive strategies to treat tuberculosis.

1.2 Key Cellular and Molecular Players Relevant to Granulomas

The immune responses induced by *M. tuberculosis* infection are myriad and complex, and even now it remains incompletely understood which responses are required for protection and which contribute to pathology [10, 11]. In truth, there is significant overlap among protective and pathologic responses, with the best outcome achieved by a balance of pro-inflammatory and anti-inflammatory responses, particularly at the level of the granuloma. It is well accepted that CD4+ and CD8+ T cells are important in defense against tuberculosis, although the mechanisms by which these cells provide protection are not completely understood. IFN-γ has been considered to be a key mechanism by which T cells activate macrophages for killing of bacilli [12, 13], although recent studies in mouse models suggest that CD4+ T cells can contribute in other ways as well [14, 15]. TNF is a cytokine that has been demonstrated to be important in preventing progression of initial infection or reactivation of latent infection in several animal models, including non-human primates [16, 17]. The use of TNF inhibitors as treatment for inflammatory diseases in humans has confirmed that TNF is a major player in the protective immune response against tuberculosis [18]. However, TNF has numerous functions in the human immune response and sorting out which are the relevant mechanisms is difficult in vivo. Although the role of B cells and antibodies in tuberculosis has not been established, some studies support the view that these cells are important contributors to protection [19].

1.3 A Systems Biology Approach to Understanding Granuloma Formation and Function

Despite decades of research on TB, our understanding of the factors that lead to active, latent, and reactivation TB remains very much incomplete. A central hypothesis to our work is that these different infection outcomes are reflected locally at the level of the granuloma and that granuloma structure and function are the result of the interplay of events at organ, tissue, cellular, and molecular scales over the time course of minutes to years. For example, the structure of a granuloma in the lung is influenced by chemokines recruiting immune cells into the lung, by antigen presentation events occurring in the LN, by the circulation of dendritic and T cells between the LN and lung granuloma, by cell–cell interactions in the developing granuloma, and by the binding of TNF to receptors. In addition, mycobacterial virulence factors are known to actively participate in granuloma formation and cell recruitment [9], supporting the view that both bacteria and host factors are relevant to this process.

We describe in this chapter a systems biology approach to understanding granuloma formation and function. In this context, systems biology is a discipline at the intersection of immunology, microbiology, mathematics, engineering, and computer science that allows us to integrate experimental and computational approaches. The power of systems biology to address complex host and bacterial responses in infections such as tuberculosis is vital to enhancing our understanding of these diseases and identifying factors to target for prevention or treatment. In the particular case of granuloma formation and function during *M. tuberculosis* infection, systems biology may help us in many ways to identify the mechanisms involved in *M. tuberculosis*–host dynamics. For example, models in which TNF–TNF receptor interactions are represented can help determine their role in influencing containment of bacteria by a granuloma, or, through the use of analysis tools, can identify potential immunological targets for immunotherapy. We describe below both animal and computational models that are being used to study granuloma formation, and then turn to the key insights these models provide.

2 Experimental and Computational Models of Tuberculosis Granulomas

2.1 Experimental Models

The TB field is extremely fortunate to have several different and well-characterized animal models for study of this disease (reviewed in [20]). The most commonly used model is the mouse. The availability of reagents and especially genetically modified animals (lacking different molecules, cytokines, or cell types) has provided invaluable information about virulence factors of the pathogen and host responses to the infection. Other advantages of mice include the relatively inexpensive cost of housing them in a Biosafety Level 3 facility, the lack of coughing (and therefore transmission), the inbred strains that limit variability among animals, and the ease of infecting, manipulating, and handling them. However, there are distinct limitations to the mouse models of TB. Mice become chronically and progressively infected with *M. tuberculosis*, and every infected mouse eventually succumbs to the disease. Thus, unlike in humans where latent infection is observed in the majority of cases, there is no true latent infection in mice. Although there are several "latency" models put forth over the years [21], these are all dependent on some type of manipulation (e.g., anti-TB drugs), and there is little indication that they resemble human latent infection. In addition, the mouse granuloma, which is best characterized as granulomatous inflammation, is substantially different from the human granuloma in terms of structure and organization, the lack of caseous necrosis, and an absence of cavity formation. Thus, studying mouse granulomas as a model for human granulomas can be problematic, since many of the features and microenvironments of human granulomas are absent in mice. There have been a few newer mouse models using genetically manipulated mice that recapitulate certain features of human granulomas and these hold some promise for granuloma studies [22]. Another rodent, the rat, has also been described as a model of TB [23].

Guinea pigs have also long been used as a model of tuberculosis, especially for vaccine studies [24, 25]. These animals are very susceptible to *M. tuberculosis* infection, and all proceed to death from tuberculosis after months of infection. The granulomas in Guinea pigs include inflammation similar to mice, but also more structured caseous and mineralized granulomas. There are several elegant studies on the pathologic features of Guinea pig granulomas [26–28]; however, immunologic manipulation of Guinea pigs is still very challenging, and the tools for studying host responses, although improving, remain limited.

Rabbits are also used in TB research, both historically and currently [29–32]. There are several interesting features of rabbits, including their relative resistance to many laboratory strains of *M. tuberculosis*, their exquisite susceptibility to *M. bovis*, the presence of caseous lesions that form and resolve in a relatively homogeneous manner, and the propensity for cavity formation. The lack of immune reagents limits this model, as does the size and containment needs of rabbits.

Zebrafish (and other fish) were developed as a model for tuberculosis over the past 15 years, and have many attractive features [33]. Zebrafish embryos are transparent, which allows one to visualize granuloma formation in various parts of the fish body. The fish form caseous granulomas, primarily composed of macrophages. The ability to manipulate the fish genetically and to examine the contributions of genes in high-throughput, forward genetic screens has provided valuable insights into the host–pathogen interaction of mycobacterial infections. However, the natural species that infects fish is *M. marinum*, not *M. tuberculosis*. Although there are substantial similarities between the two species, there are also differences that may contribute to difficulty in translating the findings in *M. marinum* models to human tuberculosis.

Non-human primates have been used for decades to address key aspects of tuberculosis. Many years ago, macaques were used in drug studies of tuberculosis. After a hiatus of a few decades, this model has reemerged as an important contributor to translational studies of tuberculosis, including drug, vaccine, pathogenesis, and immunologic studies. Low dose *M. tuberculosis* infection of cynomolgus macaques results in the full spectrum of human *M. tuberculosis* infection outcomes, from latent to active TB [7, 34]. In addition, reactivation of latent *M. tuberculosis* infection has been demonstrated after TNF neutralization [16], CD4+ T cell depletion [35], and SIV coinfection (as a model of TB and AIDS) [36, 37]. More recently, a model of latent infection in rhesus macaques has been reported, using a low-virulence strain of *M. tuberculosis* [38]. The full spectrum of human pathology is also observed in the macaque, with all varieties of granulomas observed, in both lungs and LNs, and cavity formation [5, 7]. These two features (spectrum of disease outcomes and pathology identical to humans) make this an important and useful model for studying human tuberculosis. Another feature of macaques that adds to their value as a model is the wide availability of immunologic reagents, for assessing immune responses (peripherally and in organs) and for manipulating the system. However, there are several limitations to this model system. First, there is extensive genetic variability among monkeys, requiring larger cohorts of animals to obtain statistically significant results. Second, housing macaques under Biosafety Level 3 conditions is a challenge, as these animals can cough and transmit infection to other animals and humans, and primate BSL3 facilities are not available at most institutions. Third, the cost of purchase and husbandry makes many experiments prohibitively expensive. Every effort must be made to obtain as much data as possible from each animal, and sharing of tissues and samples among other investigators allows more labs to take advantage of this resource. The cost also makes it difficult to obtain tissue samples at all the desired time points; however serial peripheral samples can be obtained from the same animal, which is ideal for matching to human studies.

A chief advantage of animal models is the ability to obtain samples from the site of disease, i.e., the lung and thoracic LNs, including granulomas. This is extremely challenging in human studies, and a model that is similar to humans provides an opportunity to assess events at the granuloma level. The types of samples and data that can be obtained include cells (numbers, phenotypes, and functions), cytokines (levels and sources), and spatial location of cell types and cytokines (using, for example, immunohistochemistry). Bronchoalveolar lavage, which allows one to sample airways serially, and blood can be obtained easily and frequently from the larger animals, as well as peripheral LNs.

More recently, imaging of live animals has been used to serially assess events during infection. Fluorescent imaging (fish and other animals), video monitoring (zebrafish), and PET/CT imaging (mice, rabbits, and monkeys) have been applied to tuberculosis models [35, 39] (Via and Barry, unpublished) and provide the unique opportunity to "watch" the infection evolve over time at the level of the granuloma, and in some cases to determine where the bacilli are during infection. These new imaging tools open up new possibilities for understanding *M. tuberculosis* infection, and in some cases can also be performed in humans (e.g., PET/CT). All of these methods can provide quantitative, serial, and spatial data for incorporation into computational models.

In addition to animal infection models, the use of in vitro granuloma models [40] and in vivo models using non-replicating agents [41, 42] provides unique insight into processes involved in granuloma formation and function. For example, mycobacterial antigen-coated beads are used to induce pulmonary granulomas with cytokine and cellular patterns that closely match those in an active mycobacterial infection [43, 44].

2.2 Computational Models of Tuberculosis Granuloma

Mathematical and computational modeling provides a unique approach to studying the behavior of complex biological systems. These methods can be used to better explore hypothesized mechanisms, generate and test new hypotheses, run virtual (in silico) experiments, interpret data, motivate particular experiments, and suggest new drug targets. A series of mathematical and computational models have been developed to investigate the host response to *M. tuberculosis* infection [45–57]. In particular, model-based analysis of the formation and function of a TB granuloma contributes to understanding the mechanisms that control the immune response to *M. tuberculosis* [45–49, 51, 57]. These models complement experimental approaches and can be used to address questions in TB that are difficult or currently impossible to approach experimentally. The high cost and time investment needed to fully explore many interacting immune factors and various outcomes involved within the *M. tuberculosis*–host interactions in an experimental setting are factors that alone should promote the use of computational models. Building computational models can also allow us to integrate data derived from experiments on different tissues, different biological scales (e.g., molecular or cellular), and different timescales into a comprehensive picture of the immune response to *M. tuberculosis*.

Differential equation (DE)-based models typically describe a deterministic relationship among several continuously varying quantities (e.g., numbers of cells and concentrations of molecules) and their rates of change in space and/or time. We have developed DE-based models for studying temporal dynamics of cytokines and effector cells during the immune response to *M. tuberculosis* [52, 53, 58, 59]. These models are based on known interactions of immune cells in the lung during *M. tuberculosis* infection. Experimental data are used to estimate parameter values.

When data are not available, uncertainty and sensitivity analyses are used to define parameter spaces. Uncertainty analysis is performed to investigate the uncertainty in the model output that results from uncertainty in input parameter values. Sensitivity analysis is then used to quantify how input uncertainty (e.g., biological variability coupled to unknown ranges of variation for model parameters) affects model outcomes and to identify critical model parameters. Once validated against experimental data, the models are used to make novel predictions about dynamics, progression of infection, and potential therapies. Examples of contributions these models have made to our understanding of TB include identifying the critical impact of delays in either DC migration to the draining LN or T cell trafficking to the site of infection on the outcome of infection [58], and identifying the key role of cytokine IL-10 in balancing macrophage phenotypes in the lung and LN [59]. DE models can also be used to examine spatial aspects of the immune response, including analysis of the process of granuloma formation and cytokine availability in a granuloma [42, 51].

In contrast to DE-based models, agent-based models (ABMs, also known as individual based models) are rule-based models that capture a variety of stochastic and discrete events occurring in the immune system. An ABM has the following components: *agents* (e.g., immune cells and bacteria), the *environment* where agents reside (e.g., a two-dimensional grid representing a section of lung tissue), the *rules* that govern the dynamics of agents, including movements, actions, and interactions between agents as well as between agents and environment, and *timescales* on which the rules are executed. In an ABM, the local, possibly stochastic interactions occurring at the level of agents lead to global, system-wide dynamics and emergent spatial and temporal patterns. Hence, ABMs are particularly useful for studying complex systems such as TB granulomas in which cell heterogeneity and spatial interactions are important.

We developed first- and second-generation ABMs to describe the immune response to *M. tuberculosis* and to identify mechanisms that control granuloma formation and function [46, 47]. Next-generation granuloma ABMs were developed in response to new biological data that indicated the importance of including additional cell types (e.g., effector CD8+ T cells and regulatory T cells), cytokines (e.g., TNF and IL-10), and chemokines (e.g., CCL2, CCL5, and CXCL9) [45, 46, 57]. The major cell types, biological activities, and interactions captured in our current granuloma ABM are listed in Table 7.1. An overview of selected ABM rules governing cellular activities and interactions on a grid representing a section of lung tissue is presented in Fig. 7.2. These models are the first to track the dynamics of formation and maintenance of a granuloma in space and time, simultaneously providing critical details regarding cellular interactions and molecular concentrations. There are no experimental methods to obtain these detailed, continuous data in primates.

A critical aspect of studying mechanisms underlying the formation and function of a granuloma during *M. tuberculosis* infection is the integration of information across multiple biological scales (molecular, cellular, tissue/organ, and host scales; see Fig. 7.2). Immunity to *M. tuberculosis* in humans and animal models has been attributed to activities of a variety of cytokines, including TNF, IFN-γ, and IL-10 (reviewed in [60]).

Table 7.1 List of agents, biological activities and interactions captured in the lung granuloma agent-based model and lymph node DE-model

Cell types and states	Activities and interactions in the lung granuloma (ABM)	Activities and interactions in the lymph node (DE model)
Macrophage	Death due to age Chemokine-dependent movement	N/A
Resting macrophage (M_r)	TNF and chemokine-dependent recruitment STAT-1 activation due to interaction with T_γ Uptake of extracellular *M. tuberculosis* (either kills *M. tuberculosis* or gets infected ($M_r \rightarrow M_i$)) Macrophage activation ($M_r \rightarrow M_a$) if both STAT-1 and NF-κB are activated and M_r is not down-regulated by T_{reg}	N/A
Infected macrophage (M_i)	Growth of intracellular *M. tuberculosis* Uptakes extracellular *M. tuberculosis* Gets chronically infected ($M_i \rightarrow M_{ci}$) with growth of intracellular *M. tuberculosis* STAT-1 activation due to interaction with T_γ Macrophage activation ($M_i \rightarrow M_a$) if both STAT-1 and NF-κB are activated and M_i is not down-regulated by T_{reg}	N/A
Chronically infected macrophage (M_{ci})	Growth of intracellular *M. tuberculosis* Bursts and disperse *M. tuberculosis* to the environment due to intracellular growth of *M. tuberculosis*	N/A
Activated macrophage (M_a)	Kills extracellular *M. tuberculosis* effectively	N/A
T cell	Death due to age Chemokine-dependent movement TNF and chemokine-dependent recruitment	Death due to age Recruitment, re-circulation and migration Differentiation
Naïve CD4+ and CD8+ T cells	None	Natural turnover and enhanced recruitment induced by DC arrival Differentiation to precursor effector cells (Th0 and T80) after interacting with DCs

Precursor effector T cells	None	Proliferation and migration to the blood Further differentiation to fully effector CD4+ and CD8+ T cells
Pro-inflammatory T cell (T_γ)	If not down-regulated, can activate STAT-1 in M_r and M_i If not down-regulated by a T_{reg}, can mediate Fas/FasL-induced apoptosis of M_i and M_{ci}	Migrate out of the LN into the blood Represented in the LN by the sum of Th1 (i.e. fully effector CD4+ T cells) and T8 (i.e. fully differentiated IFN-γ producing CD8+ T cells)
Cytotoxic T cell (T_c)	If not down-regulated by a T_{reg}, can kill M_i and M_{ci} (perforin/granulysin-mediated mechanism)	Migrate out of the LN into the blood Represented in the LN by the fully differentiated cytotoxic CD8+ T cells, labeled as CTL
Regulatory T cell (T_{reg})	Down-regulate the actions of T cells and macrophages	Migrate out of the LN into the blood Represented in the model as 10% of the sum of Th1 and T8
Bacteria (*M. tuberculosis*)	Grow inside M_i and M_{ci} and in extracellular spaces Extracellular *M. tuberculosis* can infect M_r and M_i Extracellular *M. tuberculosis* can induce NF-κB activation in M_r and M_i	None
Molecules		
TNF	Secreted from M_i, M_{ci}, M_a, NF-κB-activated M_r, T_γ and T_c with different rates Degrades and diffuses in the environment (among grid micro-compartments) Induces NF-κB activation in M_r and M_i Induces apoptosis in all cells	None
Chemokines (CCL2, CCL5, CXCL9/10/11)	Secreted from M_i, M_{ci}, M_a, and NF-κB-activated M_r with different rates Degrades and diffuses in the environment (among grid micro-compartments)	None

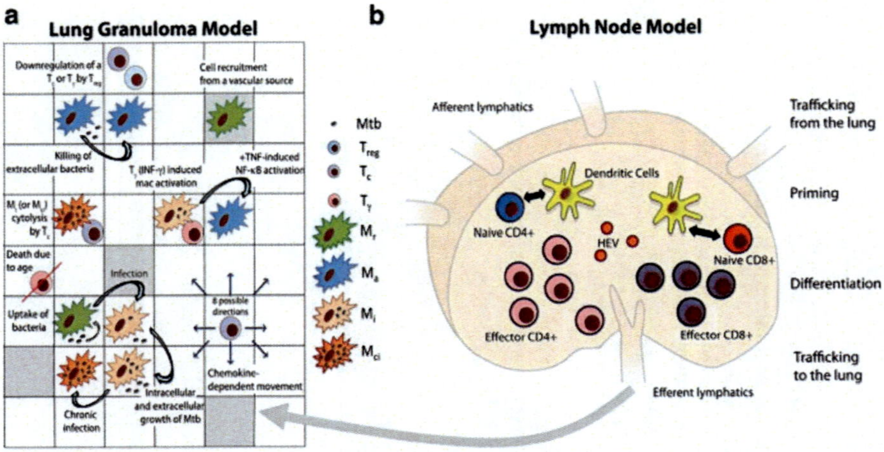

Fig. 7.2 Computational modeling of immunological processes within the lung and lymph nodes during *M. tuberculosis* infection. (**a**) An overview of selected ABM rules governing biological activities and interactions among immune cells and *M. tuberculosis* on a grid representing a section of lung tissue. Cell types and status are shown, as indicated on the *right* side of panel (**a**): M_r resting macrophage, M_i infected macrophage, M_{ci} chronically infected macrophage, M_a activated macrophage, *M. tuberculosis*: mycobacteria, T_γ pro-inflammatory IFN-γ producing T cell, T_c cytotoxic T cell, T_{reg} regulatory T cell. A complete description of all ABM rules is provided in [45]. (**b**) An overview of main mechanisms captured in the ODE-LN model. Dendritic cells migrate from the lung to the LN upon bacterial uptake and maturation. They traffic into lymphatic vessels and enter the T cell zone of the LN through afferent lymphatics. Once in the T cell zone of the LN, they interact with naïve CD4+ and CD8+ T cells [continuously circulating in the LN through high endothelial venules (HEV)], and eventually prime and activate them. Upon activation, T cells start to proliferate and differentiate into effector lymphocytes. These effector immune cells then migrate back to the lung, exiting the LN through efferent lymphatics. A complete description of all ODE-LN model rules is provided in [57]

These molecules are secreted from cellular sources in response to pathogen and host signals, interact with receptors on target cells, and trigger intracellular signaling pathways controlling cellular activities that ultimately contribute to formation of granulomas and immunologic control of *M. tuberculosis* infection. For example, TNF is secreted from infected and activated macrophages, interacts with TNF receptors (TNFRs) on the membrane of macrophages, and induces the NF-κB signaling pathway, leading to secretion of chemokines, a key process that attracts immune cells to the site of infection and influences their movement within a granuloma.

Agent-based modeling can provide a framework for describing these events. We capture cellular and tissue scale dynamics (see Fig. 7.2a) via well-described and probabilistic rules for interactions between immune cells and *M. tuberculosis*. Single-cell molecular scale processes (e.g., those controlling TNF/TNFR binding and trafficking) for each individual cell are captured by a set of ordinary DEs. Using this approach, we are able to, for example, alter a molecular property (e.g., TNF/TNFR binding affinity) and study its impact on a tissue scale outcome (e.g., size of the granuloma). As an example, we review below our recent studies on the multi-scale analysis of the role of TNF activities in controlling granuloma formation and function.

As a complex multi-scale process, granuloma models include parameters describing a large number of biological events. Hence, it is critical to understand the role that each of these parameters plays in determining how a granuloma functions. We have developed a number of useful and powerful tools to analyze these complex model systems. One approach is to perform *virtual* deletion and depletion experiments that mimic experimental gene knockout or molecule depletion studies. Loss of activity is achieved by setting relevant parameters (e.g., probabilities or rate constants) to zero or raising relevant thresholds to an unattainable level. Virtual deletion refers to the loss of activity from the beginning of simulation and virtual depletion refers to the loss of activity after establishment of a granuloma. A second approach is to use uncertainty and sensitivity analysis, which we have adapted for use in agent-based models [61]. We use uncertainty and sensitivity analysis in computational models of *M. tuberculosis* infection to analyze the impact of parameters describing events at different scales (molecular, cellular, tissue, or organ scales) on model outputs describing granuloma outcomes.

3 What Are Examples of Questions That Systems Biology Can Address?

3.1 Which Factors Influence the Ability of a Granuloma to Control Infection?

As described earlier (see Fig. 7.1), there are different types of granulomas in the lungs of non-human primates and humans with active TB [7]. Our granuloma ABM is able to recapitulate granulomas with different abilities to control infection by varying values of important model parameters. Examples of these outcomes (as shown in Fig. 7.3) include containment (control of infection within a well-circumscribed granuloma containing stable and low bacterial levels), clearance, and uncontrolled growth of bacteria.

Many immune factors are critical to a protective immune response to *M. tuberculosis* infection in animal models (reviewed in [60]). These factors include priming and activation of antigen-specific T cells [3], production of chemokines contributing to recruitment of immune cells to the site of infection [62–64], and production of cytokines such as IFN-γ and TNF [10, 17, 60, 65]. Simulations of TNF, TNF receptor 1 (TNFR1), and IFN-γ gene knockouts and deletion/depletion of T cells (described in detail in [45, 46]) lead to uncontrolled growth of *M. tuberculosis* and formation of granulomas with irregular structures that include very high numbers of extracellular bacteria, large numbers of infected macrophages, and widespread caseation. These simulations indicate that the model captures important aspects of the biology of the immune response to *M. tuberculosis*.

In addition to validating the model with experimental data, we use simulations to perform novel virtual experiments. Such studies can predict critical components of

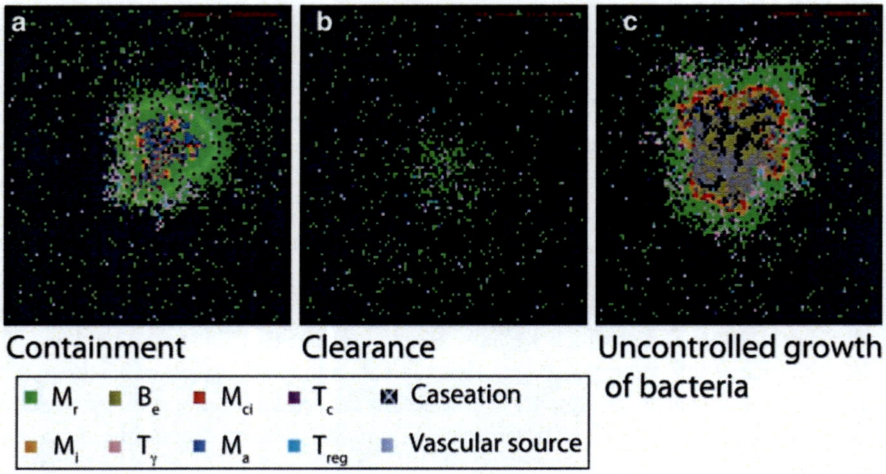

Fig. 7.3 Reproduction of possible outcomes of *M. tuberculosis* infection by the granuloma ABM by varying important model parameters. Granuloma snapshots for (**a**) a scenario of containment at 200 days post-infection, (**b**) clearance of *M. tuberculosis* in approximately 5 weeks as a result of an efficient immune response, and (**c**) a scenario of uncontrolled growth of bacteria as a result of knocking TNF out at 200 days post-infection. Cell types and status are shown by different *color squares*, as indicated at the *bottom* of the figure (M_r resting macrophage, M_i infected macrophage, M_{ci} chronically infected macrophage, M_a activated macrophage, B_e extracellular bacteria, T_γ pro-inflammatory IFN-γ producing T cell, T_c cytotoxic T cell, T_{reg} regulatory T cell). Caseation and vascular sources are also indicated

an effective immune response and can ultimately guide the design of laboratory experiments. For example, sensitivity analysis helps us predict the relative importance of various immunological processes (e.g., recruitment and movement of T cells, secretion, diffusion, and degradation of chemokines, and macrophage–T cell interactions) in immunity to *M. tuberculosis* and may suggest novel targets for control and therapy of TB. Our granuloma ABM identified cellular and tissue scale processes that significantly control bacterial numbers, caseation, and size of a granuloma: *M. tuberculosis* growth rates, activation of macrophages by IFN-γ producing T cells, and T cell movement and recruitment in lung tissue [45, 46]. Further, our model predicts molecular scale processes that have significant impact on granuloma outcomes. These processes include events at the level of TNF signaling and trafficking. Thus, we focus in the next section on our model-based findings regarding the role of TNF in granuloma formation and function.

3.2 *What Is the Role of TNF in Granuloma Formation and Function?*

The pleiotropic cytokine TNF functions as a critical part of the immune response to *M. tuberculosis* infection (see Sect. 7.1). Initial data identifying the roles of TNF include TNF knockout/neutralization and TNFR1 knockout experiments in mice [17, 62, 66]. Based on these mouse studies, TNF was long believed to be essential for formation of granulomas. However, recent studies in zebrafish and non-human primate models of TB have shown that TNF, although not required for the formation of granulomas, is necessary to restrict bacterial growth in a granuloma [16, 67]. This suggests that TNF activities *within* a granuloma determine a granuloma's ability to immunologically restrain bacteria. This is confirmed by studies in humans, where TNF neutralization leads to reactivation of latent TB; pathologic studies in a subset of humans support the view that granuloma formation is maintained in the absence of TNF, but disease exacerbation and dissemination occur, indicating a failure of the granuloma to control bacterial replication [68]. Which mechanisms control TNF availability and activities in a granuloma, and how do these activities affect granuloma function during the long-term immune response to *M. tuberculosis*? Here, we present predictions of our approach resulting from integrating experimental (animal model) data and theoretical tools to address these questions.

3.2.1 Prediction I: Establishment of a TNF Concentration Gradient Within a Granuloma

Availability of TNF within a TB granuloma has been proposed to have a key role in the protective immunity to *M. tuberculosis*, although measuring the true TNF production and consumption within a granuloma is not yet feasible. The total TNF concentration at any one time in a granuloma has been estimated by cytokine bead array technology [7], although this is simply a snapshot of the concentration at necropsy of the animal. We calculate the TNF concentration in a granuloma using two different models: a DE-based model that considers a simple representation of the spatial structure of a granuloma at steady state and the granuloma ABM described earlier. Both models explicitly include single-cell level TNF/TNFR binding and trafficking processes (i.e., synthesis, internalization, recycling, and degradation of ligands and receptors), as these processes are critical to determining TNF concentration.

Values of some model parameters, including TNF/TNFR kinetic rate constants, are estimated from the literature. Other model parameters were determined based on a simple experimental system for granuloma formation in mice. The formation of granulomas was induced in mice following injection of Sepharose beads covalently coupled to *Mycobacterium* purified protein derivative antigen [43, 69]. The cellular composition of granulomas, TNF secretion rate, and TNFR densities on different types of cells were measured for these mouse granulomas. Our experiments indicate that macrophages and DCs are the major TNF-producing immune cells within a granuloma. Further, DCs, macrophages, and B cells are found to be the major TNFR-expressing cells.

Our granuloma ABM simulations based on these data suggest that there is a TNF concentration gradient in granulomas, such that the highest concentration occurs at the center of a granuloma [42] (Fig. 7.4). This gradient results from the emergence

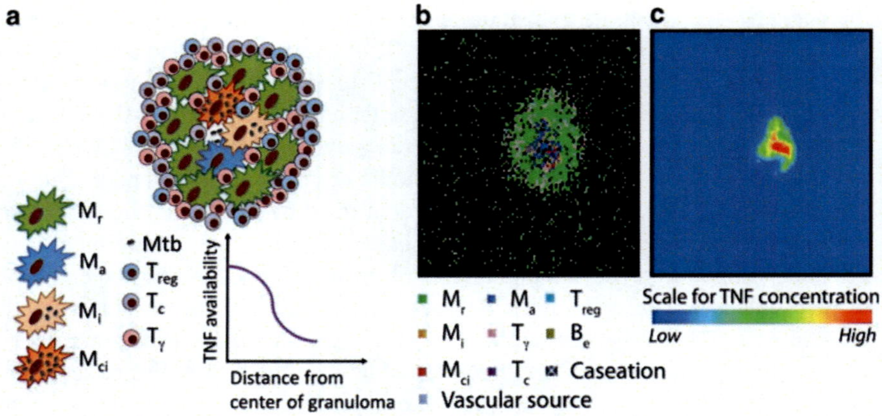

Fig. 7.4 Prediction of a TNF concentration gradient within a granuloma. This gradient, as verified by both the differential equation model [42] (**a**) and the agent-based model [45] (**b**), results from a granuloma with a specific cellular organization composed of a core of infected, activated, and resting macrophages surrounded by a ring of lymphocytes as well as TNF/TNFR binding and intracellular trafficking processes. Cell type abbreviations are as described in Fig. 7.3

of a specific organization of immune cells within a granuloma (i.e., concentration of infected macrophages at the core and concentration of lymphocytes at the periphery of the granuloma; see also Fig. 7.1) and the processes of TNF/TNFR binding and intracellular trafficking. What might the impact of this gradient be? The gradient could allow the spatial coordination of TNF-induced biological activities (i.e., activation of NF-κB and apoptotic signaling pathways) within a granuloma. Higher concentrations of TNF in the center of granuloma could induce caspase-mediated apoptotic pathway that favors antigen cross-presentation as well as the elimination of pathogen inside infected macrophages. However, very low levels of TNF at the periphery of the granuloma, although unable to induce apoptosis, are sufficient to turn on the NF-κB signaling pathway that favors cell survival and expression of pro-inflammatory genes in T cells.

3.2.2 Prediction II: A Critical Role for TNFR1 Internalization Kinetics

Experimental data suggest that TNFR1 internalization plays a key role in regulation of TNF signaling and mediates the process of TNF-induced apoptosis [70, 71]. Our simulations also predict a key role for TNFR1 internalization in control of the local TNF concentration and regulation of TNF activities during granuloma development [42, 45]. Further, the rate of TNF-induced internalization of TNFR1 regulates cell infiltration by affecting the extent and dynamics of TNF-dependent recruitment and activation of immune cells [45]. These are essential factors that control the level of inflammation in tissue. The importance of these factors to infection outcome at the level of a granuloma is demonstrated with our granuloma ABM: taken together, snapshots of model simulations (Fig. 7.5) and sensitivity

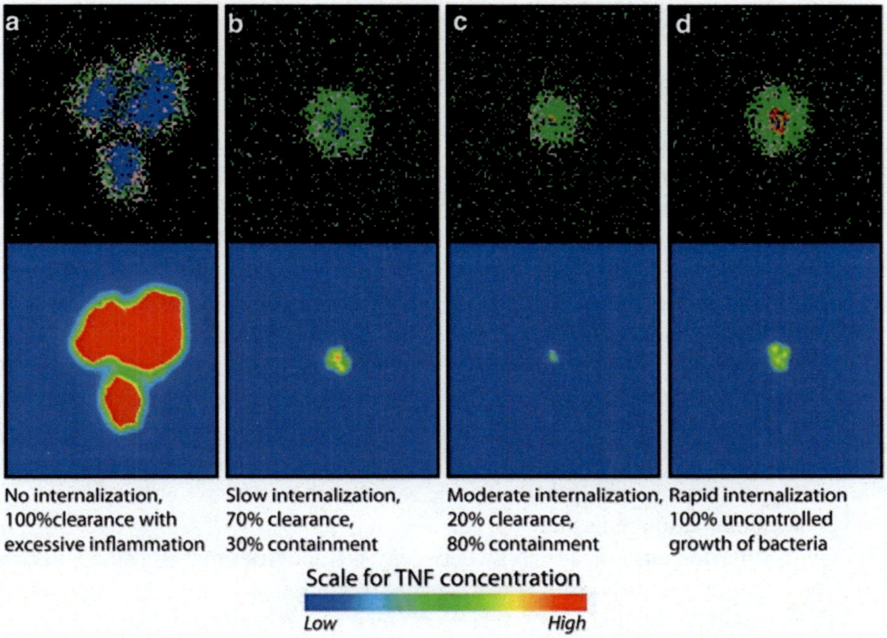

Fig. 7.5 Prediction of a key role for TNFR1 internalization kinetics in control of bacterial load and inflammation during *M. tuberculosis* infection. (**a**)–(**d**) Granuloma outcomes and tissue levels of TNF early after recruitment of T cells for varying rates of TNF-induced TNFR1 internalization. The *colors* representing cells of different type and status in granuloma snapshots are the same as those shown and defined in Fig. 7.3

analysis results demonstrate that TNF-induced TNFR1 internalization plays an important role in preventing excessive inflammation in tissue. This effect is particularly highlighted in Fig. 7.5a in which removal of the process of TNFR1 internalization leads to uncontrollably high tissue concentrations of TNF and very high rates of macrophage activation. Increasing the rate of TNFR1 internalization, however, controls the level of macrophage activation and tissue concentration of TNF (Fig. 7.5b–d).

TNFR1 internalization kinetics are also predicted to have a significant impact on bacterial numbers within a granuloma [45]. As highlighted in Fig. 7.5b–d, increasing the rate of receptor internalization reduces the rate of bacterial clearance. This effect results from reduced rates of TNF-induced activation of macrophages, diminishing their ability to kill bacteria. Overall, our results suggest the novel hypothesis that TNFR1 internalization kinetics play a role in balancing inflammation and bacterial killing within a granuloma, controlling whether there is clearance of bacteria, excessive inflammation, containment of bacteria within a stable granuloma, or uncontrolled growth of bacteria. This hypothesis can be tested in future studies investigating approaches to control and therapy of TB, as a number of ways have already been proposed to control the rate of TNFR1 internalization in vitro [70, 72, 73]

3.2.3 Prediction III: A Critical Synergy Between Individual TNF Activities

TNF has been experimentally characterized to have the following activities in *M. tuberculosis* infection: (1) macrophage activation (essential for killing of bacteria) [74, 75], (2) induction of apoptosis [76, 77], (3) induction of chemokine and cytokine production [63], and (4) regulation of cellular recruitment via transendothelial migration [78]. The use of computational modeling to describe how TNF regulates the process of granuloma formation provides an opportunity to investigate the importance of each of the TNF activities, separately or in combination, during the long-term immune response to *M. tuberculosis*. We can simulate in silico any combination of gene knockouts or deletions of biological activities by setting values of relevant parameters to either zero or very large values (infinity) in our model [46]. In particular, we simulate all 15 possible combinations in which at least one of the four TNF activities is deleted. For each case, we report the total number of bacteria and the maximal fraction of macrophages present that are activated in the granuloma (Fig. 7.6). Macrophage activation is considered here as a metric to assess the level of inflammation in tissue.

Our results demonstrate a synergy between TNF activities that contribute to control of infection within a granuloma [46]. As highlighted in Fig. 7.6, deletion of TNF-dependent activation (*act*$^{-}$), secretion (*secr*$^{-}$), or recruitment (*recr*$^{-}$) activities significantly increases bacterial levels within the granuloma. Among these activities, simulation of *act*$^{-}$ leads to significantly higher bacterial numbers. This highlights the relative importance of TNF-induced macrophage activation in control of infection compared to other TNF activities. Double and triple deletions of these activities further exacerbate infection compared to single deletion simulations. The highest level of bacteria is observed when TNF-induced activation, TNF-induced secretion of chemokines, and TNF-induced apoptosis of immune cells are all simultaneously deleted (*act*$^{-}$ *secr*$^{-}$ *apopt*$^{-}$). Interestingly, deletion of TNF-induced apoptosis activity alone (*apopt*$^{-}$) does reduce bacterial numbers. This occurs as a result of high levels of macrophage activation in lung tissue that is accompanied by high levels of TNF concentration and cell recruitment. Thus, TNF-induced apoptosis reduces inflammation by controlling the level of macrophage activation at the expense of impairing bacterial clearance. It is this type of nonintuitive result that cannot be predicted without the use of computational models in tandem with experimental models.

3.3 *What Are the Mechanisms Underlying TB Reactivation Following Anti-TNF Therapies?*

M. tuberculosis can persist for decades within the lungs of humans. This results from a latent state of infection that represents a dynamic equilibrium between host and bacteria [79]. Disturbance of this equilibrium may lead to a failure of a granuloma to contain bacteria and progression to active TB, termed reactivation.

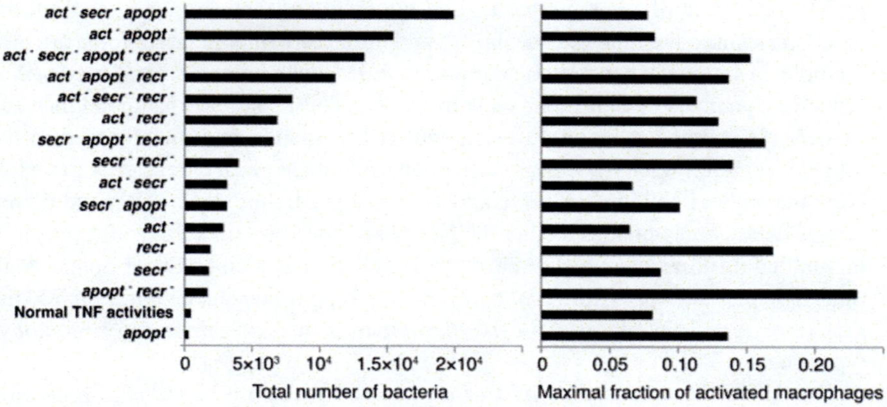

Fig. 7.6 Prediction of a synergy between four TNF-mediated biological activities: macrophage activation (*act*), inducing apoptosis (*apopt*), inducing recruitment of immune cells (*recr*), and inducing secretion of chemokines and cytokines (*secr*). Model predictions for total number of bacteria (*left*) and maximal fraction of macrophages that become activated (*right*) during granuloma development after *M. tuberculosis* infection are displayed for 16 possible scenarios with (+) or without (–) each of the TNF activities

For example, because of the inflammatory nature of TNF, treatment with TNF inhibitors (TNF-neutralizing drugs) is used in patients with inflammatory diseases such as rheumatoid arthritis and psoriasis. However, anti-TNF treatment has been recognized as one of the risk factors for reactivation of latent TB in humans. An increased incidence of TB has been reported among patients receiving treatment with TNF-neutralizing drugs [18, 80–82]. These drugs are either anti-TNF monoclonal antibodies such as infliximab, adalimumab, and certolizumab or soluble TNF receptor fusion proteins such as etanercept [83]. Although these drugs are similarly effective in treatment of some (but not all) inflammatory diseases [84, 85], the risk of TB reactivation posed by antibody-type drugs is several-fold greater than that for the soluble TNF receptor-type drugs [86–89]. A systematic and comprehensive comparison of TNF-neutralizing drugs with the aim of elucidating drug-specific reactivation mechanisms (especially in humans) has not been performed to date. Experiments required for a comprehensive analysis of the effects of drug characteristics (including TNF binding kinetics and stoichiometry, together with blood concentration and drug permeability into lung tissue, and apoptotic activities of antibody-type drugs) on the immune response to *M. tuberculosis* are at present impossible in vivo.

The granuloma ABM we described earlier, with modifications, can be used to investigate mechanisms by which TNF-neutralizing drugs interfere with granuloma function and thus immunity to *M. tuberculosis* [90]. In an earlier work, we used a DE-based model and found that the bioavailability of TNF is central to control of infection [52]; to address the mechanisms that control bioavailability during anti-TNF treatment, the ABM framework is useful. TNF-neutralizing drugs and their relevant properties can be incorporated into the ABM in a manner similar to TNF itself. The dosing of a host with drug and the ability of that drug to cross from the bloodstream into the lung (permeability) and ultimately into the granuloma are also featured in this next-generation granuloma ABM. Our computational model thus links the dynamics of molecular scale drug/TNF/TNFR interactions to cellular and tissue/scale events occurring during granuloma formation and maintenance in the lung. Using this model, we identify functional and biochemical characteristics underlying the higher likelihood of TB reactivation that occurs for some TNF-neutralizing drugs. These characteristics include TNF *binding properties* (including affinity, binding/unbinding kinetics, stoichiometry, and ability to bind membrane-bound TNF (mTNF)), *permeability* (from blood vessels into lung tissue and penetration into the granuloma), and *apoptotic and cytolytic activities* that are reported for antibody-type drugs.

Our model-based analyses lead to novel and interesting hypotheses regarding drug-induced TB reactivation at the granuloma scale (Fig. 7.7). First, we find that the ability of a drug to bind mTNF is a major factor impairing granuloma function, leading to TB reactivation. This is because the cell membrane provides a scaffold on which TNF is available at a high concentration for neutralization before it is released as soluble TNF (sTNF) and diluted in extracellular spaces. Although this is an interesting result, both the antibody-type and receptor fusion drugs bind to mTNF, so it cannot explain differences in reactivation rates observed for the two drug types. Second, our results suggest three factors: differences in blood concentrations of drugs, TNF/drug binding and unbinding kinetics, and the level of drug permeability into lung tissue can each dramatically affect the likelihood of TB reactivation. In fact, we find that these factors result in different rates of TB reactivation between antibody-type drugs (e.g., infliximab) and TNF receptor fusion proteins (etanercept). Our experimental data from a mouse model suggested that retention of drug concentration in a granuloma as well as the dissociation constant of both antibody and soluble receptors differed. In particular, the presence of high levels of TNF receptors in the granuloma competes for TNF that is temporarily not bound to drugs. Our finding that this occurs with soluble receptor drugs at a much higher level than antibody-based drugs may be involved in the differential effects of these drugs on control of established infection [91]. Finally, although there are differences in drug abilities to induce apoptosis or cytolysis in TNF-expressing key immune cells (e.g., infected and activated macrophages and T cells), our analysis suggests that these activities are not as important as other factors in driving TB reactivation. These findings suggest the characteristics of suitable anti-TNF drugs for treatment of inflammatory diseases while balancing high risks of TB reactivation.

3.4 What Is the Impact of Lymph Node Processes on Granuloma Formation and Function in the Lung?

A key step to mounting a protective immune response to most bacterial infections is effective CD4+ and CD8+ T cell priming in LNs. For TB, it remains unclear how

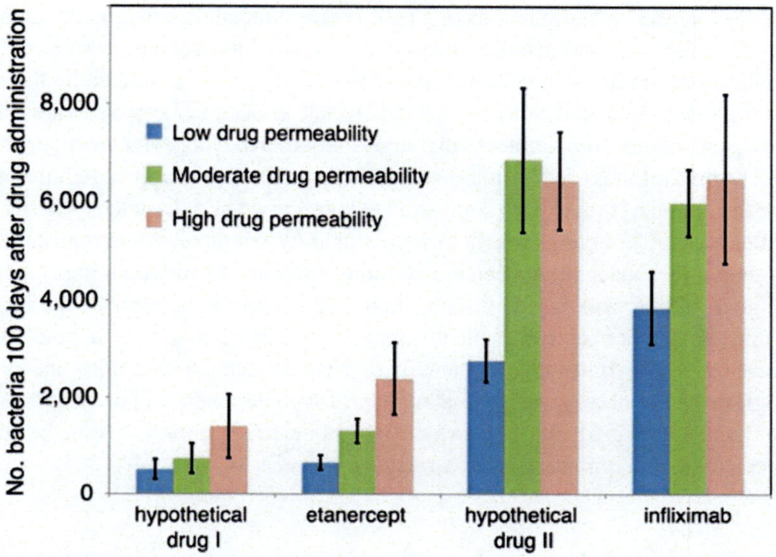

Fig. 7.7 Prediction of the impact of different types of TNF inhibitors on the outcome of *M. tuberculosis* infection at the granuloma scale. 100 days after *M. tuberculosis* infection, at which time a well-circumscribed granuloma with stable bacterial levels ($<10^3$ total bacteria) forms, the granuloma is exposed to one of the TNF-neutralizing drugs. Simulation results (bacterial levels within granulomas at 100 days after treatment with TNF inhibitor) are compared for four different drugs at different levels of drug permeability from vascular sources into lung tissue. Low, moderate, and high permeabilities represent ~10%, ~24%, and ~50% drug permeabilities into lung tissue, respectively. The hypothetical drug I is a drug defined here to only bind sTNF with a TNF:drug binding ratio of 1:1 and TNF binding/unbinding kinetics similar to infliximab. Etanercept binds both sTNF and mTNF with a TNF:drug binding ratio of 1:1. Infliximab binds both sTNF and mTNF and is assumed to have a TNF:drug binding ratio of 1:3 and can induce apoptosis and cytolysis as a result of mTNF binding and cross-linking. The hypothetical drug II is infliximab without apoptotic and cytolytic activities. TNF binding kinetic parameter values and blood concentrations of etanercept and infliximab were used as reported in [104, 105]

events occurring within LNs impact granuloma formation and maintenance. Our recent ABM studies [45–47] emphasize the critical role of T cell related mechanisms in infection progression, such as T cell movement, as well as the magnitude and timing of T cell recruitment. However, mechanistic descriptions of priming, differentiation, and recruitment of immune cells are only partially addressed in these ABM formulations, since these events occur primarily within LNs. We recently took a multi-organ (multi-compartment) approach and built onto the existing agent-based multi-scale model of the lung (described above) some of the main mechanisms of DC and T cell trafficking, as well as T cell priming and differentiation occurring in the lung-draining LN. We described the cellular dynamics occurring within a LN by a DE system, based on a simplified version of the LN compartment portion of our published two-compartmental ordinary differential equation (ODE) model [59]. Our new LN-ODE module tracks the dynamics of antigen presenting cells (APCs, defined as the sum of infected and chronically infected macrophages migrating from the lung

ABM) and several subpopulations of T cells (naïve, precursor, and effector CD4+ and CD8+ T cells). The actions and interactions included are shown in Table 7.1. The magnitude and timing of infection in the lung compartment (generated by the ABM) drive the extent of T cell priming in the LN-ODE model. Effector immune cells are generated in the LN compartment, migrate via blood to the lung, and are input onto the lung. The final result is a *hybrid* multi-compartment mathematical/computational model, where the lung (i.e., granuloma) compartment is described by a discrete/stochastic ABM module, and the LN compartment is represented by continuous/deterministic ODEs. Information is exchanged between the two compartments at every time step. One of the main goals of this work was to investigate how immune mechanisms occurring in the LN impact infection outcomes in the lung, both before and after a granuloma is established. The hybrid model recapitulates typical infection outcomes and predicts biologically relevant cell and bacterial numbers for containment and dissemination scenarios (similar to Fig. 3). Below we review two model predictions that could be relevant to vaccination and immunotherapy strategies.

3.4.1 Prediction I: Antigen Presenting Cell Migration and Immunogenicity Are Key Regulatory Mechanisms in TB Granuloma Formation and Maintenance

Whether the regulation of APC trafficking controls the nature of adaptive immune responses in the lung and in granulomatous tissue in vivo [92] is still an open question. For example, mechanisms governing pulmonary APC trafficking to LNs are still poorly understood, both at cellular and molecular scales. Another complication is that there are currently no assays that directly analyze APCs transiting through lymphatic vessels [93]. We were able to begin to address these questions using our hybrid model described above [57] and our results predicted that the rate of APC trafficking from lung to LN or T cell trafficking from LN to lung can drive the system to either clearance (both before or after a granuloma has been established; see Fig. 7.8) or dissemination and uncontrolled growth of bacteria. Enhancing APC migration is predicted to be a key regulatory mechanism that could be exploited for effective vaccination and immunotherapy strategies. Another prediction of the hybrid model is based on manipulating the efficiency of APC-T cell contacts in vivo (rather than the number of APC migrating to the LN). For example, increasing the duration of the DC–T cell interaction [94, 95], the cognate frequency of naïve T cells [96, 97] or the immunogenicity of DC [98] can all represent viable strategies to clear an infection before a granuloma is fully developed.

3.4.2 Prediction II: Differential Roles of Effector Lymphocytes in TB Containment and Clearance

Our hybrid model implementation confirms an essential role for effector T cells in a successful initial immune response to *M. tuberculosis* invasion: IFN-γ and TNF induce

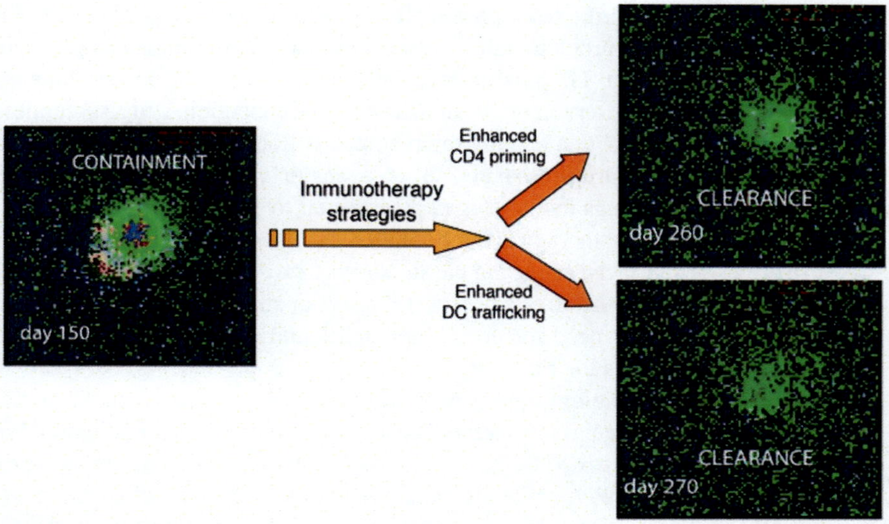

Fig. 7.8 Predictions of the hybrid lung-LN model describing granuloma development and TB infection in the presence of immunotherapy strategies. We only show the effect of enhancing CD4+ T cell priming and DC trafficking to the LN upon bacterial uptake. The snapshots capture granuloma state at different days after initiation of immunotherapy. The initial conditions used yield a typical containment scenario at day 150 post-infection. It takes approximately 10–15 days to resolve the inflammation after bacteria are cleared

macrophage activation that allows the formation of a stable granuloma (containment). In particular, we found that once a granuloma is fully formed, a viable immunotherapy strategy to clear infection in a latently infected host is to specifically enhance effector CD4+ T cell differentiation (see Fig. 7.8). We predict that effector CD8+ T cell cytotoxic activity is important to controlling the onset of infection and possibly for clearance, but has no key role when a granuloma has already been established. This follows since, once a granuloma is fully formed, it is difficult for cytotoxic T cells to reach the center of a granuloma to physically interact with infected and chronically infected macrophages due to crowding effects. Bacterial clearance is better achieved by macrophage activation, which is strictly dependent on TNF and IFN-γ secreting lymphocytes (i.e., CD+ T cells): a successful interaction between effector CD4+ T cells and resting macrophages in the outer layers of a granuloma (in the lymphocyte cuff) is a more viable strategy to combat TB once a granuloma is already established.

4 Conclusions and Future Directions

The granuloma is where the central battle in TB plays out and we believe it reflects the infection status. Thus far our systems biology approach has generated predictions and novel hypotheses regarding cellular and molecular mechanisms influencing granuloma formation and function over a time period of days to years.

Despite years of scientific research and efforts by world health organizations, TB remains a global health problem and is responsible for ~2 million deaths per year. Of great concern is that TB persists as a latent infection in ~2 billion humans worldwide, providing a reservoir of potential disease and contagion. Drug-susceptible TB can be treated only with a lengthy regimen that is fraught with compliance and drug toxicity issues. Drug-resistant TB is a major problem worldwide and development of new drugs and strategies is essential to prevent further spread of these strains. Single drug therapy is not permitted in treatment of human active TB, because drug resistance can arise, and the standard of care must be adhered to. Thus it is difficult to evaluate the effects of new TB drugs or strategies in human clinical trials. There is a critical need for novel approaches and platforms for testing and optimizing new therapies for TB.

Can we use systems biology approaches, particularly those focused on the granuloma, to identify new vaccines or therapeutic strategies for this ancient disease? We believe the field is poised to do just that. For example, combining immune modulation ("immunomodulation") with antibiotics is a potential strategy for enhancing treatment of TB [99, 100]. Several strategies have been tried in murine models (reviewed in [101, 102]) and a few in humans [101, 103], but the results are inconclusive. Appropriate delivery to granulomas and proper timing, drug combinations, and dosing are all likely to be key factors in a successful therapy, but these are difficult to study in mammalian systems due to cost, technical, and ethical issues. A computational platform such as described here could allow for development of various strategies that could then be tested in animal models.

Acknowledgements This work was supported by National Institute of Health (NIH) grants R33 HL092844 (JJL), R33 HL092853 and R01 HL106804 (DEK), and JLF was supported by grants from the NIH (R01 HL71241, R33 HL092883, HL106804) and the Bill and Melinda Gates Foundation Grand Challenges and TB Drug Accelerator Programs. R01-EB-012579 was awarded to DEK, JLL, and JLF. MF was supported by a University of Michigan Rackham Predoctoral Fellowship. We also thank Mohammed El-Kebir and Paul Wolberg for helpful discussions. The authors declare that they have no competing financial interests.

References

1. Bhatt K, Hickman SP, Salgame P (2004) Cutting edge: a new approach to modeling early lung immunity in murine tuberculosis. J Immunol 172:2748–2751
2. Wolf AJ, Desvignes L, Linas B, Banaiee N, Tamura T et al (2008) Initiation of the adaptive immune response to Mycobacterium tuberculosis depends on antigen production in the local lymph node, not the lungs. J Exp Med 205:105–115
3. Lazarevic V, Nolt D, Flynn JL (2005) Long-term control of Mycobacterium tuberculosis infection is mediated by dynamic immune responses. J Immunol 175:1107–1117
4. Chackerian AA, Alt JM, Perera TV, Dascher CC, Behar SM (2002) Dissemination of Mycobacterium tuberculosis is influenced by host factors and precedes the initiation of T-cell immunity. Infect Immun 70:4501–4509
5. Flynn JL, Klein E (2010) "Pulmonary tuberculosis in monkeys" in A color atlas of comparative pulmonary tuberculosis histopathology. In: Leong J, Dartois V, Dick T (eds) CRC, Boca Raton, pp 83–106

6. Flynn JL, Chan J, Lin PL (2011) Macrophages and control of granulomatous inflammation in tuberculosis. Mucosal Immunol 4:271–278
7. Lin PL, Rodgers M, Smith L, Bigbee M, Myers A et al (2009) Quantitative comparison of active and latent tuberculosis in the cynomolgus macaque model. Infect Immun 77:4631–4642
8. Geng E, Kreiswirth B, Burzynski J, Schluger NW (2005) Clinical and radiographic correlates of primary and reactivation tuberculosis: a molecular epidemiology study. JAMA 293:2740–2745
9. Volkman HE, Pozos TC, Zheng J, Davis JM, Rawls JF et al (2010) Tuberculous granuloma induction via interaction of a bacterial secreted protein with host epithelium. Science 327:466–469
10. Lin PL, Flynn JL (2010) Understanding latent tuberculosis: a moving target. J Immunol 185:15–22
11. Cooper AM (2009) Cell-mediated immune responses in tuberculosis. Annu Rev Immunol 27:393–422
12. Cooper AM, Dalton DK, Stewart TA, Griffin JP, Russell DG et al (1993) Disseminated tuberculosis in interferon gamma gene-disrupted mice. J Exp Med 178:2243–2247
13. Flynn JL, Chan J, Triebold KJ, Dalton DK, Stewart TA et al (1993) An essential role for interferon gamma in resistance to Mycobacterium tuberculosis infection. J Exp Med 178:2249–2254
14. Gallegos AM, van Heijst JW, Samstein M, Su X, Pamer EG et al (2011) A gamma interferon independent mechanism of CD4 T cell mediated control of M. tuberculosis infection in vivo. PLoS Pathog 7:e1002052
15. Torrado E, Cooper AM (2011) What do we really know about how CD4 T cells control mycobacterium tuberculosis? PLoS Pathog 7:e1002196
16. Lin PL, Myers A, Smith L, Bigbee C, Bigbee M et al (2010) Tumor necrosis factor neutralization results in disseminated disease in acute and latent Mycobacterium tuberculosis infection with normal granuloma structure in a cynomolgus macaque model. Arthritis Rheum 62:340–350
17. Flynn JL, Goldstein MM, Chan J, Triebold KJ, Pfeffer K et al (1995) Tumor necrosis factor-alpha is required in the protective immune response against Mycobacterium tuberculosis in mice. Immunity 2:561–572
18. Keane J, Gershon S, Wise RP, Mirabile-Levens E, Kasznica J et al (2001) Tuberculosis associated with infliximab, a tumor necrosis factor alpha-neutralizing agent. N Engl J Med 345:1098–1104
19. Maglione PJ, Chan J (2009) How B cells shape the immune response against mycobacterium tuberculosis. Eur J Immunol 39:676–686
20. Flynn JL, Tsenova L, Izzo A, Kaplan G (2008) "Experimental animal models of tuberculosis" in Handbook of tuberculosis: Immunology and cell biology, In: Kaufmann SHE, Britton WJ (eds). Wiley-VCH Vol. 2, pp 389–417
21. Scanga CA, Mohan VP, Joseph H, Yu K, Chan J et al (1999) Reactivation of latent tuberculosis: variations on the cornell murine model. Infect Immun 67:4531–4538
22. Pichugin AV, Yan BS, Sloutsky A, Kobzik L, Kramnik I (2009) Dominant role of the sst1 locus in pathogenesis of necrotizing lung granulomas during chronic tuberculosis infection and reactivation in genetically resistant hosts. Am J Pathol 174:2190–2201
23. Singhal A, Aliouat el M, Herve M, Mathys V, Kiass M et al (2011) Experimental tuberculosis in the wistar rat: a model for protective immunity and control of infection. PLoS One 6:e18632
24. Orme IM (2006) Preclinical testing of new vaccines for tuberculosis: a comprehensive review. Vaccine 24:2–19
25. Helke KL, Mankowski JL, Manabe YC (2006) Animal models of cavitation in pulmonary tuberculosis. Tuberculosis (Edinb) 86:337–348
26. Turner OC, Basaraba RJ, Orme IM (2003) Immunopathogenesis of pulmonary granulomas in the guinea pig after infection with mycobacterium tuberculosis. Infect Immun 71:864–871
27. Hoff DR, Ryan GJ, Driver ER, Ssemakulu CC, De Groote MA et al (2011) Location of intra- and extracellular M. tuberculosis populations in lungs of mice and guinea pigs during disease progression and after drug treatment. PLoS One 6:e17550

28. Basaraba RJ (2008) Experimental tuberculosis: The role of comparative pathology in the discovery of improved tuberculosis treatment strategies. Tuberculosis (Edinb) 88(suppl 1): S35–S47
29. Subbian S, Tsenova L, O'Brien P, Yang G, Koo MS et al (2011) Phosphodiesterase-4 inhibition combined with isoniazid treatment of rabbits with pulmonary tuberculosis reduces macrophage activation and lung pathology. Am J Pathol 179:289–301
30. Via LE, Lin PL, Ray SM, Carrillo J, Allen SS et al (2008) Tuberculous granulomas are hypoxic in guinea pigs, rabbits, and nonhuman primates. Infect Immun 76:2333–2340
31. Allison MJ, Zappasodi P, Lurie MB (1962) Host-parasite relationships in natively resistant and susceptible rabbits on quantitative inhalation of tubercle bacilli: their significance for the nature of genetic resistance. Am Rev Respir Dis 85:553–569
32. Dannenberg AM (1994) Rabbit model of tuberculosis. In: Bloom BR (ed) Tuberculosis: pathogenesis, protection, and control. American Society for Microbiology, Washington, DC, pp 149–156
33. Cosma CL, Swaim LE, Volkman H, Ramakrishnan L, Davis JM (2006) Zebrafish and frog models of mycobacterium marinum infection. Curr Protoc Microbiol Chapter 10:Unit 10B.2
34. Capuano SV 3rd, Croix DA, Pawar S, Zinovik A, Myers A et al (2003) Experimental Mycobacterium tuberculosis infection of cynomolgus macaques closely resembles the various manifestations of human M. tuberculosis infection. Infect Immun 71:5831–5844
35. Lin PL, Rutledge T, Green AM, Bigbee M, Fuhrman C et al (2012) CD4 T cell depletion exacerbates acute *Mycobacterium tuberculosis* while reactivation of latent infection is dependent on severity of tissue depletion in cynomolgus macaques. AIDS Res Hum Retroviruses. (in press)
36. Mattila JT, Diedrich CR, Lin PL, Phuah J, Flynn JL (2011) Simian immunodeficiency virus-induced changes in T cell cytokine responses in cynomolgus macaques with latent Mycobacterium tuberculosis infection are associated with timing of reactivation. J Immunol 186:3527–3537
37. Diedrich CR, Mattila JT, Klein E, Janssen C, Phuah J et al (2010) Reactivation of latent tuberculosis in cynomolgus macaques infected with SIV is associated with early peripheral T cell depletion and not virus load. PLoS One 5:e9611
38. Mehra S, Golden NA, Dutta NK, Midkiff CC, Alvarez X et al (2011) Reactivation of latent tuberculosis in rhesus macaques by coinfection with simian immunodeficiency virus. J Med Primatol 40:233–243
39. Davis SL, Nuermberger EL, Um PK, Vidal C, Jedynak B et al (2009) Noninvasive pulmonary [18F]-2-fluoro-deoxy-D-glucose positron emission tomography correlates with bactericidal activity of tuberculosis drug treatment. Antimicrob Agents Chemother 53:4879–4884
40. Puissegur MP, Botanch C, Duteyrat JL, Delsol G, Caratero C et al (2004) An in vitro dual model of mycobacterial granulomas to investigate the molecular interactions between mycobacteria and human host cells. Cell Microbiol 6:423–433
41. Bowdish DM, Sakamoto K, Kim MJ, Kroos M, Mukhopadhyay S et al (2009) MARCO, TLR2, and CD14 are required for macrophage cytokine responses to mycobacterial trehalose dimycolate and mycobacterium tuberculosis. PLoS Pathog 5:e1000474
42. Fallahi-Sichani M, Schaller MA, Kirschner DE, Kunkel SL, Linderman JJ (2010) Identification of key processes that control tumor necrosis factor availability in a tuberculosis granuloma. PLoS Comput Biol 6:e1000778
43. Chensue SW, Warmington K, Ruth J, Lincoln P, Kuo MC et al (1994) Cytokine responses during mycobacterial and schistosomal antigen-induced pulmonary granuloma formation. production of Th1 and Th2 cytokines and relative contribution of tumor necrosis factor. Am J Pathol 145:1105–1113
44. Chensue SW, Kunkel SL (2003) Cytokines and chemokines in granulomatous inflammation. In: Boros DL (ed) Granulomatous infections and inflammations: cellular and molecular mechanisms. ASM Press, Washington, DC, pp 29–64.

45. Fallahi-Sichani M, El-Kebir M, Marino S, Kirschner DE, Linderman JJ (2011) Multiscale computational modeling reveals a critical role for TNF-α receptor 1 dynamics in tuberculosis granuloma formation. J Immunol 186:3472–3483
46. Ray JC, Flynn JL, Kirschner DE (2009) Synergy between individual TNF-dependent functions determines granuloma performance for controlling Mycobacterium tuberculosis infection. J Immunol 182:3706–3717
47. Segovia-Juarez JL, Ganguli S, Kirschner D (2004) Identifying control mechanisms of granuloma formation during M. tuberculosis infection using an agent-based model. J Theor Biol 231:357–376
48. Warrender C, Forrest S, Koster F (2006) Modeling intercellular interactions in early mycobacterium infection. Bull Math Biol 68:2233–2261
49. Bru A, Cardona PJ (2010) Mathematical modeling of tuberculosis bacillary counts and cellular populations in the organs of infected mice. PLoS One 5:e12985
50. Ganguli S, Gammack D, Kirschner DE (2005) A metapopulation model of granuloma formation in the lung during infection with mycobacterium tuberculosis. Math Biosci Eng 2:535–560
51. Gammack D, Doering CR, Kirschner DE (2004) Macrophage response to Mycobacterium tuberculosis infection. J Math Biol 48:218–242
52. Marino S, Sud D, Plessner H, Lin PL, Chan J et al (2007) Differences in reactivation of tuberculosis induced from anti-TNF treatments are based on bioavailability in granulomatous tissue. PLoS Comput Biol 3:1909–1924
53. Sud D, Bigbee C, Flynn JL, Kirschner DE (2006) Contribution of CD8+ T cells to control of Mycobacterium tuberculosis infection. J Immunol 176:4296–4314
54. Wigginton JE, Kirschner D (2001) A model to predict cell-mediated immune regulatory mechanisms during human infection with mycobacterium tuberculosis. J Immunol 166:1951–1967
55. Magombedze G, Garira W, Mwenje E (2006) Modelling the human immune response mechanisms to Mycobacterium tuberculosis infection in the lungs. Math Biosci Eng 3:661–682
56. Day J, Friedman A, Schlesinger LS (2009) Modeling the immune rheostat of macrophages in the lung in response to infection. Proc Natl Acad Sci USA 106:11246–11251
57. Marino S, El-Kebir M, Kirschner D (2011) A hybrid multi-compartment model of granuloma formation and T cell priming in tuberculosis. J Theor Biol 280:50–62
58. Marino S, Pawar S, Fuller CL, Reinhart TA, Flynn JL et al (2004) Dendritic cell trafficking and antigen presentation in the human immune response to mycobacterium tuberculosis. J Immunol 173:494–506
59. Marino S, Myers A, Flynn JL, Kirschner DE (2010) TNF and IL-10 are major factors in modulation of the phagocytic cell environment in lung and lymph node in tuberculosis: a next-generation two-compartmental model. J Theor Biol 265:586–598
60. Flynn JL (2004) Immunology of tuberculosis and implications in vaccine development. Tuberculosis (Edinb) 84:93–101
61. Marino S, Hogue IB, Ray CJ, Kirschner DE (2008) A methodology for performing global uncertainty and sensitivity analysis in systems biology. J Theor Biol 254:178–196
62. Algood HM, Lin PL, Flynn JL (2005) Tumor necrosis factor and chemokine interactions in the formation and maintenance of granulomas in tuberculosis. Clin Infect Dis 41(suppl 3):S189–S193
63. Algood HM, Lin PL, Yankura D, Jones A, Chan J et al (2004) TNF influences chemokine expression of macrophages in vitro and that of CD11b+ cells in vivo during Mycobacterium tuberculosis infection. J Immunol 172:6846–6857
64. Algood HM, Chan J, Flynn JL (2003) Chemokines and tuberculosis. Cytokine Growth Factor Rev 14:467–477
65. Lin PL, Plessner HL, Voitenok NN, Flynn JL (2007) Tumor necrosis factor and tuberculosis. J Investig Dermatol Symp Proc 12:22–25
66. Chakravarty SD, Zhu G, Tsai MC, Mohan VP, Marino S et al (2008) Tumor necrosis factor blockade in chronic murine tuberculosis enhances granulomatous inflammation and disorganizes granulomas in the lungs. Infect Immun 76:916–926

67. Clay H, Volkman HE, Ramakrishnan L (2008) Tumor necrosis factor signaling mediates resistance to mycobacteria by inhibiting bacterial growth and macrophage death. Immunity 29:283–294
68. Iliopoulos A, Psathakis K, Aslanidis S, Skagias L, Sfikakis PP (2006) Tuberculosis and granuloma formation in patients receiving anti-TNF therapy. Int J Tuberc Lung Dis 10:588–590
69. Chensue SW, Warmington KS, Ruth JH, Lincoln P, Kunkel SL (1995) Cytokine function during mycobacterial and schistosomal antigen-induced pulmonary granuloma formation. Local and regional participation of IFN-gamma, IL-10, and TNF. J Immunol 154:5969–5976
70. Schutze S, Machleidt T, Adam D, Schwandner R, Wiegmann K et al (1999) Inhibition of receptor internalization by monodansylcadaverine selectively blocks p55 tumor necrosis factor receptor death domain signaling. J Biol Chem 274:10203–10212
71. Schneider-Brachert W, Tchikov V, Neumeyer J, Jakob M, Winoto-Morbach S et al (2004) Compartmentalization of TNF receptor 1 signaling: internalized TNF receptosomes as death signaling vesicles. Immunity 21:415–428
72. Schneider-Brachert W, Tchikov V, Merkel O, Jakob M, Hallas C et al (2006) Inhibition of TNF receptor 1 internalization by adenovirus 14.7K as a novel immune escape mechanism. J Clin Invest 116:2901–2913
73. Neumeyer J, Hallas C, Merkel O, Winoto-Morbach S, Jakob M et al (2006) TNF-receptor I defective in internalization allows for cell death through activation of neutral sphingomyelinase. Exp Cell Res 312:2142–2153
74. Harris J, Hope JC, Keane J (2008) Tumor necrosis factor blockers influence macrophage responses to mycobacterium tuberculosis. J Infect Dis 198:1842–1850
75. Gutierrez MG, Mishra BB, Jordao L, Elliott E, Anes E et al (2008) NF-kappa B activation controls phagolysosome fusion-mediated killing of mycobacteria by macrophages. J Immunol 181:2651–2663
76. Keane J, Shurtleff B, Kornfeld H (2002) TNF-dependent BALB/c murine macrophage apoptosis following Mycobacterium tuberculosis infection inhibits bacillary growth in an IFN-gamma independent manner. Tuberculosis (Edinb) 82:55–61
77. Keane J, Balcewicz-Sablinska MK, Remold HG, Chupp GL, Meek BB et al (1997) Infection by Mycobacterium tuberculosis promotes human alveolar macrophage apoptosis. Infect Immun 65:298–304
78. Zhou Z, Connell MC, MacEwan DJ (2007) TNFR1-induced NF-kappaB, but not ERK, p38MAPK or JNK activation, mediates TNF-induced ICAM-1 and VCAM-1 expression on endothelial cells. Cell Signal 19:1238–1248
79. Russell DG, Barry CE 3rd, Flynn JL (2010) Tuberculosis: what we don't know can, and does, hurt us. Science 328:852–856
80. Wallis RS, Broder M, Wong J, Lee A, Hoq L (2005) Reactivation of latent granulomatous infections by infliximab. Clin Infect Dis 41(suppl 3):S194–S198
81. Keane J (2005) TNF-blocking agents and tuberculosis: new drugs illuminate an old topic. Rheumatology (Oxford) 44:714–720
82. Winthrop KL (2006) Risk and prevention of tuberculosis and other serious opportunistic infections associated with the inhibition of tumor necrosis factor. Nat Clin Pract Rheumatol 2:602–610
83. Wallis RS (2008) Tumour necrosis factor antagonists: structure, function, and tuberculosis risks. Lancet Infect Dis 8:601–611
84. Hochberg MC, Tracy JK, Hawkins-Holt M, Flores RH (2003) Comparison of the efficacy of the tumour necrosis factor alpha blocking agents adalimumab, etanercept, and infliximab when added to methotrexate in patients with active rheumatoid arthritis. Ann Rheum Dis 62(suppl 2):ii13–ii16
85. Gladman DD (2008) Adalimumab, etanercept and infliximab are equally effective treatments for patients with psoriatic arthritis. Nat Clin Pract Rheumatol 4:510–511
86. Wallis RS (2009) Infectious complications of tumor necrosis factor blockade. Curr Opin Infect Dis 22:403–409
87. Wallis RS, Broder MS, Wong JY, Hanson ME, Beenhouwer DO (2004) Granulomatous infectious diseases associated with tumor necrosis factor antagonists. Clin Infect Dis 38:1261–1265

88. Tubach F, Salmon D, Ravaud P, Allanore Y, Goupille P et al (2009) Risk of tuberculosis is higher with anti-tumor necrosis factor monoclonal antibody therapy than with soluble tumor necrosis factor receptor therapy: the three-year prospective french research axed on tolerance of biotherapies registry. Arthritis Rheum 60:1884–1894
89. Fonseca JE, Canhao H, Silva C, Miguel C, Mediavilla MJ et al (2006) Tuberculosis in rheumatic patients treated with tumour necrosis factor alpha antagonists: the portuguese experience. Acta Reumatol Port 31:247–253
90. Fallahi-Sichani M, Flynn JL, Linderman JJ, Kirschner DE (2012) Differential risk of tuberculosis reactivation among anti-TNF therapies is due to drug binding kinetics and permeability. J Immunol 188:3169–3178
91. Plessner HL, Lin PL, Kohno T, Louie JS, Kirschner D et al (2007) Neutralization of tumor necrosis factor (TNF) by antibody but not TNF receptor fusion molecule exacerbates chronic murine tuberculosis. J Infect Dis 195:1643–1650
92. Cook DN, Bottomly K (2007) Innate immune control of pulmonary dendritic cell trafficking. Proc Am Thorac Soc 4:234–239
93. Randolph GJ, Angeli V, Swartz MA (2005) Dendritic-cell trafficking to lymph nodes through lymphatic vessels. Nat Rev Immunol 5:617–628
94. Celli S, Garcia Z, Bousso P (2005) CD4 T cells integrate signals delivered during successive DC encounters in vivo. J Exp Med 202:1271–1278
95. Celli S, Lemaitre F, Bousso P (2007) Real-time manipulation of T cell-dendritic cell interactions in vivo reveals the importance of prolonged contacts for CD4+ T cell activation. Immunity 27:625–634
96. Zheng H, Jin B, Henrickson SE, Perelson AS, von Andrian UH et al (2008) How antigen quantity and quality determine T-cell decisions in lymphoid tissue. Mol Cell Biol 28:4040–4051
97. Linderman JJ, Riggs T, Pande M, Miller M, Marino S et al (2010) Characterizing the dynamics of CD4+ T cell priming within a lymph node. J Immunol 184:2873–2885
98. Steinman RM (2001) Dendritic cells and the control of immunity: enhancing the efficiency of antigen presentation. Mt Sinai J Med 68:160–166
99. Kirschner DE, Webb GF (1998) Immunotherapy of HIV-1 infection. J Biol Syst 6:71–83
100. Kirschner D, Panetta JC (1998) Modeling immunotherapy of the tumor-immune interaction. J Math Biol 37:235–252
101. Churchyard GJ, Kaplan G, Fallows D, Wallis RS, Onyebujoh P et al (2009) Advances in immunotherapy for tuberculosis treatment. Clin Chest Med 30:769–782, ix
102. Rook GA, Lowrie DB, Hernandez-Pando R (2007) Immunotherapeutics for tuberculosis in experimental animals: is there a common pathway activated by effective protocols? J Infect Dis 196:191–198
103. Wallis RS (2005) Reconsidering adjuvant immunotherapy for tuberculosis. Clin Infect Dis 41:201–208
104. Kim MS, Lee SH, Song MY, Yoo TH, Lee BK et al (2007) Comparative analyses of complex formation and binding sites between human tumor necrosis factor-alpha and its three antagonists elucidate their different neutralizing mechanisms. J Mol Biol 374:1374–1388
105. Nestorov I (2005) Clinical pharmacokinetics of TNF antagonists: how do they differ? Semin Arthritis Rheum 34:12–18

Chapter 8
Stochastic Gene Expression in Bacterial Pathogens: A Mechanism for Persistence?

Andrea Rocco, Andrzej Kierzek, and Johnjoe McFadden

Abstract Recent experiments have shown the relevance of stochastic fluctuations to numerous biological phenomena. Intrinsic and extrinsic sources of noise existing at the subcellular level are capable of influencing the population dynamics and are believed to be responsible for the appearance of different phenotypes in clonal bacterial populations. Single cell level phenotypic diversity is a likely key factor in the emergence of persistence in *Mycobacterium tuberculosis*. Stochastic phenomena in molecular interaction networks have been first postulated in theoretical studies and later confirmed by experimental observations of individual cells and molecules. Here, we shall review the main modeling tools that can be used in this context, namely stochastic differential equations (Langevin equations) and Master Equations and their simulational counterparts, such as the Gillespie algorithm. We will distinguish between intrinsic and extrinsic noise in subcellular networks, highlighting in particular the unexpected and sometimes counterintuitive behaviors induced by extrinsic noise. We will discuss the dependence of prokaryotic gene expression noise on transcription and translation rates, as emerged from theoretical and experimental studies of stochasticity in biochemical processes. These findings have direct consequences for understanding more complex gene regulatory networks, such as catabolic repression and two-component systems. Finally we will discuss the insights into the emergence of persistence of *M. tuberculosis* resulting from our understanding of stochastic gene expression, and delineate directions of future research.

A. Rocco, Ph.D. (✉) • A. Kierzek, Ph.D. • J. McFadden, Ph.D.
Faculty of Health and Medical Sciences, University of Surrey, Guildford, Surrey GU2 7XH, UK
e-mail: a.rocco@surrey.ac.uk; a.kierzek@surrey.ac.uk; j.mcfadden@surrey.ac.uk

J. McFadden et al. (eds.), *Systems Biology of Tuberculosis*,
DOI 10.1007/978-1-4614-4966-9_8, © Springer Science+Business Media, LLC 2013

1 Persistence

When *Mycobacterium tuberculosis* is grown in vitro at mid-log phase it is relatively sensitive to standard anti-mycobacterial drugs. However, in vivo, live *drug-sensitive* bacteria may be recovered from patients for many weeks or months after initiation of chemotherapy [1]. Examination of this and similar studies indicate that killing appears to be biphasic with a rapidly killed majority population and a very slowly killed (phenotypically tolerant, PT) minority population (Fig. 8.1). The biphasic nature of the kill curve is also found in vitro with a range of antibiotics and in animal models [2] and is generally assumed to be due to non-inherited antibiotic *tolerance* of cells in a distinct genetic and/or physiological state known generally as persistence. It is these persisters that are thought to be the principal problem in TB control, since treatment has to be maintained for at least 6 months to kill the persister population leading to poor rates of compliance and consequent emergence of genetic drug resistance [3]. The phenomenon of persistence in tuberculosis is often considered to be synonymous with dormancy or latency in which the pathogen may reside asymptomatically in the host for many years or decades before resurgence to cause post-primary TB. The underlying assumption is that that population of bacteria that evades the immune response during latency (dormant cells) is functionally equivalent to the population of bacteria that survives antibiotic treatment (persisters). However, although this is plausible, it is yet to be proved. In this review, the term 'persistence' will be used to indicate an antibiotic-tolerant subpopulation of cells. The relationship between these cells and disease latency will not be addressed here but has been discussed in other recent reviews [4–6].

Persistence (the presence of phenotypically drug-tolerant subpopulation) is not confined to mycobacteria. Indeed it was described by Bigger in 1944 who observed

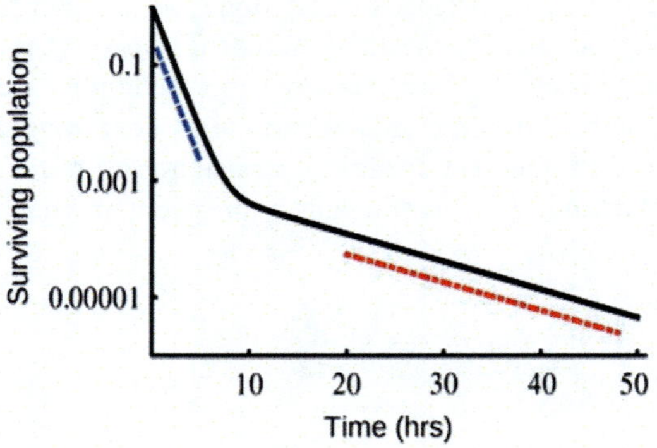

Fig. 8.1 Typical biphasic kill curve defining the phenomenon of persistence. The graph depicts the survival of a bacterial population when exposed to antibiotic. The *full curve* is the solution of (8.1), corresponding to Balaban type I persisters, with the following parameter values: $\mu = -1.0$, $\mu_p = 0$, $a = 0$, $b = 0.1$ (all units in h^{-1}). The *dot-dashed line* corresponds to the asymptotic behavior given by (8.2), and the *dashed line* is the transient killing of normal cells, also given by (8.2)

that when a genetically homogenous culture of *Staphylococcus aureus* was exposed to the bactericidal action of penicillin, a small number of cells (the persisters), which were not genetically resistant mutants, survived the treatment. Since then persistence has been described in nearly all known microbes and is considered to be largely responsible for the resistance to antibiotic therapy of many chronic bacterial infections and to the resistance of biofilms to microbiocides [7]. Although the phenomenon differs between bacterial species, certain features appear to be universal characteristics. The first is that persisters are invariably more abundant in slow-growing, stationary phase and nongrowing cultures [8], a feature that is often used to enrich for persisters. The second key characteristic is that the phenomenon of tolerance appears to be relatively general: persistent cell tends to be tolerant to a wide variety of antibiotics, stresses, and microbiocidal agents [9–12]. Lastly, of course, is the persister defining feature that phenotypic tolerance to drugs and stresses is non-heritable: it is not caused by genetic mutation. However, despite several decades of research, the underlying mechanisms responsible for persistence—the ability to survive antibiotic exposure—remain elusive.

A major difficulty in the study of PT is the low frequency of persistent cells. However, a key advance was the isolation of the *hipA7* mutant of *Escherichia coli* by Moyad and colleagues which exhibits PT at a level about 1,000-fold higher than wild-type cells [13]. A landmark study by Balaban et al. [14] examined growth of individual *hipA7* cells in a microfluidics device and demonstrated that persisters had been either slow-growing or nongrowing at the time of antibiotic administration. The experiments allowed them to identify key properties of 'type I' persisters in the *hipA7* mutant strain. These appeared to constitute a preexisting population of non-growing cells that were generated by unknown trigger events only during stationary phase growth. However, examination of another persister mutant, *hipQ*, demonstrated the existence of another type of persister, termed type II persister, which was continuously generated during exponential growth.

There are two key puzzles of persistence. The first is how cells with at least two distinct phenotypes (normal and persisters) can coexist in an apparently isogenic population. The second major puzzle is how the persister cells remain viable in the presence of antibiotic that kills their isogenic sister (normal) cells. Considerably more progress has been made with the first problem, particularly the use of mathematical equations to model the transitions between normal and persister cells. The authors of [14] developed a simple dynamic model of persistence (Fig. 8.2) in which the number of persisters and that of the normally growing cells are denoted by the symbols p and n, respectively. The cells switch from the n state to the p state either through the action of a trigger at stationary phase (type I) or with a constant rate a (type II), and from the p state to the n state with a constant rate b. These dynamics can then be translated into the following population models:

$$\frac{dn}{dt} = -an + bp + \mu_n n, \quad \frac{dp}{dt} = an - bp + \mu_p p, \tag{8.1}$$

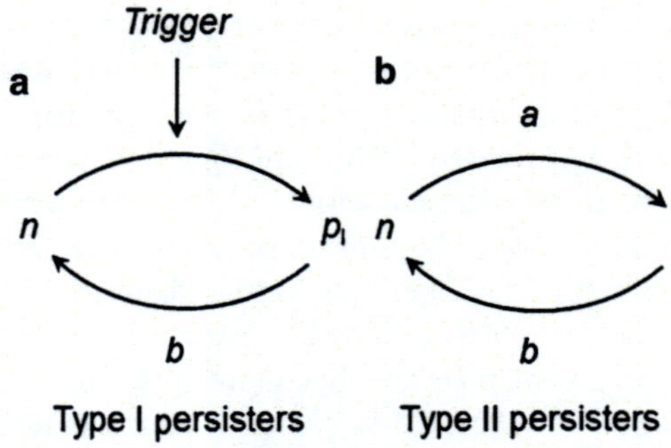

Fig. 8.2 The phenotypic switch as from [14]. In the case of type I persisters (panel **a**), these are created by some triggering event during stationary phase and have the capability of reverting to the normal cell state with a rate *b*. Type II persisters (panel **b**), in contrast, are continuously generated during log phase

where $a = 0$ defines type I persisters, and $a \neq 0$ type II persisters. Note that the parameters μ_n and μ_p represent the differences between growth and killing rates of normal and persister cells, respectively, and, for type I persisters, $\mu_p \approx 0$, whereas μ_n must always be less than zero. The solution of (8.1) has the form:

$$N(t) = n(t) + p(t) = A\mathrm{e}^{\mu_n t} + B\mathrm{e}^{(\mu_p - b)t}, \tag{8.2}$$

where *N*(*t*) is the total population of surviving cells, and the prefactors *A* and *B* can computed as

$$A = n_0 + \frac{p_0}{\mu_n - \mu_p + b} \quad \text{and} \quad B = p_0 - \frac{p_0}{\mu_n - \mu_p + b}, \tag{8.3}$$

with n_0 and p_0 the number of normal and persister cells, respectively, at the beginning of antibiotic exposure (at time t=0). The total cell population from (8.2) is depicted in Fig. 8.1, with the dashed blue and dot-dashed red lines corresponding to the two exponential components, decaying, respectively, as $\mathrm{e}^{\mu_n t}$ (fast killing) and $\mathrm{e}^{(\mu_p - b)t}$ (slow killing).

Parameters of the persister populations for the *hip* strains and wild-type cells were obtained by fitting plots of growth and antibiotic killing to the solutions of the population equations to obtain the typical biphasic kill curves that are characteristic of persistence. Note that in this model the resistance of persisters to antibiotics is a fitting parameter and not derived from the model; that is, the model provides mechanistic underpinning of the first problem of persistence (the existence of distinct cells) but not the second problem (mechanism of drug tolerance).

A plausible molecular mechanism for generation of persisters, at least in the *hipA* strain, emerged when it was discovered that HipA is a toxin component of a member of the toxin–antitoxin (TA) system. TA modules are widespread in bacteria and consist of a toxin, capable of killing or slowing growth of the cell, and an antitoxin that is able to neutralize the action of the toxin [15–19]. In normal conditions both toxin and antitoxin are expressed so that the toxin is neutralized. However, the toxin is more stable than the antitoxin so if the expression of both proteins is slowed, particularly during conditions of stress, then the toxin component will tend to persist and kill the cell. Balaban's group demonstrated that stochastic fluctuations in expression of HipA levels beyond a threshold (set by the capacity of the HipB antitoxin to neutralize the toxin) cause switching to a state of transient growth arrest and associated antibiotic tolerance in *E. coli* [20].

It seems likely that TA modules are involved in persistence in at least some systems. However, TA-based mechanism seems unlikely to account for persistence in all bacteria since a plethora of genes and mechanisms are known to enhance or depress PT levels [21–23]. Several alternative or complementary models of persistence have been published by various authors [8, 10, 21, 22, 24–27] involving toxin systems, cell senescence, metabolite stress, and other mechanisms. A common feature of many of these models is phenotypic heterogeneity that is maintained by some sort of switching process to generate a bistable or multistable population [28, 29]. However, it is worth noting that, in contrast to other bistable systems, it has not so far been possible to engineer (e.g., by modification of the switch) or isolate mutant strains that exist solely in either a pure non-persister or persister state.

2 Persistence in *M. tuberculosis*

Much less is known about persistence in *M. tuberculosis*, but what little is known is consistent with the phenomenon being broadly similar to persistence described in other bacteria. Because of the difficulty of examining persistence in vivo and the absence of *hip*-like mutant strains of *M. tuberculosis* with high rates of persisters, much of the research has focused on characterizing mycobacteria in various in vitro systems that enrich the population of *M. tuberculosis* for drug-tolerant persisters. By far the most popular of these is the 'Wayne model', developed by Larry Wayne, in which tubercle bacilli are grown in sealed tubes with slow stirring until the bacteria cease replicating when oxygen concentrations decrease to the microaerobic or hypoxic levels. *M. tuberculosis* then enters a non-replicating persistent state and becomes more generally tolerant to antibiotics [30–32], although, interestingly, the cells become more sensitive to the drug metronidazole [32]. The trigger for the high persister state is unclear in this model as the cells are in the stationary phase of growth and are therefore in an undefined environment with changing oxygen concentrations and nutrients combined with the buildup of potentially toxic waste products. Many other in vitro models of persistence have been developed, such as treatment with nitric oxide [33], extended stationary phase, and the extended hypoxia model [34–36]. An alternative route towards enriching for *M. tuberculosis* persisters was recently developed via treatment with antibiotic (to kill sensitive cells) and differential centrifugation to isolate surviving persisters [37]. A mutant screen in mice was also recently used to identify mutants with either impaired or enhanced 'persistence' in mice treated with isoniazid [38]. However, the impairment or enhancement of survival of the mutant strains appeared to be specific to isoniazid, and therefore the relevance of the study to the in vitro phenomenon of broad-spectrum antibiotic tolerance (persistence as defined here) is currently unclear.

Transcriptome studies demonstrated that expression of very large number of genes is perturbed when *M. tuberculosis* is grown in any of the in vitro persistence models [36, 39–41]. The predominant signal is downregulation of most genes particularly those involved in central metabolism and biosynthetic and ribosomal synthesis. However, some genes are upregulated in each of the models. In the hypoxia model, many of the regulated genes form part of a regulon of approximately 50 transcripts, called the dormancy survival (dos) regulon [40] which includes many genes involved in energy and carbon source acquisition. An additional, larger, set of genes have recently been identified to form the 'enduring hypoxia response' [42]. The dos regulon is controlled by a two-component response regulator composed of the sensor kinases DosT and DosS which activate the transcriptional regulator DosR in response to hypoxia, as well as nitric oxide and carbon monoxide exposure [43]. Disruption of DosR resulted in loss of the ability of BCG to adapt to survival in the Wayne hypoxia model [44] and defects in maintaining ATP and NADH balance in the cell [45]. More recently, another regulatory system controlled by the redox sensor, WhiB3, has also been implicated in the hypoxia persistence model and was also shown to be essential for survival during nutrient starvation [46, 47]. Transcriptional analysis of *M. tuberculosis* persisters recovered by the antibiotic treatment and differential centrifugation method demonstrated similar features to the other in vitro methods: downregulation of genes involved in central metabolism and biosynthetic pathways [37]. However, an interesting feature of this analysis was the upregulation of several of the 88 toxin–antitoxin (TA) systems that are encoded in the *M. tuberculosis* genome [48]. Overexpression of TA toxins slows growth of *M. tuberculosis* and confers increased tolerance to antibiotics, intriguingly, in a toxin-specific manner such that the expression of different toxins confers tolerance to different sets of drugs [49].

Overall, it appears that persistence in *M. tuberculosis* involves a variety of molecular adaptations [6, 12, 38, 50–53], but the mechanisms that are central to persistence remain unclear.

3 Noise

As discussed above, the first problem of persistence is the existence of multiple cell types in an isogenic population. In models such as the Balaban model discussed above, normal cells transform into persister cells and persister cells transform into normal cells at rates that are fitted to the equations rather than mechanistically generated. But why do isogenic cells transit between different states? One of the most intriguing possibilities to have emerged in recent years is that noise, at the molecular level, provides the driving force for stochastic phenotypic transitions. We now examine whether an understanding of molecular noise can contribute to our understanding of persistence.

Stochastic processes and noise have been studied extensively for decades in many different fields of Physics (both theoretical and experimental) and Mathematics, as well as more in applied sciences such as Engineering. More recently, stochasticity has gained a renewed attention in biology as well, where it appears to play a major role at many different scales, ranging from the dynamics taking place in subcellular molecular networks to the behavior of entire organisms. In particular, mathematical modeling of gene expression processes has predicted large variances in the numbers of protein molecules produced by a gene, and has indicated that randomness in gene expression outcomes is an important factor determining cellular behavior [54–56]. The advent of fluorescent reporter protein strains enabled the observation of protein amounts in single cells. A large body of experimental evidence has been collected to validate the predictions of mathematical models, and shows that gene expression is indeed stochastic. The level of detail of recent observations is astonishing. The expression of the *lacY* gene in *E. coli* for instance has been observed with single molecule resolution, which has allowed the direct recording of random gene expression events occurring in single cells. It is interesting to note that stochasticity in gene expression was first hypothesized by theoreticians and later validated experimentally. The discovery of the role of gene expression noise in molecular biology is an example of the applicability of an exact science approach to biology, and one of the major successes in the field of systems biology.

It is now well established that stochastic processes play a role in the emergence of different phenotypes within the known genetic circuits regulating the functional behaviors of different networks, cell, or organisms. A widespread view is that subcellular networks have evolved so as to allow for multiple different operating modes (states), with noise allowing the system to access all of them, and switch among them. When the genetic circuit is engineered so that transitions among these states are slow, the system will appear "stuck" in the same state for a long time, possibly larger than the observation time. These states are naturally identified as distinct phenotypes, and their occurrence within an isogenic population gives rise to so-called epigenetic population heterogeneity. In other words, the phenotypic state of the cell is not solely determined by environmental signals activating regulatory processes, but can also be the result of random switching between different states, coexisting in the same environment. The randomness inherent to the processes which express genetic information in response to environmental changes will produce different, unpredictable outcomes under the same conditions.

Noise is usually assumed detrimental to normal biological function, implying that organisms have evolved so as to minimize its effect, by buffering against it. We shall show below that this might not always be the case, and in fact noise could be beneficial to the normal physiological function. In this respect it may well be that organisms have actually evolved so as to utilize noise to attain operating points which would not be easily accessible if the dynamics were completely deterministic. In this view, a trade-off is evolutionary sought in order to exploit the advantageous properties of noise, at the same time avoiding its possibly destructive effects [57].

Furthermore, when scaled up to the entire population, noise-induced heterogeneity can provide the organism with a relevant protection mechanism. Variability in cell populations clearly represents an asset at the moment of taking advantage of specific environmental conditions, or to resist external threats. Bacterial pathogens are under particularly strong selective pressure to allow for noisy gene expression, as this population diversity offers a selective advantage when confronting host immune defenses and antibiotic treatment. Drug tolerance/persistence is a clear example of this. While different drug tolerance phenotypes can be associated with cleverly engineered gene circuits, it may also be that stochastic fluctuations originating at the gene expression level become (quasi) static heterogeneities at the population level, and as such protect the whole population against antibiotics or more generally any type of stress.

But what is the source of noise in molecular networks? In his book *What is Life?* [58], Erwin Schrödinger introduced in 1944 for the first time the so-called $1/\sqrt{n}$ rule. His argument, based on assuming equilibrium for the intracellular reactions, thereby implying Poisson statistics, states that if a molecular species is present with copy number *n*, then fluctuations with intensity of the order of $1/\sqrt{n}$ should be expected. It is clear then that in a macroscopic sample of substance, with an Avogadro number of molecules, of the order of 10^{23}, noise is safely negligible, and we can rely on a deterministic description of the system. However, if we have a much lower number of molecules, for instance 10 or 100, then fluctuations are not negligible, and stochastic processes need to be described. This source of stochastic effects is usually referred to as *intrinsic noise*, and occurs typically in weakly expressed genes.

Another source of stochasticity is so-called *extrinsic noise*, associated with fluctuations affecting either control parameters of the system, such as temperature or pH, or otherwise molecular species assumed constant in the system, such as, for instance, abundances of RNA polymerases, or ribosome concentrations. In contrast to intrinsic noise, which is gene specific, extrinsic noise is global, in that it affects in the same way the expression of all genes in the cell [59]. An important feature of extrinsic noise is that it can produce drastic modifications of the dynamics of the system, including shifts of operating points of the network, oscillating behaviors, or even appearance of multiple steady states, what is generally referred to as noise-induced transitions [60].

3.1 Langevin Equations

A useful framework to describe stochastic processes is the so-called Langevin approach. In the Langevin approach, the effect of a stochastic process acting on a dynamical variable *x* (corresponding for instance to the expression level of gene *X*) is phenomenologically described by decomposing the total dynamics into two contributions. The first contribution is a collective effect, which can be thought as an average dynamics, which Langevin proposes to treat deterministically. The second contribution describes the fluctuations about this average. This is the stochastic component, or noise, which is treated in a probabilistic manner. In mathematical terms, for a system composed by only one molecular species,

$$\frac{dx}{dt} = f(x) + \xi(t), \tag{8.4}$$

where $f(x)$ is the deterministic (average) dynamics, and $\xi(t)$ is the noise. The noise term is defined in terms of its statistical properties and besides being Gaussian (normally distributed), it is usually assumed to be zero average, and delta correlated over time (so-called white noise which is not correlated over time),

$$\begin{cases} \xi(t) = 0, \\ \xi(t)\xi(t') = \Gamma\delta(t - t'), \end{cases} \tag{8.5}$$

with the brackets indicating ensemble averages, Γ being the noise intensity, and $\delta(t)$ indicating the Dirac delta function. These assumptions are typical, but, of course, are an idealization of the real system, and ultimately a modeling choice. Gaussianity is usually assumed because of the Central Limit Theorem, but different statistics can also be considered. Furthermore, the assumption of the noise being zero average is not restrictive. In fact, if it is not, the average can be formally subtracted out, and recast into a redefinition of the deterministic part of the equation, leading again to the same structure of (8.4) and (8.5). Finally, the choice of the noise being delta correlated over time corresponds to assuming that the process is memory-less, namely Markov. This assumption is justified in the case when the real fluctuations have a correlation time smaller than any other timescale of the system. If this is not the case, the finite correlation time of the noise needs to be considered (Ornstein–Uhlenbeck or colored noises) [61]. A noticeable example of the application of this simple scheme appears for instance in [62], where a set of two coupled Langevin equations, respectively, for mRNA and protein concentrations, is used to account for the observed positive correlation of stochastic fluctuations and translation rates in a single gene system.

When extrinsic noise is considered, the Langevin equation acquires a different structure. Let a be a control parameter for the system, representing either environmental factors, such as pH levels, or fixed intracellular elements, such as amount of RNAP, or number of ribosomes. Let us assume that a fluctuates over time, and let us represent its dynamics as

$$a \to a(t) = a_0 + \varepsilon^{1/2}\xi(t), \tag{8.6}$$

where a_0 is the average value of a, $\xi(t)$ is the noise, which for simplicity we again assume Gaussian, zero average, and white, and ε is the noise intensity. The exponent ½ accounts for the dimensions of the noise intensity as from (8.5). The reaction dynamics $f(x, a)$ can be linearized in the noise intensity for sufficiently small ε, and the corresponding Langevin equation becomes:

$$\frac{dx}{dt} = f(x, a_0) + \varepsilon^{1/2} g(x, a_0)\xi(t), \tag{8.7}$$

with

$$g(x, a_0) = \left.\frac{\partial f(x, a)}{\partial a}\right|_{a=a_0}. \tag{8.8}$$

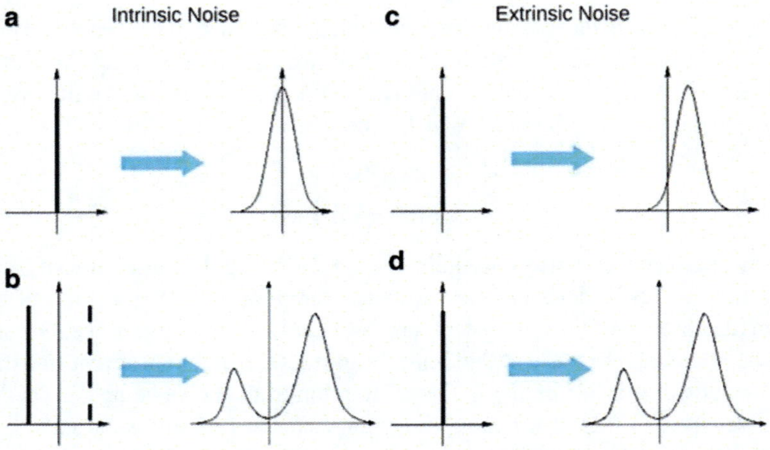

Fig. 8.3 Schematic representation of possible effects of intrinsic and extrinsic noise. Panels (**a**) and (**b**) represent the broadening of probability distribution when passing (*arrows*) from deterministic to stochastic dynamics when only intrinsic noise is present. The *thick line* indicates the degenerate (zero width) probability distribution associated with the existence of a (stable) steady state of the deterministic dynamics. In panel (**b**), the deterministic dynamics allows already for two stable steady states (indicated with the alternative *full* and *dashed thick lines*), and the system becomes free to hop from one to the other as a result of noise. Panels (**c**) and (**d**) show instead the counter-intuitive effects that may appear due to extrinsic noise. In panel (**c**) a shift of the maximum of the probability distribution is shown, while in panel (**d**) a system which is deterministically monostable acquires a further stable state

Equation (8.7) is a multiplicative noise stochastic differential equation, whose solution can be difficult to compute. In particular it needs to be supplemented with a specific prescription on how to evaluate the associated stochastic integral. In the case when fluctuations happen on a shorter timescale than any other process in the system, it is possible to show [61] that this prescription is the so-called Stratonovich prescription. Without entering into the related mathematical analysis, it is important to notice that this prescription implies interesting consequences for the system. In particular it is possible to show that the mode(s) of the probability distribution associated with the stochastic process defined by the equation above, and supplemented with the Stratonovich prescription, are determined by the equation:

$$f(x,a_0) - g(x,a_0)g'(x,a_0) = 0, \tag{8.9}$$

where, with the prime, we indicate the derivative with respect to x. As a result, the position of the mode can change with respect to the deterministic solution (identified by $f(x,a_0) = 0$), its stability properties can change, and new modes may appear (Fig. 8.3).

The consequences of extrinsic stochastic fluctuations have been analyzed in molecular networks in recent literature. For instance noise-induced oscillations have been predicted in enzymatic futile cycles [63], and the corresponding "deviant" effects have been catalogued in [64]. Also, the possibility of using extrinsic noise as a control mechanism in metabolic pathways has been explored in [57]. The bottom line of these studies is that extrinsic noise can in fact change the macroscopic behavior of molecular networks, and therefore can act so as to allow the cell to explore regions of the phase space otherwise unreachable if the dynamics were only deterministic. In this sense, this begs the question to what extent stochastic dynamics can be evolutionary selected for because of dynamic benefit [57].

The solution of the Langevin equations above, (8.4) or (8.7), is the time evolution of the dynamical variable x for a sequence in time of realizations of the stochastic process ξ. It produces a single trajectory in the x phase space. Collecting many trajectories leads to the construction of an ensemble of dynamical realization of the system, over which the time evolution of the probability distribution can be measured. As we shall see in the next section, an alternative, statistically equivalent approach is to construct dynamical equations for the probability distributions directly.

3.2 *Master Equations*

As an alternative to the single trajectory Langevin approach, it is always possible to describe the system in terms of probability distributions. In the case of a memory-less Markov processes, this can be done by using the so-called Master equation formalism. In its more general form, the time evolution of the probability distribution reads

$$\frac{\partial p(x,t)}{\partial t} = \int \left\{ W(x \mid x')p(x',t) - W(x' \mid x)p(x,t) \right\} dx'. \tag{8.10}$$

The derivation of this equation proceeds from the Chapman–Kolmogorov equation, which is the most general equation obeyed by any Markov process. We will not present this derivation here, and refer to [65] for the related mathematical details. We highlight the aspects of the approach relevant to stochastic gene expression.

In (8.10), $p(x,t)$ is the probability that the stochastic variable takes up the value x at time t, and $W(x \mid y)$ is the transition probability that the stochastic variable makes a jump from the value y to the value x in the unit time. The first term corresponds to the gain of probability of finding the system in the state x due to transitions from any other state. The minus term represents the loss of probability associated with the system being in the state x and making a transition to any other state. Because of this interpretation of the structure of the equation, the master equation is also referred to as gain–loss equation, or birth–death master equation.

The formalism of master equations has been recently used to compute analytically the stationary distribution of protein within the two-stage model of gene expression:

$$\begin{array}{ccccc} \text{DNA} & \xrightarrow{k_1} & \text{mRNA} & \xrightarrow{k_2} & \text{protein} \\ & & \gamma_1 \big\downarrow & & \gamma_2 \big\downarrow \\ & & \varnothing & & \varnothing \end{array}$$

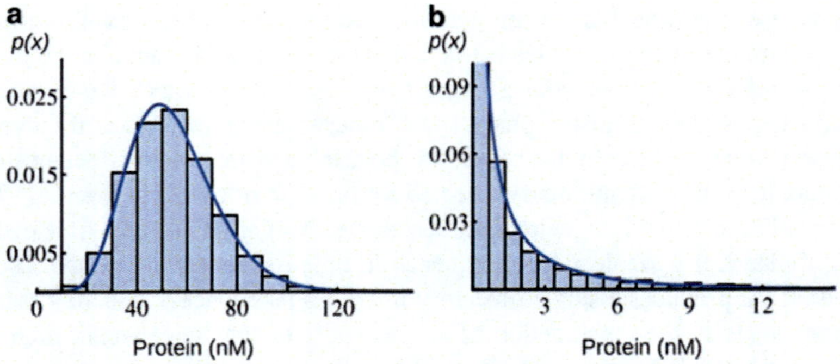

Fig. 8.4 The two regimes of Gamma distributions. Histograms are obtained by direct simulation of the two-stage gene expression model presented above. The *solid curves* represent the Gamma distribution as given by (8.12) with parameters a and b estimated as explained in the text. *Left panel*: $k_1=0.1$ (s^{-1}), $k_2=0.4$ (s^{-1}), $\gamma_1=0.05$ (s^{-1}), $\gamma_2=0.01$(s^{-1}), $T_0=2{,}100$ (s), resulting in $a=9.7$ and $b=8\log 2$. *Right panel*: $k_1=10^{-4}$ (s^{-1}), $k_2=0.4$ (s^{-1}), $\gamma_1=0.05$ (s^{-1}), $\gamma_2=10^{-3}$ (s^{-1}), $T_0=2{,}100$ (s), resulting in $a=0.075$ and $b=8\log 2$

Here k_1 and k_2 are transcription and translation rates, respectively, while γ_1 and γ_2 correspond to mRNA and protein degradation. For this process, the following equation has been proposed [66]:

$$\frac{\partial p(x,t)}{\partial t}=\left(\gamma_2+\frac{\ln 2}{T_0}\right)\frac{\partial}{\partial x}\left[xp(x,t)\right]-k_1\int w(x-x')p(x',t)\mathrm{d}x'. \tag{8.11}$$

This equation has the gain–loss structure mentioned before, with the first term corresponding to protein dilution due to protein degradation and cell division (T_0 being the division time), and the second term to protein production. The production kernel reproduces the observed exponential bursting of proteins, namely, $w(x)=(1/b)\exp(-(x/b))-\delta(x)$, where b measures the average burst size, and the Dirac delta function accounts for transitions away of x. Equation (8.11) can be solved analytically in stationary conditions, and the solution reads

$$p(x)=\frac{1}{b^a\,\Gamma(a)}x^{a-1}e^{-x/b}, \tag{8.12}$$

which is a Gamma distribution, with $a=k_1/(\gamma_2+\ln 2/T_0)$ and $b=k_2/\gamma_1$. It is interesting to note that Gamma distributions are associated with distinct regimes, identified by $a>1$ and $a<1$, respectively. An example is given in Fig. 8.4, where a direct stochastic simulation is performed by using the Gillespie algorithm (see next section), and compared with the Gamma distribution as defined by (8.12). The two regimes correspond to qualitatively different situations, in which the majority of cells present some amount of protein ($a>1$, graph on left), or the majority of cells do not contain any protein at all ($a<1$, graph on right). This distribution has been confirmed experimentally [67] and is thought to be the relevant distribution in prokaryotic gene expression. It is noticeable that the same distribution holds when extrinsic noise is also considered [68].

Another example of the application of the Chemical Master Equation approach is the work of Komorowki and coworkers [69], where the general model of a two-gene regulatory network has been studied. The study compares gene expression noise of protein or RNA repressors acting on transcription or translation initiation and resulting in the same effective transcription and translation initiation rates of a regulated gene. In this comparison translational repression introduces more noise than repression on promoter activity for a wide range of biologically relevant parameters. The relative magnitude of the noise introduced by protein and RNA repressors also depends on the protein and mRNA degradation rates.

3.3 Exact Stochastic Simulation

In most cases, however, both the Langevin and the Master equations are difficult to solve analytically, and it is necessary to rely on direct stochastic simulations of the system. One way which has become extremely popular is to use the so-called Gillespie algorithm [70], and more recent modifications. In this chapter we shall review the original formulation of the algorithm, and some of its extensions that have proven extremely useful in simulating molecular systems.

The Gillespie algorithm applies to a set of N chemical species in a volume V, each characterized by a number of molecules X_i, with $i = 1,\ldots,N$, and interacting through a set of M reactions R_μ, with $\mu = 1,\ldots,M$. The system is considered well stirred. At the heart of the algorithm is the definition of the so-called propensity function, a_μ, which defines the probability for the reaction R_μ to happen between t and $t + \mathrm{d}t$ as

$$a_\mu \,\mathrm{d}t = \text{probability for the } \mu^{\text{th}} \text{ reaction to occur in } \mathrm{d}t.$$

The propensity in turn can be written as the product of the stochastic reaction constants and the number of distinct molecular combinations of reactants participating in R_μ. Once the propensities are known, it is possible to use them to compute exactly the probability of reaction R_μ to occur within a finite interval of time τ:

$$P(\tau,\mu) = a_\mu \exp[-a_0\tau] \quad \text{with} \quad a_0 = \sum_{\nu=1}^{M} a_\nu \tag{8.13}$$

This simple equation is the key to simulating the time evolution of a chemically reacting system. The algorithm is based on drawing two random numbers, say r_1 and r_2, uniformly from the unit interval [0,1], and setting

$$\tau = \left(\frac{1}{a_0}\right)\ln\left(\frac{1}{r_1}\right), \tag{8.14}$$

and identifying μ in such a way that

$$\sum_{\nu=1}^{\mu-1} a_\nu < r_2 a_0 \le \sum_{\nu=1}^{\mu} a_\nu \tag{8.15}$$

These two equations allow for the computation of the waiting time for a reaction to occur, and for the identification of which reaction actually occurred. Iteration of the process, which includes updating the molecular species, produces an explicit time evolution of the system.

This algorithm, usually referred to as the direct method, has the advantage of being exact, but at the same time can be computationally intensive because of the numerous calls to the random number generator. Gillespie proposed also an approximated version of the algorithm, the so-called τ-leap method [71], where the number of reaction happening in a time interval τ is assumed to be distributed as a Poisson distribution. Therefore τ does not correspond to the minimal waiting time for a reaction to occur, but is a time over which a number of reactions are lumped together. This results in a significant gain in the simulation speed, particularly when all species are characterized by a large number of molecules. The typical τ can be chosen of the order of the waiting time of the reaction with the smallest number of molecules. Therefore when all species are present in high copy numbers, this can be a much larger time than the waiting time of the exact Gillespie algorithm. In contrast if the number of reactant molecules is small even for just one species, the τ-leap method can still be applied, but τ will have to be similar to the typical waiting time of the exact algorithm, and τ-leaping does not confer any efficiency advantage to the simulation.

These difficulties naturally lead to distinguish between "fast" and "slow" reactions, classified according to propensity values. This is particularly relevant to systems whose number of molecules or more in general the propensities vary over several orders of magnitude. Ideally it would be desirable to have a multiscale algorithm that is capable of treating differently fast and slow reactions, and apply the exact Gillespie scheme to the slow ones, and the approximate τ-leap method to the fast ones. Such an algorithm, named the Maximal Time Step Method, has been in fact developed [72], and is based on a dynamical partitioning of fast and slow reactions according to the evaluation of their propensities. This algorithm has been applied to perform direct stochastic simulations of glucose, lactose, and glycerol metabolism in *E. coli*, and has allowed the study of the propagation of gene regulation noise onto the level of metabolic processes [72].

4 Emergence of Distinct Functional Phenotypes

As an application of the formalism illustrated above, we discuss now the role played by stochastic fluctuations in gene regulatory networks, and in particular in the emergence of distinct phenotypes.

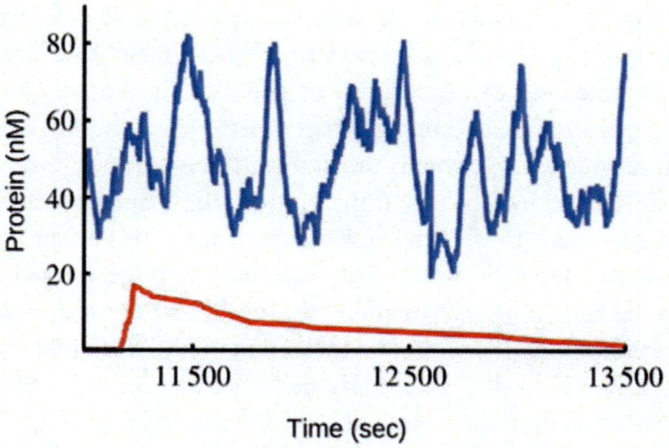

Fig. 8.5 Stiffness of protein fluctuations over time. Two representative trajectories of protein concentrations as a function of time are shown. The *blue* (respectively *red*) *curve* corresponds to the left (*right*) panel and set of parameters as given in Fig. 8.4. Notice that at high transcription and high protein degradation (*blue curve*), the protein fluctuations are much "wilder" than in the low transcription/degradation case (*red curve*). Slowly evolving changes in protein concentrations can be associated with specific phenotypes, at least over timescales shorter or at most comparable to the division time

We analyze first the effect of transcription and translation rates on the stochastic fluctuations in the single gene expression of prokaryotic cells. In [73], a detailed kinetic model for the expression of *LacZ* was built, and simulated by using the Gillespie algorithm. The main result of the study was that high rates of transcription produce proteins at uniform rate and minimize variation in the number of molecules (stochastic fluctuations). Conversely, imposing production of the same average amount of protein but lowering gene transcription causes fluctuations to increase, with the protein produced in bursts. Finally, tuning down translation efficiency, whilst keeping the same amount of protein, requires a higher transcription rate, and therefore does not lead to noisy production patterns.

It is interesting to note that these results can be interpreted within the scheme presented before and leading to the Gamma distribution equation (8.12) as the stationary protein distribution solution of the Master equation (8.11). For the Gamma distribution, we have the protein average $\langle p \rangle = ab \sim k_1 k_2$ and the variance $\sigma^2 = ab^2 \sim k_1 k_2^2$. Therefore the analysis as from [73] can be naturally accounted for within this scheme. An increase of translation rate is the predominant factor that increases fluctuations in gene expression. Notice in fact the expression of the Fano factor (which is a measure of noise to signal ratio), $F = \sigma^2 / \langle p \rangle = b \sim k_2$, or $F = \langle p \rangle / a \sim 1/k_1$, meaning that the randomness of the process correlates positively with translation, and negatively with transcription (at constant protein production).

Another interesting result presented in [73] is the dependency of protein fluctuations timescales on transcription rates. This property can be visualized as the "stiffness" of the trajectory, meaning its capability of conserving its profile over times longer or at least comparable to the observational time (Fig. 8.5). The more the transcription rate is reduced, the more the majority of cells express very low protein concentrations. This is in agreement with the behavior of the Gamma distribution in the regime $a<1$ discussed above. The more interesting observation, however, is that all trajectories, and in particular those corresponding to the outlier cells, presenting large number of protein molecules, fluctuate over time very slowly. In other words, in the regime of low transcription, high translation, and low protein degradation, cells seem to be stuck for a long time with having a protein amount which came out of a burst, or which they inherited from the mother cell. If this time is longer than the time of the experiment, these cells will appear as different phenotypes in the populations. However, it is important to highlight that this heterogeneity is only transient (protein will eventually degrade, new bursts will occur, and cells will divide), and its source is dynamical, in that it depends on temporal noise in gene expression. This is in contrast with other explanations of the emergence of distinct phenotypes, based either on preexisting and fixed heterogeneities in the cellular environment (for instance number of ribosomes), or on the presence of specifically engineered gene switches and circuits, or on a combination of these.

In fact, in [74] a detailed stochastic kinetic model of Two-Component System (TCS) signaling has been developed. TCSs are the most prevalent signal transduction and gene regulation mechanisms in bacteria, responsible for innumerable adaptive responses to environmental signals and host–pathogen interactions. The model has been validated by comparison with flow cytometry data and used to study the stochastic switching of a reporter gene controlled by the TCS as a function of extracellular signal strength. The model shows that, depending on kinetic parameters, the two-component system can exhibit all-or-none, graded, or mixed mode responses. The positively autoregulated TCS exhibits all-or-none response in accordance with other studies. Surprisingly, the TCS lacking a positive feedback loop can also exhibit behavior leading to coexistence of two cellular populations. In this mixed mode, variation of the signal strength changes the level of gene expression in induced cells while the regulated gene continues to be expressed at the basal level in a substantial fraction of cells. It is an intriguing observation that graded response of the TCS can be changed to a mixed mode response by the increase of the translation initiation rate of the histidine kinase gene. It shows that a TCS is an evolvable design pattern capable of implementing deterministic regulation and stochastic switches associated with both graded and threshold responses.

5 Noise and Persistence: Conclusions

Biochemical processes occurring in living cells frequently involve very small numbers of reactant molecules. This results in substantial random fluctuations in the outcomes of these processes, which propagate to the level of cell physiology and may lead to phenotypic heterogeneity in isogenic bacterial populations. We have introduced here the mathematical modeling and computer simulation approaches which have been used extensively to study these phenomena. In fact, many theoretical studies have drawn the attention of biologists to the underlying stochasticity of cellular processes, and motivated the development of single cell, quantitative experimental methods, capable of validating the theoretical predictions. The integration of modeling approaches and experimental studies not only confirmed the existence of phenotypic diversity at the single cell level but also demonstrated its importance for resisting antibiotic challenge.

The functional phenotypic states available to cells within a clonal population are likely to be determined by a combination of the deterministic dynamics associated with the cascade of gene regulatory interactions, and intrinsic and extrinsic noise sources. We have shown here that extrinsic noise can contribute dynamically to building up multimodality, even in systems which are deterministically monostable. In any case noise remains the mechanism for which transitions among different states become possible.

However, partitioning of the cell population into different subpopulations cannot be taken yet as a manifestation of the emergence of distinct phenotypes. Wild fluctuations of proteins levels, and corresponding wild transitions among different cellular states, cannot be associated with observable cellular phenotypes. For cells to be identified as phenotypically different, we need to find mechanisms for those cells to maintain their state for a time longer or at least comparable either to our observational time, or to the other timescales over which other relevant physiological processes happen, for instance, the killing by antibiotic drugs. The interplay between transcription and translation rates discussed in Sect. 8.4 suggests that a dynamical regime exists where protein levels remain correlated over long times. However, cell division may provide an efficient mixing mechanism, and may limit phenotypic identity across generations within cell lines.

The search for mechanisms for persistence, and the emergence of phenotypic diversity more in general, becomes then the search for mechanisms capable of slowing down all stochastic components. Extrinsic noise, again, is a natural candidate to look for sources of slow fluctuations, and it could be directly interpreted as responsible for the observation of different functional states. Cells, whose only difference is in the number of ribosomes, for instance, would be naturally classified as functionally distinct, even though their genetic material is identical. Whether extrinsic noise plays a role in the emergence of persistence remains to be seen. Likewise, highly complex relaxational dynamics might also play a fundamental role in slowing down fluctuations.

In this chapter we propose a scenario in which the phenotypic heterogeneity leading to persistence emerges because of a dynamical slowing down of stochastic fluctuations in gene expression levels. In this scenario gene expression noise generates variation in the survival characteristics of individual bacterial cells when exposed to antibiotics. Antibiotic exposure will thereby select those minority cells that are tolerant to antibiotic, due to noisy fluctuation in their internal dynamics, and thereby generate the phenomenon of persistence. Note that this mechanism does not a priori require any specific mechanisms of switching between normal and persistent populations states, nor any specific gene controlling the dynamics of persister formation. The only assumption is that molecular noise impacts on the factors responsible for conferring those properties to cause relatively long-lived cell-to-cell variation in their phenotypic expression. Thereafter, evolution may act on the system to tune the level of noise (by, for instance, modifying gene expression levels) so as to optimize the population structure (balance of persisters to normal growing cells) in order to survive antibiotic exposure and/or other stresses (such as the host immune response). In this way bacterial persistence can be considered as social trait, expressed at the level of population [24] but caused by dynamical molecular noise.

Most of the studies on noise in gene regulation have been done for the model bacterial pathogen *E. coli*. Understanding the role of noise in the persistence of *M. tuberculosis* requires a twofold effort. On the theoretical side, more work is needed to identify mechanisms for slowing down dynamically stochastic fluctuations, and to understand their relationship with the emergence of distinct phenotypes. On the experimental side we need to develop single cell level quantitative experimental approaches for mycobacterial species. This is likely to involve the use of microfluidic devices coupled with automated microscopy and quantitative measurement of fluorescent reporter proteins. These are challenging tasks due to the slow and branched cell growth of *Mycobacteria*. We believe that the results already obtained from a combined theoretical–experimental approach on model organisms provide further motivation for this effort. As quantitative single cell level data become available, the quantitative theoretical approaches reviewed in this chapter will provide new insights into the origin of persistence in *M. tuberculosis*.

References

1. Jindani A, Aber VR, Edwards EA, Mitchison DA (1980) The early bactericidal activity of drugs in patients with pulmonary tuberculosis. Am Rev Respir Dis 121:939–949
2. Ahmad Z, Klinkenberg LG, Pinn ML, Fraig MM, Peloquin CA, Bishai WR, Nuermberger EL, Grosset JH, Karakousis PC (2009) Biphasic kill curve of isoniazid reveals the presence of drug-tolerant, not drug-resistant, Mycobacterium tuberculosis in the guinea pig. J Infect Dis 200:1136–1143
3. Mitchison DA (1998) How drug resistance emerges as a result of poor compliance during short course chemotherapy for tuberculosis. Int J Tuberc Lung Dis 2:10–15
4. Stewart GR, Robertson BD, Young DB (2003) Tuberculosis: a problem with persistence. Nat Rev Microbiol 1:97–105
5. Barry CE III, Boshoff HI, Dartois V, Dick T, Ehrt S, Flynn J, Schnappinger D, Wilkinson RJ, Young D (2009) The spectrum of latent tuberculosis: rethinking the biology and intervention strategies. Nat Rev Microbiol 7:845–855
6. Chao MC, Rubin EJ (2010) Letting sleeping dos lie: does dormancy play a role in tuberculosis? Annu Rev Microbiol 64:293–311
7. Stewart PS, Costerton JW (2001) Antibiotic resistance of bacteria in biofilms. Lancet 358: 135–138
8. Jayaraman R (2008) Bacterial persistence: some new insights into an old phenomenon. J Biosci 33:795–805
9. Keren I, Kaldalu N, Spoering A, Wang Y, Lewis K (2004) Persister cells and tolerance to antimicrobials. FEMS Microbiol Lett 230:13–18
10. Keren I, Shah D, Spoering A, Kaldalu N, Lewis K (2004) Specialized persister cells and the mechanism of multidrug tolerance in Escherichia coli. J Bacteriol 186:8172–8180

11. Tuomanen E, Durack DT, Tomasz A (1986) Antibiotic tolerance among clinical isolates of bacteria. Antimicrob Agents Chemother 30:521–527
12. Dhar N, McKinney JD (2007) Microbial phenotypic heterogeneity and antibiotic tolerance. Curr Opin Microbiol 10:30–38
13. Moyed HS, Bertrand KP (1983) hipA, a newly recognized gene of Escherichia coli K-12 that affects frequency of persistence after inhibition of murein synthesis. J Bacteriol 155:768–775
14. Balaban NQ, Merrin J, Chait R, Kowalik L, Leibler S (2004) Bacterial persistence as a phenotypic switch. Science 305:1622–1625
15. Gerdes K, Christensen SK, Lobner-Olesen A (2005) Prokaryotic toxin-antitoxin stress response loci. Nat Rev Microbiol 3:371–382
16. Gotfredsen M, Gerdes K (1998) The Escherichia coli relBE genes belong to a new toxin-antitoxin gene family. Mol Microbiol 29:1065–1076
17. Gronlund H, Gerdes K (1999) Toxin-antitoxin systems homologous with relBE of Escherichia coli plasmid P307 are ubiquitous in prokaryotes. J Mol Biol 285:1401–1415
18. Pandey DP, Gerdes K (2005) Toxin-antitoxin loci are highly abundant in free-living but lost from host-associated prokaryotes. Nucleic Acids Res 33:966–976
19. Pedersen K, Christensen SK, Gerdes K (2002) Rapid induction and reversal of a bacteriostatic condition by controlled expression of toxins and antitoxins. Mol Microbiol 45:501–510
20. Rotem E, Loinger A, Ronin I, Levin-Reisman I, Gabay C, Shoresh N, Biham O, Balaban NQ (2010) Regulation of phenotypic variability by a threshold-based mechanism underlies bacterial persistence. Proc Natl Acad Sci USA 107:12541–12546
21. Hansen S, Lewis K, Vulic M (2008) Role of global regulators and nucleotide metabolism in antibiotic tolerance in Escherichia coli. Antimicrob Agents Chemother 52:2718–2726
22. Vazquez-Laslop N, Lee H, Neyfakh AA (2006) Increased persistence in Escherichia coli caused by controlled expression of toxins or other unrelated proteins. J Bacteriol 188:3494–3497
23. Korch SB, Hill TM (2006) Ectopic overexpression of wild-type and mutant hipA genes in Escherichia coli: effects on macromolecular synthesis and persister formation. J Bacteriol 188:3826–3836
24. Gardner A, West SA, Griffin AS (2007) Is bacterial persistence a social trait? PLoS One 2:e752
25. De Leenheer P, Cogan NG (2009) Failure of antibiotic treatment in microbial populations. J Math Biol 59:563–579
26. Klapper I, Gilbert P, Ayati BP, Dockery J, Stewart PS (2007) Senescence can explain microbial persistence. Microbiology 153:3623–3630
27. Korch SB, Henderson TA, Hill TM (2003) Characterization of the hipA7 allele of Escherichia coli and evidence that high persistence is governed by (p)ppGpp synthesis. Mol Microbiol 50:1199–1213
28. Avery SV (2006) Microbial cell individuality and the underlying sources of heterogeneity. Nat Rev Microbiol 4:577–587
29. Dubnau D, Losick R (2006) Bistability in bacteria. Mol Microbiol 61:564–572
30. Wayne LG, Hayes LJ (1996) An *in vitro* model for sequential analysis of shiftdown of *Mycobacterium tuberculosis* through two stages of non-replicating persistence. Infect Immun 64:2062–2069
31. Wayne LG, Sohaskey CD (2001) Nonreplicating persistence of mycobacterium tuberculosis. Annu Rev Microbiol 55:139–163
32. Wayne LG, Sramek HA (1994) Metronidazole is bactericidal to dormant cells of Mycobacterium tuberculosis. Antimicrob Agents Chemother 38:2054–2058
33. Bryk R, Gold B, Venugopal A, Singh J, Samy R, Pupek K, Cao H, Popescu C, Gurney M, Hotha S et al (2008) Selective killing of nonreplicating mycobacteria. Cell Host Microbe 3:137–145
34. Beste DJ, Espasa M, Bonde B, Kierzek AM, Stewart GR, McFadden J (2009) The genetic requirements for fast and slow growth in mycobacteria. PLoS One 4:e5349
35. Betts JC, Lukey PT, Robb LC, McAdam RA, Duncan K (2002) Evaluation of a nutrient starvation model of Mycobacterium tuberculosis persistence by gene and protein expression profiling. Mol Microbiol 43:717–731

36. Bacon J, James BW, Wernisch L, Williams A, Morley KA, Hatch GJ, Mangan JA, Hinds J, Stoker NG, Butcher PD et al (2004) The influence of reduced oxygen availability on pathogenicity and gene expression in Mycobacterium tuberculosis. Tuberculosis (Edinb) 84:205–217
37. Keren I, Minami S, Rubin E, Lewis K (2011) Characterization and transcriptome analysis of Mycobacterium tuberculosis persisters. MBio 2:e00100–e00111
38. Dhar N, McKinney JD (2010) Mycobacterium tuberculosis persistence mutants identified by screening in isoniazid-treated mice. Proc Natl Acad Sci USA 107:12275–12280
39. Hampshire T, Soneji S, Bacon J, James BW, Hinds J, Laing K, Stabler RA, Marsh PD, Butcher PD (2004) Stationary phase gene expression of Mycobacterium tuberculosis following a progressive nutrient depletion: a model for persistent organisms? Tuberculosis (Edinb) 84:228–238
40. Voskuil MI, Visconti KC, Schoolnik GK (2004) Mycobacterium tuberculosis gene expression during adaptation to stationary phase and low-oxygen dormancy. Tuberculosis (Edinb) 84:218–227
41. Beste DJ, Laing E, Bonde B, Avignone-Rossa C, Bushell ME, McFadden JJ (2007) Transcriptomic analysis identifies growth rate modulation as a component of the adaptation of mycobacteria to survival inside the macrophage. J Bacteriol 189:3969–3976
42. Rustad TR, Harrell MI, Liao R, Sherman DR (2008) The enduring hypoxic response of Mycobacterium tuberculosis. PLoS One 3:e1502
43. Kumar A, Toledo JC, Patel RP, Lancaster JR Jr, Steyn AJ (2007) Mycobacterium tuberculosis DosS is a redox sensor and DosT is a hypoxia sensor. Proc Natl Acad Sci USA 104:11568–11573
44. Boon C, Dick T (2002) Mycobacterium bovis BCG response regulator essential for hypoxic dormancy. J Bacteriol 184:6760–6767
45. Leistikow RL, Morton RA, Bartek IL, Frimpong I, Wagner K, Voskuil MI (2010) The Mycobacterium tuberculosis DosR regulon assists in metabolic homeostasis and enables rapid recovery from nonrespiring dormancy. J Bacteriol 192:1662–1670
46. Singh A, Crossman DK, Mai D, Guidry L, Voskuil MI, Renfrow MB, Steyn AJ (2009) Mycobacterium tuberculosis WhiB3 maintains redox homeostasis by regulating virulence lipid anabolism to modulate macrophage response. PLoS Pathog 5:e1000545
47. Singh A, Guidry L, Narasimhulu KV, Mai D, Trombley J, Redding KE, Giles GI, Lancaster JR Jr, Steyn AJ (2007) Mycobacterium tuberculosis WhiB3 responds to O2 and nitric oxide via its [4Fe-4S] cluster and is essential for nutrient starvation survival. Proc Natl Acad Sci USA 104:11562–11567
48. Ramage HR, Connolly LE, Cox JS (2009) Comprehensive functional analysis of Mycobacterium tuberculosis toxin-antitoxin systems: implications for pathogenesis, stress responses, and evolution. PLoS Genet 5:e1000767
49. Singh R, Barry CE III, Boshoff HI (2010) The three RelE homologs of Mycobacterium tuberculosis have individual, drug-specific effects on bacterial antibiotic tolerance. J Bacteriol 192:1279–1291
50. McKinney JD, Honer zu Bentrup K, Munoz-Elias EJ, Miczak ACB, Chan W-T, Senson D, Sacchettini JC, Jacobs WR Jr, Russell DG (2000) Persistance of *Mycobacterium tuberculosis* in macrophages and mice requires the glyoxylate shunt enzyme isocitrate lyase. Nature 406:735–738
51. Wang X, Wang H, Xie J (2011) Genes and regulatory networks involved in persistence of Mycobacterium tuberculosis. Sci China Life Sci 54:300–310
52. Rifat D, Bishai WR, Karakousis PC (2009) Phosphate depletion: a novel trigger for Mycobacterium tuberculosis persistence. J Infect Dis 200:1126–1135
53. Farhana A, Guidry L, Srivastava A, Singh A, Hondalus MK, Steyn AJ (2010) Reductive stress in microbes: implications for understanding Mycobacterium tuberculosis disease and persistence. Adv Microb Physiol 57:43–117
54. Rigney DR, Schieve WC (1977) Stochastic model of linear, continuous protein synthesis in bacterial populations. J Theor Biol 69:761–766
55. Berg OG (1978) A model for the statistical fluctuations of protein numbers in a microbial population. J Theor Biol 71:587–603

56. McAdams HH, Arkin A (1997) Stochastic mechanisms in gene expression. Proc Natl Acad Sci USA 94:814–819
57. Rocco A (2009) Stochastic control of metabolic pathways. Phys Biol 6:016002
58. Schrödinger E (1944) What is life? Cambridge University Press, London
59. Elowitz MB, Levine AJ, Siggia ED, Swain PS (2002) Stochastic gene expression in a single cell. Science 297:1183–1186
60. Horsthemke W, Lefever R (1984) Noise-induced transitions theory and applications in physics, chemistry, and biology. Springer-Verlag, Berlin
61. Gardiner CW (1985) Handbook of stochastic methods for physics, chemistry, and the natural sciences. Springer-Verlag, Berlin
62. Ozbudak EM, Thattai M, Kurtser I, Grossman AD, van Oudenaarden A (2002) Regulation of noise in the expression of a single gene. Nat Genet 31:69–73
63. Samoilov M, Plyasunov S, Arkin AP (2005) Stochastic amplification and signaling in enzymatic futile cycles through noise-induced bistability with oscillations. Proc Natl Acad Sci USA 102:2310–2315
64. Samoilov MS, Arkin AP (2006) Deviant effects in molecular reaction pathways. Nat Biotechnol 24:1235–1240
65. Van Kampen NG (2007) Stochastic processes in physics and chemistry. Elsevier, Amsterdam
66. Friedman N, Cai L, Xie XS (2006) Linking stochastic dynamics to population distribution: an analytical framework of gene expression. Phys Rev Lett 97:168302
67. Cai L, Friedman N, Xie XS (2006) Stochastic protein expression in individual cells at the single molecule level. Nature 440:358–362
68. Taniguchi Y, Choi PJ, Li GW, Chen H, Babu M, Hearn J, Emili A, Xie XS (2010) Quantifying E. coli proteome and transcriptome with single-molecule sensitivity in single cells. Science 329:533–538
69. Komorowski M, Miekisz J, Kierzek AM (2009) Translational repression contributes greater noise to gene expression than transcriptional repression. Biophys J 96:372–384
70. Gillespie DT (1977) Exact stochastic simulation of coupled chemical-reactions. J Phys Chem 81:2340–2361
71. Gillespie DT (2001) Approximate accelerated stochastic simulation of chemically reacting systems. J Chem Phys 115:1716–1733
72. Puchalka J, Kierzek AM (2004) Bridging the gap between stochastic and deterministic regimes in the kinetic simulations of the biochemical reaction networks. Biophys J 86:1357–1372
73. Kierzek AM, Zaim J, Zielenkiewicz P (2001) The effect of transcription and translation initiation frequencies on the stochastic fluctuations in prokaryotic gene expression. J Biol Chem 276:8165–8172
74. Kierzek AM, Zhou L, Wanner BL (2010) Stochastic kinetic model of two component system signalling reveals all-or-none, graded and mixed mode stochastic switching responses. Mol Biosyst 6:531–542

Chapter 9
Drug Discovery

Nagasuma Chandra

Abstract Despite availability of several drugs, a vaccine, decades of research and concerted efforts from medical and allied communities to manage tuberculosis, it is clear that the *Mycobacterium tuberculosis* has been successful in defying these efforts and continues to pose a major threat to mankind. Newer approaches, in particular, newer strategies for drug discovery are therefore urgently required. The science of drug discovery has witnessed multiple paradigm shifts in the past few decades, from a predominantly ligand-centric approach to a target-centric approach and now recently leaning towards a systems-based approach. The shifts can be attributed to several factors such as availability of publicly accessible databases containing genome sequences, functional and structural data of macromolecules, high-throughput experimental profiling, protein–protein interactions and pathway models, as well as adaptation and application of computational methods for efficient data mining and modeling. Several *omics-scale* experimental and *in silico* approaches have emerged recently to systematically address important questions in biology, with an obvious impact on drug discovery.

A systems view enables a broad understanding of the system as a whole, providing significant insights at multiple stages in the drug discovery pipeline, from target identification, understanding pharmacokinetics and pharmacodynamics, to personalized medicine. Of the systems approaches for drug discovery, modeling metabolism in the causative agent has received some attention. Flux balance analysis, and metabolic control analysis that can simulate the relative reaction fluxes under a variety of conditions, have provided lists of predicted essential proteins and hence potential drug targets. Perturbations such as gene knock-outs, drug inhibitions, double and triple knock-outs, exposure to different chemical environments can all be modeled through this approach. Interactomes capturing structural and functional protein–protein linkages have been useful in identifying proteins strategically

N. Chandra (✉)
Department of Biochemistry, Indian Institute of Science, Bangalore, 560 012, India
e-mail: nchandra@biochem.iisc.ernet.in

J. McFadden et al. (eds.), *Systems Biology of Tuberculosis*,
DOI 10.1007/978-1-4614-4966-9_9, © Springer Science+Business Media, LLC 2013

located in the network, which when inhibited would perturb the network significantly. There have also been examples of rule-based or logic-based modeling studies that will help in identifying the effect of different scenarios of host–pathogen interactions and adaptations within each, thereby identifying optimal strategies for therapeutic intervention. The models themselves are increasingly being enriched with experimental information, as more and more genomics and proteomics data is becoming available. The potential of these methods that still remains to be tapped in drug discovery programs are discussed. The stage seems set for the integration and application of skills from mathematics, computer science, and engineering disciplines, to address complex problems in biology and drug discovery, in a big way.

1 Introduction

Tuberculosis, one of the oldest infectious diseases known to man, continues to top the list of killer infectious diseases [1]. Available statistics indicate two million deaths every year globally or one death approximately every 17 s [2]. Decades of efforts of tackling tuberculosis through chemotherapeutic agents or the BCG vaccine, have clearly not been sufficient. Besides the deadly pact of the causative agent *Mycobacterium tuberculosis* (Mtb) with HIV, its ability to thrive in macrophages, hibernate for decades, and resurface at opportunistic moments, only emphasize the urgent need for new drugs and perhaps new approaches to discover them [3]. Confronting tuberculosis thus serves as a good example to highlight the need for a more holistic view of the pathogen and its interaction with the host.

For several decades, drug discovery has been ligand-centric with medicinal chemistry and traditional pharmacology as the main components in its toolkit, from which the paradigm shift to target-centric approaches backed by the power of molecular biology has marked an important milestone, several years ago [4]. In this post-genomic era with the added power of new-age biology armaments, the science of drug discovery has been witnessing another major paradigm shift, marking another significant milestone in the evolutionary trajectory, a shift in focus from studying single molecules with a reductionist philosophy to using knowledge from holistic "systems" behavior [5–7].

Systems-thinking is not new to pharmacology and drug discovery. Heavy reliance on whole animal models, in vivo assays versus in vitro studies of the same molecules in a test tube, and more importantly clinical trials that are carried out in different phases, all stand as testimony to such thinking. However, there is a major difference between conventional physiology-based approaches and the currently emerging practice of systems biology [8]. Conventional approaches have surely benefitted from systems philosophy, but appreciation of the "system" is at best only implicit, in fact more of a "black box," which only facilitates a systems output as a "readout," but does not tell us why or how, such an output results. The discipline of systems biology on the other hand, seeks to reconstruct the system brick by brick so as to facilitate an understanding of why and how an event takes place, automatically leading to "what if" type of questions, and hence enabling predictions. With systems

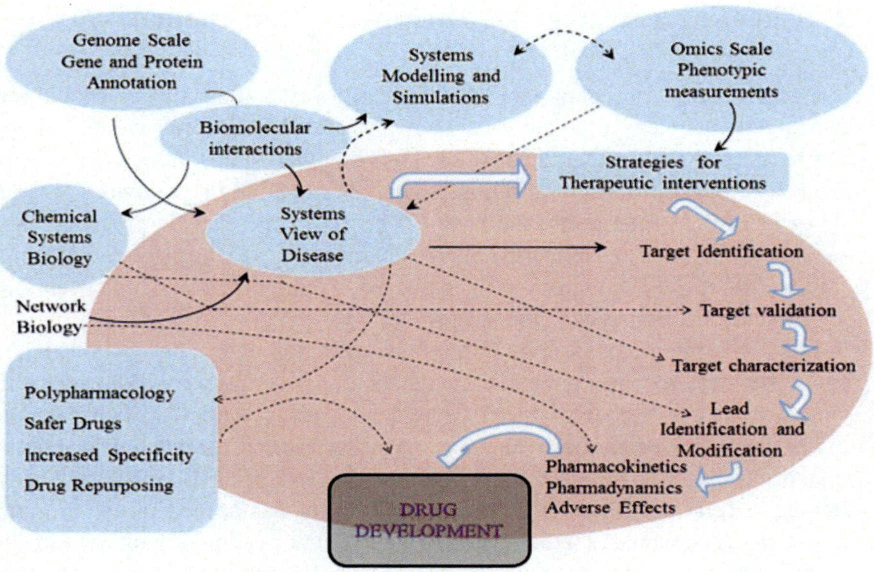

Fig. 9.1 A systems biology-based drug discovery pipeline that starts from a "System map of the disease" and has an additional step of identifying the right strategy for therapeutic intervention, as compared to the standard discovery pipeline. Various aspects that can be addressed through this pipeline are shown in *boxes* as also the methods and techniques that will be useful for that goal

biology becoming both feasible and a well-appreciated discipline, its influence on drug discovery is almost inevitable. The tight orchestration of a biological system involving thousands of molecular components and a tenfold higher number of interactions among them, the interactions themselves spanning across the levels of hierarchical organization of the cellular constituents, necessitates systems approaches for their comprehension. Systems biology also differs from the "spherical cow" type of highly simplified abstractions [9]. Instead it seeks to reconstruct a system in detail capturing the complex real life phenomena, by using "omics" scale data on various fronts that facilitate such realistic modeling.

A systems view of pathophysiology enables a broad understanding of the system as a whole [10], providing significant insights at multiple stages in the drug discovery pipeline, by aiding in identifying best strategies for therapeutic intervention, target identification, understanding pharmacokinetics and pharmacodynamics, adverse effects, and even personalized medicine. Systems biology signals a departure from the viewpoint of "*single target per drug, lone therapeutic indication.*" Obtaining systems perspectives of different diseases also has the potential to enable a comparison of pathogenesis of different diseases as well as their differential manifestations and enables exploration of targeting common pathways leading to "polypharmacology" type of approaches [11]. The area is rapidly evolving and promises to transform the level at which we understand biology, automatically leap-frogging advances in drug discovery. Figure 9.1 illustrates a new pipeline for drug discovery incorporating the systems approach. The following sections give an account of how systems approaches are being applied for antitubercular drug discovery.

2 Choosing a Strategy for Therapeutic Intervention

In a general sense, any important process or pathway in the pathogen should serve as a targeting strategy, but then some pathways and some proteins are more important for the survival and virulence of the pathogen than others. A logical strategy to tackle a pathogen would be to inhibit in part or full one of the following processes (1) entry of the bacteria into the host, (2) host-immune evasion processes in the bacterium, (3) processes rendering unique characteristics to the pathogen, (4), nutrient absorption and metabolism in the pathogen, (5) important signaling events, and (6) virulence determinants in the bacterium.

First, it is important to review the mode of action of existing antituberculous drugs. We must of course remember that most of these drugs were discovered either serendipitously or through large-scale screening using whole cell cultures. Their mechanism of action has been unraveled only much later. Current clinical antitubercular drugs target the following (a) mycolic acid biosynthesis (isoniazid, and ethionamide), (b) cell wall biosynthesis through inhibition of alanine racemase (cycloserine) or arabinosyl transferases (ethambutol), (c) DNA gyrases (fluroquinolone) and DNA synthesis via thymidylate synthase (aminosalicylic acid), (d) RNA polymerase (rifampicin), (e) ribosomes (streptomycin), and (f) trans-translation (pyrazinamide) [12–15].

The mechanism of action of each drug varies considerably but most appear to target processes essential for bacterial growth and/or virulence. Isoniazid, a prodrug that gets converted to an isonicotinic acid adduct by KatG, a catalase peroxidase enzyme, binds tightly and thus inhibits the enoyl-acyl carrier protein reductase known as InhA. InhA is a key enzyme of the type-II fatty acid synthase system and essential for biosynthesis of mycolic acids, and thus for the formation of the waxy coat of the mycobacterial cell wall [16, 17]. Rifampicin targets the beta subunit of RNA polymerase, thereby inhibiting transcript elongation, a critical process for the cell survival [18]. Ethambutol inhibits mycobacterial arabinosyl transferases (encoded by the embCAB operon), an essential cell wall component involved in the polymerization of D-arabinofuranose to arabinoglycan. Pyrazinoic acid derived from pyrazinamide is reported to target the essential ribosomal protein S1 (RpsA), involved in protein translation and the ribosome-sparing process of trans-translation [15]. Thus, viewing them from a reverse pharmacological perspective, these targets are seen to pass target validation criteria in terms of essentiality.

Reconstruction of large scale models of pathogen and host (as described in Chap. 1) will be extremely valuable in identifying best strategies for intervention. However, most genome-scale models built so far have incorporated only a single cellular level, for instance metabolism or gene regulation. Methodologies for modeling at multiple levels to build "virtual cells" are still under development but hold a lot of promise for development of optimal strategies for killing the bacterium whilst minimizing damage to the host.

3 Target Identification

In order to pursue strategies that would be useful for killing the bacterium, appropriate proteins must be identified as drug targets. First, it is important to analyze what makes a good drug target. Criteria that can be used either to evaluate or to predict an antibacterial drug target are, whether:

(a) The target is directly or indirectly a cause of the disease/symptoms (relevance to the given pathophysiology).
(b) The target is essential to the system responsible for the pathophysiology and its knock-down will have sufficient impact on the bacterial cell (essentiality and impact).
(c) The target is specific to the disease process or state (process/condition specificity).
(d) The target is specific to the pathogen species or other closely related organisms (species/family specificity).
(e) The target's function can be manipulated by an appropriate small molecule (druggability or chemical tractability).
(f) The lead molecule is available in sufficient quantities in vivo and suitable methods are available to test the function of the target and thereby study the extent of inhibition or activation by candidate lead molecules (biological tractability and assayability).

Several of these criteria have already been described above for current drugs. Species-specificity can be addressed through bioinformatics analysis of gene and protein sequences in a given species and through comparison of genome sequences to identify unique proteins and any other unique features. It can also be addressed through structural analyses by comparing three-dimensional structures of the target protein(s) with other proteins particularly with those from host cells [19–21]. Druggability can be inferred from structural analysis, by, for instance, characterization of functional regions such as the binding sites in a protein molecule and perhaps through ligand binding and docking studies in conjunction with approaches now well established for structure-based drug discovery [22–24]. Biological tractability as well as condition specificity can be addressed through genomics data to understand expression patterns and assayability is typically addressed by studying the individual molecular species [25].

Two important criteria are "relevance to the given pathology" and "essentiality"; which are both inherently systems' properties and cannot generally be addressed by studying proteins individually. For example, a protein that might seem to be a good drug target for a given disease may not actually be critical or essential, when viewed in the context of the entire metabolism in the cell. Systems-level studies may reveal the presence of an alternate pathway and mechanisms that may naturally exist to compensate for the absence of that protein. Systems level models hence provide a basis to assess criticality of potential targets. The conventional method of studying a single protein at a time, even at the highest level of detail, are no doubt extremely

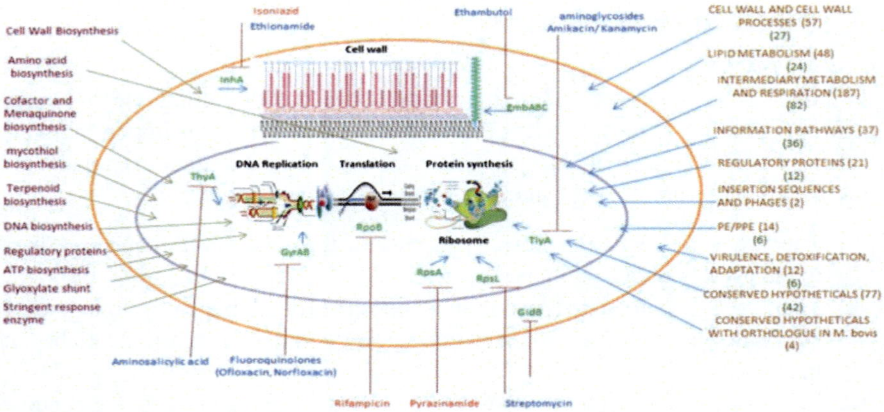

Fig. 9.2 The known antitubercular drugs and their targets. Targets (in *green*) involved in the cellular processes of the first-line drugs (in *red*) and second-line drugs (in *blue*) and their Tuberculist classes are shown. Strategies for targetting as put together by Global Alliance for TB Drug Development are depicted in *purple* on the *left*. The number of high confidence targets in different Tuberculist classes as predicted by targetTB pipeline are shown in *brown* on the *right*

important to characterize individual capabilities and limits of that molecule, but are clearly insufficient to obtain a perspective of its larger context and hence may not provide the right answers to questions such as druggability. However, where possible, it would of course be most insightful to integrate systems level studies with those at sequence, structural, and biochemical levels.

Established and possible targets for antitubercular drugs have been compiled from experimental target validation methods such as knock-out studies [26] (Fig. 9.2). The analysis indicates that at least 32 different protein molecules that are essential can be explored as possible drug targets [26]. Another study identified unique pathways in Mtb based on sequence comparisons of metabolic enzymes [27]. Hasan et al. [28] uses various features such as metabolic choke-points to identify drug targets. Enzymes involved in those reactions that uniquely consume a substrate or produce a product in a metabolic network, are described as choke points [28]. In this study, the mycolic acid biosynthetic pathway was reconstructed involving 218 reactions, 197 metabolites, and 28 different proteins. Application of FBA (Chap. 2) led to the identification of key points in the pathway and the delineation of potential drug targets through systematic gene knock-outs as well simulation of enzyme inhibition through hypothetical drugs [29]. In another study, the two available genome-scale reconstructions of Mtb were used to identify essential genes and hard-coupled reaction sets (HCRs, which are groups of adjacent reactions involving metabolites with one-to-one connectivity) and thereby identify potential drug targets [30, 31]. Both these studies reconstruct genome-scale metabolic networks of Mtb and utilize a constraint-based approach called flux balance analysis (FBA) to obtain insights about metabolism as a whole in this pathogen. Through FBA of *in-silico* gene knock-out conditions, predictions of gene essentiality were mad.

Most of the known drug targets to be essential. In addition, the known drug targets mapped to 25 of the hard-coupled reaction sets identified through this approach. It has further been proposed that other enzymes in these 25 sets could be explored for their potential as alternate but metabolically equivalent drug targets to some of those that are currently being studied [26].

4 The TargetTB Pipeline

The *targetTB* target identification pipeline [19] highlights the promise that systems approaches have for drug discovery. The pipeline uses a set of multilevel filters to evaluate and retain only those proteins that pass the filters at each stage. It starts with a systems level analysis to address the essentiality criteria of a drug target and indirectly also the biological relevance to disease criteria as well. The next filter involves a comparative genomics analysis at the sequence level, which evaluates aspects of specificity, particularly with respect to the human genome. The next filter carries out a structural assessment of targetability using a novel scheme of pocketome characterizations and large scale comparisons. Pocketome refers to a comprehensive set of small molecule binding pockets in the entire set of proteins coded by the genome [32]. This filter evaluates both specificity at a higher resolution than that is possible from a sequence level and simultaneously evaluates druggability and amenability of the protein to be manipulated by a small molecule. Further filters prune the shortlisted target candidates based on expression profiles using available expression data, addressing biological tractability of suggested target proteins. A comparison at the sequence level with gut flora proteins to rule out all those that have homologues in the latter species, addresses aspects of adverse effects and drug safety, while a similar analysis combined with phylogenetic profiling against several common pathogen genomes, addresses aspects of designing broad-spectrum antibiotic targets, while also identifying those specific to tuberculosis.

The pipeline resulted in identifying a set of 451 high confidence drug targets, belonging to several functional classes. There are more than 100 enzymes, some transcription factors, some transporters, which pass all the filters and are seen to make it to the final list. Here again, established targets for drugs currently used in the clinic were identified, lending credence to the approach. The study indicates that clearly there are many more targets in the proteome that could be explored, than those currently targeted (Fig. 9.2).

A comparison of targetTB hits with previously suggested targets indicates how the integrated approach could also be used to evaluate different criteria in target candidates. For instance, Anishetty and coworkers [27] compared enzymes in metabolic pathways between human and Mtb and proposed 186 proteins unique in Mtb as possible drug targets. Fifty-one of them were found in the *targetTB* pipeline, the rest were found to be nonessential by the systems or uniqueness criteria and hence were eliminated. In another study, Hasan and coworkers [28] report a scheme for prioritizing and ranking all proteins in the Mtb genome as possible drug targets, based on consideration of metabolic choke-points, in vitro essentiality for growth and druggability as judged by sequence similarity to proteins capable of binding small molecule ligands. One hundred and forty-six in the top 500 ranks of this study were also in the *targetTB* shortlist, while the rest were again eliminated due to systems and expression criteria. Different types of analyses carried out independently, lead to different shortlists of putative drug targets, since they use different criteria for the identification. Such lists can be compared or combined to identify those proteins which satisfy multiple criteria applied for evaluating drug targets and hence appear in multiple lists, increasing their chance of being successful drug targets.

5 Polypharmacology, Combination Targets, and Drug Repurposing

Another concept that is receiving much attention in the recent years is that of polypharmacology, facilitated through the study of drug-target networks [11]. Polypharmacology can be defined to encompass both situations of one drug binding to multiple targets and multiple drugs binding to different targets within a network. A systems view of the pathogen and reconstruction of comprehensive networks at various levels of hierarchy in the cell (enabled through integration of "*omics*" level experimental data) render it feasible to address polypharmacology questions. Indeed, it has previously been claimed that the success of crude drug preparations from plant or animal origin may be due to their ability to act on multiple targets with multiple mechanisms [33]. Various drug-target databases such as DrugBank [34], the Therapeutic Targets Database (TTD) [35], and World Molecular Bioactivity (WOMBAT) [36] are now available, which are highly useful for identifying multiple targets. Indeed drug-target networks have been constructed and analyzed in several ways [37].

Multiple drugs at multiple targets, which is best described as "drug combinations with different mechanisms of action" is already used extensively in the clinic, with examples from many diseases, including of course tuberculosis. However, achieving polypharmacology through one drug is less common, most probably due to lack of enabling technologies until recently. With the drug-target networks and development of newer and more efficient algorithms to model three-dimensional structures and to compare their functional sites at a genomics-scale is making it possible to identify such possibilities. In the future, it can be envisaged that rational design protocols can take advantage of such capabilities and incorporate these criteria at the lead design stage itself.

An example in this direction is the discovery of a common structural motif in fatty acyl-AMP ligases and acyl-CoA-synthesizing fatty acyl-CoA ligases in Mtb. These are proteins involved in generating functional versatility in the activation of fatty acids to acyl-adenylates [38]. Since the acyl-AMP ligases are crucial nodes in biosynthetic network of virulent lipids, inhibitors of these proteins are likely to have a multipronged approach of simultaneously disrupting several pathways. The group of bi-substrate acyl-sulfamoyl analogues was shown to inhibit both proteins and had a marked effect on the cell surface architecture of Mtb [38]. These are being further explored as promising polypharmacology lead compounds.

Current combination therapy for tuberculosis was developed more than 50 years ago on empirical rather than rational grounds. Despite its obvious success, development of resistance is still a problem [39], which is likely due to the pathogen's intrinsic robustness mechanisms which allow it to survive in the presence of drugs for long enough to develop resistance. One way to counter this would be to attack the pathogen at multiple points which are rationally designed to achieve more rapid killing. To address this, a recent study analyzed metabolome and corresponding protein interaction networks to identify a combination of targets such that multi-modal destruction to bacterial metabolism could be optimally achieved [40]. This study computed metabolic disruptability indices to shortlist efficient disruption strategies by identifying pairs, triplets, and quadruplets that wielded the largest influence on the metabolism.

In another study, Kinnings et al., combined molecular modeling, structural bioinformatics, and systems biology approaches to construct a large-scale drug-target network for Mtb [20]. They constructed a network which they termed the TB-drugome, from a dataset of 274 drugs approved for human use for any condition and for which structural data was available through crystallographic studies. Similarities of predicted binding sites of Mtb proteins with known drug binding sites was used as a basis for associating Mtb proteins with known drugs (the premise being that the two proteins have the possibility of binding the same drug if they share similarity in their binding sites). Using a similar approach to computationally detect cross-reactivity between different drug target families, the drugs entacapone and tolcapone prescribed for the treatment of Parkinson's disease were identified as potential TB drugs and shown to inhibit InhA, thus leading to a possibility of drug repurposing [38].

6 Addressing Drug Resistance

As indicated earlier, drug resistance is perhaps the most important problem plaguing chemotherapy in general. The emergence of drug resistant varieties of several bacteria and viruses has been on the rise and requires urgent attention. In tuberculosis too, the MDR and XDR varieties are increasing in numbers and has been a cause for concern [41–43]. Although, several measures are being taken to counter the problem of drug resistance, success is hampered by lack of full understanding of how resistance emerges. Mechanisms include activation of alternate pathways; manipulation of the drug; altering its bioavailability; or modification of the target; modulating expression of the target; mutation of the target that hinders drug binding; and activation of pumps and transporters to efflux out the drug [44]. A comprehensive resource, TBDReaMDB that contains drug resistance mutations in Mtb has been developed [45]. Experimentally characterized drug resistance mutations were curated from diverse sources of existing literature. The database serves as a valuable resource to map mutations to the resistance phenotypes.

A systems-based approach was recently applied to identify likely resistance mechanisms, particularly to probe if there were possible pathways in the cell from the drug target(s) to trigger one or more proteins that comprise the resistance machinery. A large scale protein–protein interaction network was first reconstructed, based on experimentally observed and predicted protein–protein interactions capturing both direct structural complexes as well as functional linkages among pairs of proteins [46]. This network was probed using graph theoretical methods to identify possible resistance paths, which resulted in finding a few such paths and some proteins that were as central to many resistance pathways. Moreover, different targets and hence different drugs were predicted to exhibit different propensities for triggering emergence of resistance [47]. It was suggested that such hub-proteins could serve as "co-targets," so that their inhibition simultaneously with their primary targets might lead to reducing rates of drug resistance.

Pharmacokinetics and pharmacodynamics of drugs, both existing ones as well as new ones in the discovery pipeline, have not received sufficient attention, especially from a systems perspective. Having a rational handle to understanding their bioavailability, controlling release of active drug ingredients from the administered drug formulation, drug metabolism, tissue distribution as well as unintended drug interactions with other target molecules, is essential to design the right drug combinations, dosage regimens, and better ways of clinical management. Although not sufficiently explored at this time, systems level models incorporating key molecular components should be immensely useful for addressing these issues.

7 Targeting Host–Pathogen Interactions and Critical Host Factors

It is quite clear that the complex web of interactions between the host immune system and the pathogen determines the outcome of any infection, including tuberculosis. A systems perspective is important to understand how certain bacterial factors interact with various host factors to evade host immune response, divert host nutrients into the bacterium, and induce other toxic effects. Knowledge of how various host and pathogen molecules play their role will provide additional handles to identify novel drug targets. For instance, using a genome-wide siRNA screen, host factors that are important for survival of the pathogen inside the host cell have been identified [48]. It appears that Mtb, an intracellular pathogen functions as a type of hub inside the host network, interacting and influencing simultaneously several of the constituent nodes of the host system [49]. Molecules identified as essential for maintaining infection and hence as host susceptibility factors through these siRNA screens should serve as potential targets for controlling Mtb growth in vivo. As a proof of concept, when two such molecules TGFbRI and CSNK1d were simultaneously inhibited using a pharmacological agent, Mtb survival in infected mice was seen to be severely compromised.

A host–pathogen interactions model containing 75 nodes corresponding to host and pathogen molecules, cells, cellular states, or processes has been built using Boolean rules [40]. Simulations were used to explore vaccination effects, clearance efficiencies due to drugs on the bacterial growth rate, and hence the outcome of exposure to Mtb. Simulations indicate a high propensity of the pathogen to persist under different conditions. Another study reporting a rule-based modeling of iron homeostasis in tuberculosis was able to identify key factors in maintaining the iron balance and suggest possible strategies for controlling bacterial growth, such as inhibition of mycobactin biosynthesis [50]. A human alveolar macrophage genome-scale metabolic reconstruction and its interaction with Mtb has also been recently reported. Using flux balance analysis, the authors show that metabolic changes as well as gene essentiality predicted from the model depict the infection scenario [51] and hence enable identification of drug targets using a model of the metabolic state resembling that of active infection.

8 Future Perspectives

While there is still tremendous scope for improvement, the promise of systems biology leapfrogging drug discovery is evident. The approaches described here clearly demonstrate the potential of systems thinking in drug discovery. A goal of systems biology is to transform biomedical research and drug discovery into a more predictive science. A key to success perhaps lies in the modeling of complete systems, accounting for all component reactions, the localization of these components and their interactions. The interaction between the reconstructed system and its immediate external environment and ultimately the physical world becomes more relevant in the final layers of hierarchy. The predictive power is important because it reflects the extent of our collective understanding of that system through all available approaches and at any level of detail—physiological, cellular, biochemical, molecular, or biophysical levels. Towards that goal, it is increasingly being appreciated that collaborative efforts are essential to make rapid progress, given the scale and complexity of such experiments. In that light, it is encouraging that consortia such as the Open Source Drug Discovery consortium [52] and the NM4TB programs are already well under way, as also are the TBDB and the collaborative chemistry databases [53, 54].

The promise of predictions and data-driven computations has long demonstrated in product development and safety testing in various engineering disciplines, such as aerospace engineering and electronic circuit design. For drug discovery too, it can be envisaged that such opportunities, are expected to increase significantly in the coming years, both due to comprehensive data gathering through high-throughput leading-edge technologies as well cutting-edge concepts and methods to comprehend and interpret that data. Thus, it seems to be an answer to several problems in drug discovery. Given the extent of *omics* data and a large number of groups working on the tuberculosis pathogen, many components and modules are increasingly being well characterized, making the systems view more and more complete.

Several promising target candidates have been identified. Besides lead molecules for individual target proteins, examples of lead compounds binding to more than one target have also begun to emerge. It is also likely that the systems perspective of the pathogen and in future the host cell as well, will have a high impact on pharmacokinetic and pharmacodynamic models and characterizations as well. It may no longer be mere fiction to envision that in the future, in a clinical setting, a disease could get diagnosed and characterized at the systems level with precise genotype and phenotype definitions, both by phenome-typing the pathogen, as well as the host, leading all the way up to predictive quantitative titrations of the available remedies and finally personalized prescriptions.

Acknowledgments The author thanks Deepika Ajit Sakorey and Priyanka Baloni for assistance in preparation of this article.

References

1. Wirth T, Hildebrand F, Allix-Beguec C, Wolbeling F, Kubica T, Kremer K, van Soolingen D, Rusch-Gerdes S, Locht C, Brisse S, Meyer A, Supply P, Niemann S (2008) Origin, spread and demography of the Mycobacterium tuberculosis complex. PLoS Pathog 4(9):e1000160. doi:10.1371/journal.ppat.1000160
2. World Health Organisation (2008) World Health Organisation
3. Triccas JA, Davenport MP (2008) Infectious diseases: too little, too late for tuberculosis. Immunol Cell Biol 86(4):293–294. doi:icb20087[pii]10.1038/icb.2008.7
4. Drews J (2000) Drug discovery: a historical perspective. Science 287(5460):1960–1964. doi:8361[pii]
5. Butcher EC, Berg EL, Kunkel EJ (2004) Systems biology in drug discovery. Nat Biotechnol 22(10):1253–1259. doi:nbt1017[pii]10.1038/nbt1017
6. Hood L, Perlmutter RM (2004) The impact of systems approaches on biological problems in drug discovery. Nat Biotechnol 22(10):1215–1217. doi:nbt1004-1215[pii]10.1038/nbt1004-1215
7. Chandra N (2009) Computational systems approach for drug target discovery. Expert Opin Drug Discov 4(12):1221–1236. doi:10.1517/17460440903380422
8. Kitano H (2002) Systems biology: a brief overview. Science 295(5560):1662–1664. doi:10.1126/science.1069492 295/5560/1662[pii]
9. Doyle J (2001) Computational biology. Beyond the spherical cow. Nature 411(6834):151–152. doi:10.1038/35075703 35075703[pii]
10. Kohl P, Noble D (2009) Systems biology and the virtual physiological human. Mol Syst Biol 5:292. doi:msb200951[pii]10.1038/msb.2009.51
11. Boran AD, Iyengar R (2010) Systems approaches to polypharmacology and drug discovery. Curr Opin Drug Discov Devel 13(3):297–309
12. Zhang Y (2005) The magic bullets and tuberculosis drug targets. Annu Rev Pharmacol Toxicol 45:529–564. doi:10.1146/annurev.pharmtox.45.120403.100120
13. Rengarajan J, Sassetti CM, Naroditskaya V, Sloutsky A, Bloom BR, Rubin EJ (2004) The folate pathway is a target for resistance to the drug para-aminosalicylic acid (PAS) in mycobacteria. Mol Microbiol 53(1):275–282. doi:10.1111/j.1365-2958.2004.04120.xMMI4120[pii]
14. Lamichhane G (2011) Novel targets in M. tuberculosis: search for new drugs. Trends Mol Med. doi:S1471-4914(10)00150-4[pii]10.1016/j.molmed.2010.10.004
15. Shi W, Zhang X, Jiang X, Yuan H, Lee JS, Barry CE 3rd, Wang H, Zhang W, Zhang Y (2011) Pyrazinamide inhibits trans-translation in Mycobacterium tuberculosis. Science 333(6049):1630–1632. doi:science.1208813[pii]10.1126/science.1208813

16. Banerjee A, Dubnau E, Quemard A, Balasubramanian V, Um KS, Wilson T, Collins D, de Lisle G, Jacobs WR Jr (1994) inhA, a gene encoding a target for isoniazid and ethionamide in Mycobacterium tuberculosis. Science 263(5144):227–230
17. Larsen MH, Vilcheze C, Kremer L, Besra GS, Parsons L, Salfinger M, Heifets L, Hazbon MH, Alland D, Sacchettini JC, Jacobs WR Jr (2002) Overexpression of inhA, but not kasA, confers resistance to isoniazid and ethionamide in Mycobacterium smegmatis, M. bovis BCG and M. tuberculosis. Mol Microbiol 46(2):453–466. doi:3162[pii]
18. Campbell EA, Korzheva N, Mustaev A, Murakami K, Nair S, Goldfarb A, Darst SA (2001) Structural mechanism for rifampicin inhibition of bacterial RNA polymerase. Cell 104(6):901–912. doi:S0092-8674(01)00286-0[pii]
19. Raman K, Yeturu K, Chandra N (2008) targetTB: a target identification pipeline for Mycobacterium tuberculosis through an interactome, reactome and genome-scale structural analysis. BMC Syst Biol 2:109. doi:1752-0509-2-109[pii]10.1186/1752-0509-2-109
20. Kinnings SL, Liu N, Buchmeier N, Tonge PJ, Xie L, Bourne PE (2009) Drug discovery using chemical systems biology: repositioning the safe medicine Comtan to treat multi-drug and extensively drug resistant tuberculosis. PLoS Comput Biol 5(7):e1000423. doi:10.1371/journal.pcbi.1000423
21. Yamanishi Y, Kotera M, Kanehisa M, Goto S (2010) Drug-target interaction prediction from chemical, genomic and pharmacological data in an integrated framework. Bioinformatics 26(12):i246–i254
22. Congreve M, Murray CW, Blundell TL (2005) Keynote review: Structural biology and drug discovery. Drug Discov Today 10:895–907
23. Keller TH, Pichota A, Yin Z (2006) A practical view of 'druggability'. Curr Opin Chem Biol 10(4):357–361
24. Vajda S, Guarnieri F (2006) Characterization of protein-ligand interaction sites using experimental and computational methods. Curr Opin Drug Discov Devel 9:354–362
25. Bunnage ME (2011) Getting pharmaceutical R&D back on target. Nat Chem Biol 7:335–339
26. Mdluli K, Spigelman M (2006) Novel targets for tuberculosis drug discovery. Curr Opin Pharmacol 6(5):459–467
27. Anishetty S, Pulimi M, Pennathur G (2005) Potential drug targets in Mycobacterium tuberculosis through metabolic pathway analysis. Comput Biol Chem 29(5):368–378. doi:S1476-9271(05)00076-9[pii]10.1016/j.compbiolchem.2005.07.001
28. Hasan S, Daugelat S, Rao PS, Schreiber M (2006) Prioritizing genomic drug targets in pathogens: application to Mycobacterium tuberculosis. PLoS Comput Biol 2(6):e61. doi:05-PLCB-RA-0376R3[pii]10.1371/journal.pcbi.0020061
29. Raman K, Rajagopalan P, Chandra N (2005) Flux balance analysis of mycolic acid pathway: targets for anti-tubercular drugs. PLoS Comput Biol 1(5):e46. doi:10.1371/journal.pcbi.0010046
30. Beste DJ, Hooper T, Stewart G, Bonde B, Avignone-Rossa C, Bushell ME, Wheeler P, Klamt S, Kierzek AM, McFadden J (2007) GSMN-TB: a web-based genome-scale network model of Mycobacterium tuberculosis metabolism. Genome Biol 8(5):R89. doi:gb-2007-8-5-r89[pii]10.1186/gb-2007-8-5-r89
31. Jamshidi N, Palsson BO (2007) Investigating the metabolic capabilities of Mycobacterium tuberculosis H37Rv using the in silico strain iNJ661 and proposing alternative drug targets. BMC Syst Biol 1:26. doi:1752-0509-1-26[pii]10.1186/1752-0509-1-26
32. An J, Totrov M, Abagyan R (2005) Pocketome via comprehensive identification and classification of ligand binding envelopes. Mol Cell Proteomics 4:752–761
33. Csermely P, Agoston V, Pongor S (2005) The efficiency of multi-target drugs: the network approach might help drug design. Trends Pharmacol Sci 26(4):178–182. doi:S0165-6147(05)00055-6[pii]10.1016/j.tips.2005.02.007
34. Knox C, Law V, Jewison T, Liu P, Ly S, Frolkis A, Pon A, Banco K, Mak C, Neveu V, Djoumbou Y, Eisner R, Guo AC, Wishart DS (2011) DrugBank 3.0: a comprehensive resource for 'omics' research on drugs. Nucleic Acids Res 39:D1035–D1041. doi:gkq1126[pii]10.1093/nar/gkq1126
35. Chen X, Ji ZL, Chen YZ (2002) TTD: therapeutic target database. Nucleic Acids Res 30(1):412–415

36. Olah M, Bologa C, Oprea TI (2004) An automated PLS search for biologically relevant QSAR descriptors. J Comput Aided Mol Des 18(7–9):437–449
37. Yildirim MA, Goh KI, Cusick ME, Barabasi AL, Vidal M (2007) Drug-target network. Nat Biotechnol 25(10):1119–1126. doi:nbt1338[pii]10.1038/nbt1338
38. Arora P, Goyal A, Natarajan VT, Rajakumara E, Verma P, Gupta R, Yousuf M, Trivedi OA, Mohanty D, Tyagi A, Sankaranarayanan R, Gokhale RS (2009) Mechanistic and functional insights into fatty acid activation in Mycobacterium tuberculosis. Nat Chem Biol 5(3):166–173. doi:nchembio.143[pii]10.1038/nchembio.143
39. du Toit LC, Pillay V, Danckwerts MP (2006) Tuberculosis chemotherapy: current drug delivery approaches. Respir Res 7:118. doi:1465-9921-7-118[pii]10.1186/1465-9921-7-118
40. Raman K, Bhat AG, Chandra N (2010) A systems perspective of host-pathogen interactions: predicting disease outcome in tuberculosis. Mol Biosyst 6(3):516–530. doi:10.1039/b912129c
41. Johnson R, Streicher EM, Louw GE, Warren RM, van Helden PD, Victor TC (2006) Drug resistance in Mycobacterium tuberculosis. Curr Issues Mol Biol 8(2):97–111
42. Argyrou A, Jin L, Siconilfi-Baez L, Angeletti RH, Blanchard JS (2006) Proteome-wide profiling of isoniazid targets in Mycobacterium tuberculosis. Biochemistry 45(47):13947–13953. doi:10.1021/bi061874m
43. Tan YT, Tillett DJ, McKay IA (2000) Molecular strategies for overcoming antibiotic resistance in bacteria. Mol Med Today 6(8):309–314. doi:S1357-4310(00)01739-1[pii]
44. Nguyen L, Thompson CJ (2006) Foundations of antibiotic resistance in bacterial physiology: the mycobacterial paradigm. Trends Microbiol 14(7):304–312. doi:S0966-842X(06)00125-9[pii]10.1016/j.tim.2006.05.005
45. Sandgren A, Strong M, Muthukrishnan P, Weiner BK, Church GM, Murray MB (2009) Tuberculosis drug resistance mutation database. PLoS Med 6(2):e2. doi:08-PLME-HIA-2556[pii]10.1371/journal.pmed.1000002
46. Raman K, Chandra N (2008) Mycobacterium tuberculosis interactome analysis unravels potential pathways to drug resistance. BMC Microbiol 8:234. doi:1471-2180-8-234[pii]10.1186/1471-2180-8-234
47. Padiadpu J, Vashisht R, Chandra N (2011) Protein-protein interaction networks suggest different targets have different propensities for triggering drug resistance. Syst Synth Biol 4(4):311–322. doi:10.1007/s11693-011-9076-5
48. Kumar D, Nath L, Kamal MA, Varshney A, Jain A, Singh S, Rao KV (2010) Genome-wide analysis of the host intracellular network that regulates survival of Mycobacterium tuberculosis. Cell 140(5):731–743. doi:S0092-8674(10)00127-3[pii]10.1016/j.cell.2010.02.012
49. Chandra N, Kumar D, Rao K (2011) Systems biology of tuberculosis. Tuberculosis (Edinb) 91(5):487–496. doi:S1472-9792(11)00031-X[pii]10.1016/j.tube.2011.02.008
50. Ghosh S, Prasad KV, Vishveshwara S, Chandra N (2011) Rule-based modelling of iron homeostasis in tuberculosis. Mol Biosyst 7(10):2750–2768. doi:10.1039/c1mb05093a
51. Bordbar A, Lewis NE, Schellenberger J, Palsson BO, Jamshidi N (2010) Insight into human alveolar macrophage and M. tuberculosis interactions via metabolic reconstructions. Mol Syst Biol 6:422. doi:msb201068[pii]10.1038/msb.2010.68
52. Bhardwaj A, Scaria V, Raghava GP, Lynn AM, Chandra N, Banerjee S, Raghunandanan MV, Pandey V, Taneja B, Yadav J, Dash D, Bhattacharya J, Misra A, Kumar A, Ramachandran S, Thomas Z, Brahmachari SK (2011) Open source drug discovery—a new paradigm of collaborative research in tuberculosis drug development. Tuberculosis (Edinb) 91(5):479–486. doi:S1472-9792(11)00102-8[pii]10.1016/j.tube.2011.06.004
53. Ekins S, Kaneko T, Lipinski CA, Bradford J, Dole K, Spektor A, Gregory K, Blondeau D, Ernst S, Yang J, Goncharoff N, Hohman MM, Bunin BA (2010) Analysis and hit filtering of a very large library of compounds screened against Mycobacterium tuberculosis. Mol Biosyst 6(11):2316–2324. doi:10.1039/c0mb00104j
54. Galagan JE, Sisk P, Stolte C, Weiner B, Koehrsen M, Wymore F, Reddy TB, Zucker JD, Engels R, Gellesch M, Hubble J, Jin H, Larson L, Mao M, Nitzberg M, White J, Zachariah ZK, Sherlock G, Ball CA, Schoolnik GK (2010) TB database 2010: overview and update. Tuberculosis (Edinb) 90(4):225–235. doi:S1472-9792(10)00041-7[pii]10.1016/j.tube.2010.03.010

Chapter 10
Immunological Biomarkers for Tuberculosis: Potential for a Combinatorial Approach

Richard Pine, Yuri Bushkin, and Maria Laura Gennaro

Abstract Host responses to *Mycobacterium tuberculosis* infection provide a basis for diagnosis of latent infection and active disease. T cell responses have been the mainstay for diagnosis of latent infection and may contribute to diagnosis of active disease. Recent advances in characterizing humoral responses will likely contribute to improved diagnosis of active disease. However, these measures fail to distinguish the continuum of infection states. Moving to a systems approach to biomarker discovery may provide the resolution that current methods of diagnosis lack. The chapter evaluates the current use of T cell and B cell responses for diagnosis and the limitations of applying them separately. The possibility that macrophage or monocyte activation may serve as a biomarker is also addressed. We consider whether methodologies that combine (a) multifunctional T cell responses and T cell types, (b) monocyte/macrophage characteristics that reveal response to infection, and (c) dominant B cell responses to *M. tuberculosis* growth-phase-specific antigens can further contribute to a systems approach for biomarker discovery that can distinguish among infection states.

List of Abbreviations

AM	Alveolar macrophage
APC	Antigen presenting cell
BAL	Bronchoalveolar lavage
CD	Cluster of differentiation
DC	Dendritic cell
IFN	Interferon

R. Pine • Y. Bushkin • M.L. Gennaro (✉)
Public Health Research Institute, New Jersey Medical School,
225 Warren Street, Newark, NJ 07103, USA
e-mail: gennarma@umdnj.edu

J. McFadden et al. (eds.), *Systems Biology of Tuberculosis*,
DOI 10.1007/978-1-4614-4966-9_10, © Springer Science+Business Media, LLC 2013

IL	Interleukin
LTBI	Latent *Mycobacterium tuberculosis* infection
MHC	Major histocompatibility complex
PBmo	Peripheral blood monocyte
TB	Tuberculosis
TCR	T cell receptor
TNF	Tumor necrosis factor

Systems biology approaches, to which this volume is dedicated, can be applied to biomarker discovery, an area of tuberculosis (TB) research with critical consequences for TB control effectiveness. Biomarkers are needed to diagnose active TB when detection of tubercle bacilli alone is insufficient, slow, or impractical (reviewed in [1–3]) and to discriminate stable forms of latent *Mycobacterium tuberculosis* infection (LTBI) from recent or progressive infection to target the latter for treatment [4, 5]. Biomarkers are also actively sought for vaccine research. Due to the typically slow evolution and low frequency of TB activation in immunocompetent individuals (5–10% of infected individuals develop active TB during their lifetime), testing vaccine efficacy in clinical trials is extremely cumbersome and expensive. Finding biomarkers of vaccine efficacy is therefore imperative. Similarly, developing new drugs or drug regimens against TB would be greatly facilitated by acquiring biomarkers of treatment outcome that can be assessed more easily and rapidly than culturing *M. tuberculosis*. For some of the above applications, biomarker discovery research may overlap. For example, a biomarker that is used to diagnose asymptomatic stable infection could also work as a surrogate marker of vaccine effectiveness. Moreover, biomarkers of bacillary burden may in principle be employed as indicators of disease progression and drug treatment outcome. These multiple applications may allow research integration.

Even when biomarker discovery does not extend to multiple areas of application at once, integrated approaches should be favored because *M. tuberculosis* infection presents with a spectrum of multiple, often poorly separated, clinical conditions. It has been increasingly recognized that LTBI covers a spectrum of "subconditions" [6]: it may be a stable state, it may be associated with high risk of progression to disease, or it may represent a preclinical stage of disease. Also active disease may present with various clinical characteristics: it may be minimal (i.e., accompanied by low-grade symptoms) or it may exhibit various degrees of severity in terms of symptoms, bacterial burden, and tissue damage. Thus, it is unlikely that a single marker (or marker set) can reflect the complexity of the clinical forms of TB.

A need for integrated biomarker research and application also derives from the complexity of the cellular and molecular events occurring during infection. The granuloma, which is the histopathological hallmark of tuberculous infection, is a multicellular, dynamic structure. The host–pathogen interactions at the granuloma lead to dynamic changes of tubercle bacilli, of the phenotypes of the host immune cells, and of the levels of the immune mediators they produce. Since immune cells recirculate between local immune compartments and the periphery, systemic levels of immune markers and released soluble factors also vary in association with infection outcome. The "sign" of the variation (up or down) derives from whether marker expression is directly linked to immune function, for example, protective or suppressive, and its regulation, whether it merely reflects changes in relative bacterial antigen burden during infection, or whether it is affected by both. Thus, biomarker discovery could involve any combination of host responses.

An additional level of complexity is introduced by genetic and epigenetic sources of variability among individuals. Antigen processing, presentation, and recognition are tightly linked to the immunogenetic background of the host, such as histocompatibility and T and B cell repertoires. Moreover, different types of granulomatous lesions can be found in the same individual [6, 7], further complicating biomarker discovery. Comorbidities affecting immune competence, such as HIV, or host cellular metabolism, such as diabetes, further compound interindividual variation. Also the pathogen contributes to variation, as different clinical strains of *M. tuberculosis* vary in relative gene expression [8, 9]. Moreover, any alteration of the host-derived microenvironments due to inter-lesion or inter-patient differences may be reflected in the stress response of the pathogen. This will result in bacterial gene expression differences, which in turn will affect the profiles of the antigen-specific immune responses. Recognizing that the association between biomarkers and infection stages is at the intersection of multiple host- and pathogen-derived covariates, the argument for systems approaches to biomarker discovery is all but compelling.

In this chapter, we discuss effector and regulatory T cell responses, monocyte and macrophage properties, and B cell responses in relation to the dynamics of host–pathogen interactions in *M. tuberculosis* infection. We conclude by considering how an integrated approach to biomarker discovery can advance our ability to track responses to infection, vaccination, or pharmacological treatment.

1 T Cell Responses

1.1 Overview

Airborne tubercle bacilli infect their host by reaching the respiratory mucosa and lung alveoli where they are engulfed by myeloid dendritic cells (DC) and resident macrophages. DC containing viable *M. tuberculosis* or mycobacterial components circulate from the respiratory mucosa to the draining lymph nodes, where they prime naïve T cells. The resulting effector T cells can return to the lung and exert control of infection. Endogenously processed antigens and those captured from the extracellular milieu can be presented to B cells or to T cells in the context of MHC class II, class I, and MHC class I-like molecules expressed on the surface of DC. Conventional CD4+ and CD8+ T cells and unconventional T cells, all expressing antigen-specific T cell receptors (TCR), recognize a vast array of protein and lipid antigens, respectively, and play a major role in adaptive immunity against TB [10, 11]. Adaptive immunity mounted by T cells that utilize antigen-specific TCR usually requires days, a time frame that allows for priming of naïve lymphocytes and clonal expansion. Ultimately, effector T cells are generated; these can further differentiate and establish immunological memory. Conventional CD4+

and CD8+ T cells are central to the anti-mycobacterial immune response, while both conventional and unconventional T cell populations are involved in granuloma formation at the site of infection [11–15]. Finally, regulatory T (Treg) cells are an important part of the immune network, as they maintain a fine balance between anti-mycobacterial pro-inflammation responses and immunopathological consequences of excessive inflammation on tissues [16]. Below we briefly review critical concepts related to these cell types and their main features, such as antigen specificity, cytokine production, and function, that are relevant to biomarker research.

1.2 Natural Memory Immunity

Natural memory immunity is established in the early stages of infection, prior to expression of adaptive immunity mediated by conventional T cells (reviewed in [10]). These T cells express either αβ or γδ TCR, recognize a broad scope of both self and foreign lipid antigens presented in the context of CD1 molecules, and exhibit rapid responses [17–19]. The kinetics of response varies among cell populations: the αβ-expressing CD1a-, b-, and c-restricted T cells respond rapidly and robustly to secondary antigen challenge, while the γδ-expressing CD1d-restricted invariant natural killer T (iNKT) cells respond readily to primary antigen exposure. Despite a limited life span [20, 21], the iNKT subset appears to be essential for optimal protection against infection, apparently through interleukin (IL)-17 production and participation of the liver X receptor (LXR) [11, 22, 23]. LXR, a transcription factor mainly known for its role in lipid homeostasis, is broadly expressed in T cells [24]. The IL-23/IL-17 cytokine pathway associated with differentiation of naïve CD4+ T cells into IL-17-secreting T helper 17 (Th17) cells may play a crucial role in protective immunity against mycobacteria and other intracellular pathogens by regulating both innate and adaptive immune responses [15, 25, 26]. Moreover, IL-17 appears to be essential for granuloma development [27]. Differentiation of naïve CD4+ T cells into Th17 cells can be negatively regulated by LXR [11, 22, 23]. It is possible, therefore, that early responses to infection involve the LXR-dependent IL-17 response mediated by iNKT cells. Functional characteristics of iNKT cells, such as the constitutive presence of cytokine-encoding mRNA transcripts in their cytoplasm [28, 29], also point to their role in mediating protective immune responses at early stages of infection.

In addition to iNKT, another subset of CD1d-restricted NKT cells carrying a structurally distinct, more diverse TCR (dNKT) has been recently identified [18, 19]. Both CD1d-restricted dNKT and iNKT subsets have a potential for rapid cytokine secretion and cytolytic activity, but they may differ in their chemokine and integrin expression profiles [30]. A clear distinction in immunological functions mediated by these two NKT subsets may also be related to differences in their respective TCR structures enabling dNKT cells to recognize a broader array of antigens and undergo clonal expansion in response to TCR stimulation [19].

1.3 Adaptive Immunity

Th1 responses mediated by conventional interferon gamma (IFNγ)-producing CD4+ T cells protect against acute and primary mycobacterial infection. In contrast, CD8+ T cells play a role in resistance to and control of infection due to IFNγ production and cytotoxicity [10]. These T cell populations are important mediators of granuloma formation and maturation [11, 14]. Intracellular signaling and cell–cell communications that are involved in generation of adaptive immunity effector and memory CD4+ and CD8+ T cells principally begin with the recognition of an antigenic peptide by the appropriate TCR. Two defined features of these conventional T cells expressing antigen-specific TCR, therefore, are important in the search for biomarkers: the antigen specificity and the cytokine production phenotype. More than two decades' effort has been devoted to identifying the *M. tuberculosis*-derived antigens that could elicit protective responses against infection. Due to the lack of established parameters that could help identify such targets, the search and selection criteria most commonly adopted to identify candidates have been immunogenicity in humans and animals and physicochemical characteristics [31, 32]. More recently, consideration has been given to the apparent segregation of some mycobacterial antigen-specific responses with LTBI or active TB, and with severity of disease. In a major shift in antigen selection strategy, studies have focused on genes and corresponding antigens that are expressed predominantly at a particular stage of infection [33–37]. Thus, in addition to immunodominant proteins such as ESAT6, TB10.4, and Ag85 [38–40], these recent studies have described immune responses against antigens associated with *M. tuberculosis* dormancy (DosR-regulated), resuscitation, and reactivation (Rpf group-associated), and antigens from the starvation-stimulon group [41–44]. By comparing antigen-specific CD4+ and CD8+ T cell-mediated Th1 responses in active TB and LTBI, immune responses directed to the proteins expressed predominantly in non-replicating bacilli were found with higher frequencies in persons with latent infection. These results are in agreement with earlier hypotheses that actively replicating and dormant mycobacteria differ in antigen composition [34, 45], and this may be reflected in the repertoire of immune responses at distinct stages of infection.

Dynamic changes in the immune repertoire during infection are also demonstrated by the presence of antigen-specific T cell subsets with distinct phenotypic and functional signatures in different groups of TB patients. The levels of the main Th1 cytokines, i.e., IFNγ, IL-2, and tumor necrosis factor alpha (TNFα), produced by the same cell were shown to define the phenotype and functionality of T cell subsets that are essential in anti-viral and anti-mycobacterial immune responses [46–48]. Therefore, characterization of these key subsets in terms of the proportion of effector T cells secreting mainly IFNγ only, effector memory T cells secreting both IFNγ and IL-2, and central memory T cells secreting only IL-2 in different patients' cohorts may accurately distinguish among the various stages of infection [49–53]. The cytokine-based multifunctionality of Th1 responses could also be associated with long-term memory and correlate with efficient protective immunity [54–56].

1.4 Regulation of Immunity

Inflammatory responses to infection caused by intracellular pathogens often induce excessive acute and chronic inflammation that may lead to various immune pathologies. Usually, pro-inflammatory responses are controlled at the cellular level by Treg cells. Naturally occurring thymus-derived Treg cells expressing CD4, CD25, and the forkhead winged-helix family transcriptional repressor P3 (FoxP3) are the main cell type suppressing the activity of pathogen-specific effector CD4+ and CD8+ T cells [16, 57–59]. FoxP3 represses IL-2 and IFNγ expression and interacts with nuclear transcription factors of activated T cells, resulting in poor cytokine production and impaired proliferation [60]. Mycobacteria, through their multiple Toll-receptor-like agonists, could stimulate proliferation of Treg cells and stimulate their expansion over effector T cells at the site of infection [61, 62].

The role of CD4+ CD25+ FoxP3+ Treg cells has been primarily investigated in mouse models of *M. tuberculosis* infection and in humans [63–66]. Depletion of Treg cells increased the frequency of IFNγ-producing cells in the lung but resulted in reduced bacterial burden only at early stages of infection [63, 67]. Thus alternative mechanisms related to the progressive development of other T cell populations may contribute to containment of inflammation at later stages and during chronic infection. The defining feature of Treg cells is the ability to inhibit T cell proliferation and IFNγ production through cell–cell contact and/or through production of immune suppressive cytokines such as transforming growth factor beta 1 (TGFβ1) and IL-10 [68]. These properties may explain delayed priming and/or proliferation of newly activated T cells [62]. It is thought that resident DC can receive the antigenic cargo from infected macrophages that undergo apoptosis in the lung. Since Treg cells interact with DC, they may be responsible for negative modulation of DC motility to lymph-draining lymph nodes and impede transfer of antigen to T cell priming sites [69, 70].

The human Treg population is heterogeneous and comprises multiple sets. Adaptive Treg exist as Tr1 and Th3 subsets that secrete high levels of IL-10 and TGFβ, respectively [71]. CD4+ CD25+ CD39+ Treg cells distinguishable from classical FoxP3+ Treg cells were recently identified in TB patients [72]. While FoxP3+ Treg cells produce IFNγ, CD39+ Treg cells appear to produce TGFβ but not IFNγ [72]. In addition to CD4+ Treg, a subset of CD8+ CD25+ FoxP3+ Treg cells has been found in human lymph nodes infected with mycobacteria. These CD8+ Treg cells characteristically express CD223 (an MHC class II binding CD4 homologue) and glucocorticoid-induced TNF receptor. Moreover, they exert suppressive activity, in part, by releasing chemokine (CC-motif) ligand 4, which interferes with TCR signaling [73]. In addition, circulating CD8+ CD28- Treg cells were also found in patients at levels correlating with TB progression and apparently complementing classical FoxP3+ Treg cells [74]. The mechanistic basis for suppressive activity of this Treg subset was not investigated directly, but association with higher levels of TNFα, IL-4, and IL-10 in peripheral blood was observed.

1.5 T Cell Biomarkers

Two requirements appear to be important for successful improvement of T cell-based TB diagnostics. First, novel diagnostic schemes must take into account the multiple characteristics of the various cellular populations mediating the immune response. Among the features that are readily amenable to quantitative evaluation, one should consider the frequencies of antigen-specific cells with a defined specificity (see below) and the frequencies of cells with certain cytokine production profiles, or T cell signatures, characteristic of infection and/or its stages. Second, the latest technological advances must be implemented to increase the sensitivity of methods used to detect intracellular cytokines and to identify antigenic targets recognized by TCR.

MHC class I molecules transport specific antigens, i.e., peptides originated from the degradation of proteins in cytosol or other intracellular compartments, for presentation to TCR on the surface of all nucleated cells [75]. Therefore, analysis of the peptides bound to MHC class I has been utilized to learn about the protein synthesis and degradation pathways in normal and malignant or otherwise transformed cells [76–79]. This principle can be extended to infectious diseases caused by intracellular pathogens. In the context of mycobacterial infection, direct access to the cells harboring mycobacteria is difficult. However, soluble MHC class I (sMHC-I) proteins are elevated in serum of patients and have prognostic value in malignancy and some infections [80–82]. Importantly, it has been shown that sMHC-I proteins are stable and retain a repertoire of bound peptides similar to that of the membrane-bound MHC-I [79, 83]. The sMHC-I forms, which are lacking either the cytoplasmic and transmembrane domains or the transmembrane domain only, are released by metalloproteinases [84] or by alternative splicing [85], and comprise the majority of all sMHC-I found in serum [79, 86, 87]. Thus, similar to cancer cells that release sMHC-I with peptides characteristic for the type of malignancy and/or stage of progression, serum of TB patients can be expected to contain sMHC-I with *M. tuberculosis*-derived peptides that may be characteristic of the clinical stage of infection. Based on these considerations, isolation of sMHC-I from serum of TB patients by immunoaffinity purification and subsequent identification of bound *M. tuberculosis*-specific peptides by capillary chromatography and mass spectrometry and bioinformatics analysis should be feasible.

Peptide signatures specific for the TB clinical stage obtained by this analysis could be further explored in TB diagnostics. As an example, one can use a diagnostic scheme based on detection of antigen-specific CD8+ T cells. Patients with active and latent forms of *M. tuberculosis* infection have mycobacteria-specific effector and/or memory CD8+ T cells in peripheral blood with frequencies ranging between 0.004% and 1.0% [35, 49]. Detection and analysis of small numbers of T cells are difficult, while conventional methods of T cell expansion by stimulation with autologous APC are cumbersome and require weeks. However, a recently developed methodology that uses artificial APC (aAPC) might solve this problem. The aAPC is a synthetic bead-based platform that contains dimeric MHC class I-Ig molecules that can be easily loaded with a specific peptide (or mixture) to target appropriate TCR on T cells and may also contain co-stimulatory molecules on the same bead [88]. This platform effectively allows for the expansion and easy detection of antigen-specific T cells within days rather than weeks [89]. Sensitivity might be increased and time to detection further reduced (from days to hours) by replacing flow cytometry measurements of key cytokine levels, e.g., IFNγ, IL-2, and TNFα, with measurements of the corresponding mRNAs detected following fluorescence in situ hybridization [90]. Similar receptor signaling- and activation-based approaches could be used with other T and non-T cell populations as their role in *M. tuberculosis*-induced responses is elucidated.

2 Properties of Antigen Presenting Cells

2.1 *Overview*

Including the analysis of antigen presenting cells (APC) along with the well-established consideration of T cell responses (above) and serology (below) may provide a new opportunity to distinguish among *M. tuberculosis* infection stages, assess disease progression, evaluate vaccine response, and monitor treatment efficacy. APC products that might be secreted following antigen stimulation of whole blood in vitro have been considered as the basis for improved biomarker outputs [91]. Moreover, characteristics that are detectable without ex vivo manipulation have been investigated for the ability to discriminate infection stage or response to treatment (see below). However, prior efforts at biomarker identification have largely failed to consider APC properties [92–94]. Before considering integrating data from APC with other sources of evidence, we will discuss what might be learned from APC alone and how it might be learned. We limit consideration to the spontaneous properties of monocytes and macrophages, since DC are likely impractical as a source of diagnostic or prognostic information, and B cells are represented by proxy in serological studies, which are addressed below.

A practical approach would rely preferentially on lung macrophages that are present in sputum, since they have the most favorable combination of relevance and accessibility. Alveolar macrophages (AM) are intuitively more sensitive indicators, but far less accessible outside advanced care settings. Thus, they might best be studied to identify factors that should be further assessed under different infection stages for sputum macrophages. Indeed, taking accessibility as a key criterion, peripheral blood monocytes (PBmo) are next after sputum macrophages in order of desirability as a possible avenue of investigation for establishing biomarkers.

2.2 *Sputum Macrophages*

Differences in sputum macrophages that are specific for infection stage, progression, and treatment would be powerful tools for prognostic and diagnostic purposes, regardless of whether they mirvror results from AM obtained by bronchoalveolar lavage (BAL). An immunohistochemical analysis of cell surface molecules involved in antigen presentation demonstrated clear differences between cells from TB patients and cells from a cohort of donors having no or other lung disease [95]. TB was associated with lower levels of HLA-DR and of CD86 on a per-cell basis. While HLA-DR was expressed on a similar percentage of cells from each group, CD86 was present on a smaller proportion of macrophages from TB patients. Additionally, lack of radiographic indication of disease was associated with a lower level of CD86 than that found with cavitary disease. Thus, CD86 might be useful as an indicator of transition from LTBI to active TB, or it might distinguish between the two states.

Additional possibilities can be derived from the evaluation of changes in immune mediator gene expression during treatment of TB [96]. Two key trends were reported. First, IL-10, IRAK-M, SOCS-1, SOCS-3, and TGFβRII, which limit immune activation, were found to be higher in sputum cells from TB patients than from cases of other infectious lung disease and clinically well healthcare workers. All of these markers but SOCS1 declined during treatment. Second, IFNγ, IL-12p40, IL-12p35, and IL-23p19, which are genes coupled to the Th1 immune response profile associated with TB, were also elevated before initiation of anti-TB therapy. Of these, only IL-12p40 responded to treatment with a sustained increase. The other genes either did not change or only transiently increased expression levels. Thus, the Th1 response profile is maintained with treatment as the immunosuppressive response declines. These changes in gene expression likely occurred in macrophages, which constituted the majority of the sputum cells.

Soluble mediators found in sputum suggest additional macrophage properties that might be investigated. Active TB is associated with elevated concentrations of IFNγ and IL-10 in sputum in comparison to levels found with LTBI cases, healthy controls, or other infectious lung disease cases [96, 97]. Conflicting results have been reported as to whether one or the other decreases with treatment. However, when present, IFNγ and IL-10 would activate the transcription factors STAT1 and STAT3 [98–100]. Interrogating the level of activated STAT transcription factors could integrate the effects of changes in the levels of these cytokines. While STAT activation declines when the activating stimuli are present for prolonged periods, a residual steady-state level of response may be detectable. The pattern of activation could reflect the combination of IFNγ and IL-10, since the profile of STAT activation differs between them. Crosstalk further fine-tunes the overall response. Taken together, these studies indicate that the properties of sputum macrophages, including the influence of soluble mediators in sputum, should be further investigated for potential biomarkers.

2.3 Macrophages in Bronchoalveolar Lavage

Although recovery of sputum is quite simple, more has been learned about infection-stage-specific differences using BAL. The first inherent property considered here is a straightforward analysis of AM density, a biophysical measure integrating several cell properties such as extent of maturation/differentiation and cell viability. A marked increase in hypodense AM correlates with disease severity [101]. Differences in activation and antigen presentation have been shown for hypodense compared to higher density AM in other investigations [102–106]. Whether LTBI or disease progression would result in density phenotypes distinct from uninfected and advanced disease remains to be elucidated.

In contrast to the numerous physiological features that determine AM density, a single parameter that may be informative as a biomarker is the presence of inducible nitric oxide synthase (iNOS, NOS2) or its correlate, production of NO. NO may be a critical determinant of host defense against *M. tuberculosis* in humans, as it clearly is in mice [107, 108]. In one notable study, the presence of NOS2 and its activity was demonstrated in AM from all of eleven donors having TB. On average, 65% of cells were positive. In contrast, NOS2 was detected, on average, in 10% of cells from non-TB donors [109]. Another report described significant increases in exhaled NO, AM NOS2 levels, and NOS2-dependent nitrite production by AM from TB cases in comparison to healthy controls [110]. Moreover, AM NOS2 levels significantly correlated with AM nitrite production and with exhaled NO, which decreased with anti-TB treatment. A third study also described elevated, NOS2-dependent production of NO by cultured BAL cells from TB cases compared to healthy controls [111]. These studies indicate that measurement of either NO production or intracellular NOS2 levels would be feasible. Key questions are whether the presence of NOS2 and production of NO distinguish LTBI from TB or whether they might be host responses that indicate progression from LTBI to TB.

Since NOS2 expression is a response to the cytokine milieu, the above results point to the potential value of further investigating correlations between AM responses and BAL fluid cytokines during infection, as discussed above for sputum. Elevated IL-6 in BAL fluid from TB cases was reported in comparison to healthy controls [112, 113]. Additionally, in accord with sputum analyses, IFNγ and IL-10 levels were elevated in BAL fluid from TB patients compared to other infectious lung disease cases and to healthy volunteers [114–116]. A decrease of IFNγ levels in BAL fluid occurred with treatment [116], as noted above for sputum [97]. Since IL-6, IFNγ, and IL-10 activate both STAT1 and STAT3, these two transcription factors could be appropriate readouts for the state of the AM. Analysis of the SMAD2 and SMAD3 transcription factors that are activated by TGFβ, and which associate with SMAD4 to mediate the response, might also be informative, since active TGFβ was also elevated in comparison to healthy and other infectious lung disease control groups [114]. Other mediators, including IL-1β, TNFα, CCL2, CCL5, and CXCL8 (IL-8), which have also been reported to differ in BAL fluid from TB patients relative to LTBI cases or healthy controls [111–113, 117–120], may not induce responses that could serve as biomarkers. The macrophage responses to IL-1β and TNFα may not distinguish PTB from other respiratory disease because these cytokines are induced by many inflammatory signals. In addition, chemokine receptors generally interact with more than one chemokine, which limits the utility of the response to chemokines as a biomarker. In summary, monocyte/macrophage response(s) to a subset of the mediators present in BAL fluid may be diagnostic of infection stage.

2.4 Peripheral Blood Monocytes

Three lines of investigation might acertain PBmo characteristics that would help determine infection stage, progression, and response to treatment. The first is to identify monocyte properties that differ between active TB and LTBI or between active TB patients and noninfected donors. However, the direct characterization of PBmo to date has been surprisingly limited. The second is to characterize PBmo properties that could reflect differences identified in serum. Among these, properties that are known consequences of response to a particular stimulus would serve best. The third is to ascertain whether the proportions of recently described PBmo subtypes would change in ways that distinguish among the categories of interest. In light of the evidence that TB involves substantial recruitment of monocytes from the circulation and replacement from the bone marrow [121–123], such an effect should be considered. Additional studies are required to extend and integrate the various measures that are described below.

In three comparisons of PBmo from TB and LTBI donors, effects of active disease have been described for cell surface markers, spontaneous cytokine release, and intracellular activities. First, a higher percentage of CD14+ cells was present in the blood mononuclear cell fraction, but the percentage of CD14+ CD36+ and CD14+ HLA-DR+ double-positive cells was decreased [124]. Those cells were restored during treatment. The overall expression of all three markers was also decreased in active disease and was restored during treatment. Second, increased spontaneous cytokine release was reported for IL-1 [125]. Third, higher intracellular activity (and extracellular release) was described for two lysosomal enzymes: beta-glucuronidase and N-acetylglucosaminidase [126]. The enzyme activities declined with treatment to control levels. Whether differences exist between healthy controls and TB or LTBI donors should be ascertained for these properties.

Several reports have described differences in comparisons between TB patients and healthy controls. In one study, active disease led to a lower percentage of phagocytic PBmo and reduced phagocytic activity [127]. In a separate study, an increase in cell surface Fcγ-RI and Fcγ-RIII was observed [128]. Recently, transcriptomic analysis of whole blood was interpreted to conclude that a functional network connected to Fcγ-RI distinguishes TB cases from LTBI and uninfected donors [129]. In a study of spontaneous IL-1 release that did not include comparison to LTBI, an increase relative to healthy control and a decrease with treatment were noted [130]. However, a study that did include comparison to LTBI also reported an increase in comparison to healthy controls [125]. Thus, IL-1 release may not be useful in discriminating between LTBI and TB. Moreover, increased release of IL-1 also occurs with other lung infections and with inflammatory conditions [130], limiting its utility as a biomarker for TB. Increased spontaneous release of IL-6 and

TNFα has also been described; for both, an increase with LTBI compared to noninfected has also been reported [131]. The increase noted above for lysosomal enzymes also occurred in comparison to healthy controls [126]. In a comparison only with healthy controls, another intracellular difference was found in reduced levels of IκB-α and constitutive activation of NFκB [132]. The effects on signaling to NFκB did not occur with non-tuberculous pulmonary conditions. One transcriptome analysis of whole blood from TB, LTBI, and noninfected donors identified infection-stage differences in a small number of monocyte/macrophage-specific genes [133]. These results further suggest that the PBmo population could be used for biomarker discovery, either alone or as part of a combinatorial approach.

Another approach would be to evaluate responses characteristic of exposure to peripheral stimuli that differ with infection stage, as described for sputum and BAL fluid. Indeed, as for BAL fluid, IFNγ, IL-6, and IL-10 were increased in serum from TB compared to LTBI donors [134–136] and in comparison to noninfected controls [113, 134, 136–139]. A multiplex assay that measured IFNγ, IL-6, IL-10, and 27 other analytes corroborated the elevated level of IL-6, but not the increased levels of IFNγ or IL-10 [140]. Higher serum concentration of IL-10 correlated with more advanced disease in some studies [134, 138, 141] but not in others [137]. Higher concentrations of IFNγ and IL-6 also correlated with more advanced disease in some reports [134, 138] but not in others [141]; in yet another study, advanced disease correlated with some IFNγ decrease [139]. In response to treatment, decreased levels of IFNγ but not of IL-6 [134], a decrease for both [137], and decreased levels for IFNγ and IL-10 [136] have been reported. Such disparities may limit the value of pursuing effects of these cytokines on PBmo as markers for TB severity or effects of therapy. In contrast, since increased levels of IFNγ, IL-6, and IL-10 are reliable serum indicators of active TB, investigation of PBmo responses to these cytokines as potential biomarkers seems warranted.

A third possibility for assessment based on PBmo would be to determine whether the proportions of subsets recently identified in healthy individuals are altered in LTBI or in active TB. In one example, CD14+ cells include a minor subpopulation that is relatively high in cell surface ganglioside GM1 [142], which defines the ordered membrane domains called lipid rafts. Lipid rafts are sites for cell surface receptors that mediate much of the cell's response to its environment [143, 144]. Cells having a higher level of GM1 show increased phagocytic activity and increased infection with *M. bovis* BCG compared to the monocytes having lower GM1. The CD14+ GM1high cells overlapped substantially (95%) with another recently described minor subset of PBmo, CD14+ CD16+. A recent study of PBmo in patients with stable coronary artery disease revealed that this inflammatory condition greatly increases the proportion of the minor subset, and changes expression of other cell surface markers on both subsets [145]. This result provides a proof of principle for considering changes in these PBmo subtypes as a possible biomarker for infection stage or response to treatment. Inasmuch as AM are also CD14+ CD16+, and active TB induces monocytopoiesis [122], monitoring the presence and properties of this monocyte subset in peripheral blood may be informative.

2.5 *Biomarkers Based on Monocytes and Macrophages*

Taken together, the reports described above suggest that monocyte and macrophage properties should be included in biomarker discovery. However, discrepancies in the literature call for revisiting some of the most promising possibilities, such as cell surface markers and responses characteristic of the cytokine environment. As noted for biomarkers based on T cell properties, sufficiently sensitive technology is essential. However, for monocytes and macrophages, potential measures of cell surface or intracellular markers can be accomplished with well-established flow cytometry approaches. With the development of user-friendly and relatively inexpensive flow cytometers, such assays have become feasible in clinical settings.

3 Humoral Responses

3.1 *Overview*

Ever since it was recognized that patients with active TB produced antibodies against *M. tuberculosis* that were detectable in serum [146], the antibody response has been a prime area of biomarker research in TB. Over the decades, the antibody response has been explored in relation to many antigenic targets. Most of the initial research utilized a complex immunogen of protein and non-protein nature derived from fractionation of *M. tuberculosis* cultures (purified protein derivative, culture supernatant, glycolipids, among others). With advances in protein purification and, later, with production of proteins by recombinant methods, it became increasingly more common to investigate the response to particular antigens of *M. tuberculosis*, such as the extracellular proteins (among the earlier examples are [147, 148]). In 2010, the utilization of high-throughput protein production and screening methods [149] made it possible to interrogate the entire proteome of *M. tuberculosis* (~4,000 proteins) with hundreds of sera [150]. Below we review the results of proteome-scale serological work and that of longitudinal studies investigating antibody responses during progression of the infection. Together, these results make it possible to define the characteristics of the antibody response to *M. tuberculosis* infection, and the potential value and shortcomings of circulating antibodies as TB biomarkers.

3.2 *Antibody Profiles*

Interrogation of chips arrayed with the *M. tuberculosis* proteome with hundreds of sera from TB suspects from TB-endemic countries identified approximately 10% of the bacterial proteome as reactive with human serum antibodies [150]. The reactive component of the proteome, termed the immunoproteome, contains predominantly membrane-associated and secreted proteins, in keeping with the preferential targeting of surface/external antigens by the antibody response. Active TB was associated with antibody recognition of a much smaller pool of proteins (<1% of the proteome), which were predominantly secreted proteins. These conclusions agree with much of the earlier serological work utilizing culture filtrates and purified secreted proteins (e.g., [151–154]). The immunoproteome data strongly suggest that membrane-associated proteins (which may derive from low numbers of live bacilli, dead bacilli, or macrophage-secreted exosomes) are occasionally targeted during LTBI or paucibacillary disease. In either condition, the extracellular proteins are underrepresented, either because dormant bacilli do not secrete (LTBI), or because the numbers of metabolically active mycobacteria are low (paucibacillary disease). During active disease, metabolically active bacilli secrete proteins. These become the favored targets. Thus, the specificity of antibody responses correlates with the metabolic state of tubercle bacilli during infection.

The above results lead to the conclusion that antibody responses correlate with antigen production by tubercle bacilli because those responses are exquisitely sensitive to antigen burden. Indeed, it is well established that antibody responses tend to be much stronger in sputum smear-positive than in smear-negative pulmonary TB [38, 152, 155,]. A positive correlation between antibody levels and bacillary load was seen also at the immunoproteome level. Thus, the sensitivity to burden is a general characteristic of the antibody response in TB.

The correlation between antibody levels and antigen burden suggests that antibody profiles may also reflect the progression of disease. The transition from LTBI to active TB involves resumed bacillary growth with consequent changes in bacterial metabolism and physiology, relative antigen production, and bacillary numbers. In macaques, which respond to experimental *M. tuberculosis* infection with approximately equal probability of asymptomatic infection and active disease [156], antibody profiles drastically differ with infection outcome. While antibody levels remained at pre-infection levels or increased only transiently in infected, asymptomatic animals, antibody responses to the *M. tuberculosis* proteome increased in animals exhibiting active disease [157]. The rise of antibody levels followed the temporal dynamics of disease manifestations. Additionally, the number of seroreactive proteins increased with overall antibody levels in active disease, indicating that the number of antigens reaching threshold levels for immune activation increases with antigen load. Longitudinal studies of the human antibody response to TB have only been conducted in HIV-infected cohorts, where the levels of some antibodies were seen to increase prior to the diagnosis of active TB [158–160]. Antibody levels tend to be lower in TB patients co-infected with HIV than in those who are HIV-negative [161]. Thus, these observations suggest that similar behaviors should also be observed in HIV-negative patients, where longitudinal studies of predictive biomarkers have not been conducted due to the logistical complexities associated with the low frequency of reactivation in immunocompetent individuals.

One puzzling aspect of the antibody response to TB has been the tremendous variability of antibody specificities from one individual to another. Even immunodominant targets of the antibody response, such as the 38 kDa lipoprotein (Rv0934), are recognized by only a fraction of the TB patient population [162–164], and antibody profiles are highly diverse from one patient to another. Host factors are almost certainly at play in determining serological diversity, since macaques infected by the same route with the same number of tubercle bacilli from the same strain exhibit varied antibody profiles [157]. In humans, levels of some antibodies have been reported to be associated with HLA type [165, 166]. An additional element of variation is introduced by the different antigen composition observed among clinical isolates of *M. tuberculosis* bacterial strain. Such variation presumably results from the level of expression of immunodominant antigens rather than from genetic differences [9, 167], since little diversity exists in genes encoding antigenic targets [168] and most antigenic epitopes are hyperconserved [169]. We have previously suggested [150] that relative antigen burden can be viewed as a main source of antibody variability, since relative antigen burden (a consequence of growth-state-associated relative antigen composition and bacillary load) likely varies among patients at the time of testing. Consequently, the relative frequency at which the antibody response "samples" each immunodominant antigen may vary from one patient to another. The effect of antigen load on the frequency of sampling will also vary from one antigen to another due to the relative antigen immunodominance. As bacillary burden increases during disease, more antigens reach threshold levels. Indeed, at high bacillary counts in patients' sputa, antibody profiles tend to become more homogeneous [150].

3.3 *Antibodies as Biomarkers*

The usefulness of antibodies as biomarkers of active TB is the result of the characteristics reviewed above. On the one hand, antibodies can differentiate active TB from asymptomatic infection, since the bacillary burden associated with stable asymptomatic infection is low. Moreover, seroconversion could be used as an early indication of disease progression, which is presumably accompanied by increased bacillary load. On the other hand, however, the low bacterial burden associated with paucibacillary forms of pulmonary TB and extrapulmonary TB makes antibodies poorly suited as markers of the hard-to-diagnose forms of active TB [154, 162–164, 170]. Additionally, an intriguing correlation has been repeatedly seen between past TB and seropositivity to *M. tuberculosis* antigens [150, 171, 172]. Subjects with a history of past TB may harbor a larger (or metabolically more active) bacterial population than the general latently infected population, as suggested by the association of past TB with increased risk of TB reactivation [173, 174]. The lack of longitudinal studies noted above has impeded determining whether increased antibody levels in past TB cases can be explained by ongoing disease reactivation, which would make antibody detection of important prognostic value. In the absence of this information, the occasional seropositivity seen in the past TB group poses another challenge to serodiagnosis of active TB.

The objective limitations of the antibody response as biomarker—poor detection of paucibacillary TB cases, antibody profile diversity among patients, and ill-understood confounding factors such as history of past TB—together with the irresponsible commercial release and utilization of low-quality serological tests (http://www.who.int/mediacentre/news/releases/2011/tb_20110720/en/index.html) make TB serodiagnostic development a complex area of investigation. However, it is important to recognize that antibody detection assays are ideal for user-friendly, point-of-care diagnosis, which is the next priority in TB diagnostics development [175]. Therefore, it is worthwhile considering how combinatorial, multimarker approaches could help overcome the shortcomings of the antibody response by enhancing diagnosis of paucibacillary forms of TB and by excluding from among the seropositive responses those associated with past TB rather than with active TB. For the latter limitation, multiple possibilities exist. For example, in the context of an antibody assay, relative ratios of IgG isotypes, which reflect T cell help and the cytokine environment, may skew the diagnostic decision towards the presence or absence of an active disease process. Moreover, in a multimarker approach, concurrent detection of markers of ongoing inflammation based on monocytes/macrophage properties could help discriminate seropositivity caused by active TB from that associated with history of past TB. Furthermore, it is well known that particular cellular responses are reduced in active TB patients due to immunosuppression associated with active disease. Thus, the diagnostic association of antibody to active disease might be strengthened by the concurrent absence of one such cellular response.

4 Conclusions

The yin and yang of immune biomarkers for TB are clear. On the one hand, no biomarker should be better suited than those of immunological nature to distinguish among the various stages of *M. tuberculosis* infection since immunological events are at the core of TB pathogenesis. This means that the spectrum of the clinical manifestations is associated with corresponding, interrelated spectra of tissue damage and infection control, which both result from immunological events [176]. Thus, each particular stage of the *M. tuberculosis* infection is associated with specific T cell phenotypes, monocyte/macrophage characteristics, and antibody profiles. These are all potential stage-specific biomarkers. On the other hand, due to the multifactorial nature of the immunological events and (host- and pathogen-derived) inter-patient variation, no single marker or marker set has yet identified a particular *M. tuberculosis* infection stage with adequate diagnostic accuracy (the shortcomings of TB immunodiagnostics have been extensively reviewed [170, 177–179]). Thus, translating biomarker discovery into tests with diagnostic and prognostic value requires integrated approaches to data collection, analysis, and interpretation.

For the reasons stated above, the immune signature of each particular infection stage most likely comprises multiple biomarkers. So far, the search for multiple markers has been applied to markers of the same kind. For example, to increase diagnostic sensitivity, serodiagnostic research has been oriented towards multi-antigen tests while IFNγ release assays (IGRA) for the diagnosis of LTBI have included two or three antigens. However, multi-antigen serology has yet to provide accurate diagnostics for active TB, while the current IGRAs do not distinguish between stable and progressive LTBI (only the latter requires antibiotic treatment). The limitations of the current methods raise the possibility that the immunological signature of each *M. tuberculosis* infection stage rests on a combination of immune markers of different types. For example, the need for multiple antibodies to boost sensitivity reduces diagnostic specificity due to accumulating positive results in the non-TB-diseased population. As noted above, concurrent testing for markers indicative of an active disease process might increase the diagnostic specificity of antibody-based assays. Moreover, cellular response detected by the current IGRAs often wanes in TB patients, due to the immunosuppression associated with active disease. Thus the concurrent detection of specific antibody and reduced IGRA responses might be better associated with active disease than either marker alone. Additional diagnostic insight should result from the antigen specificity of T cells, and from monocyte or macrophage markers associated with contained infection rather than with failed immunity. Concomitant detection of biomarkers that define stable LTBI and active TB may be the signature of early reactivation. Certain biomarker combinations may be informative as correlates of effective vaccination, or even of effective pharmacological treatment (e.g., treatment-associated decline in biomarker combinations specific for active TB). Thus, a systems approach to biomarker identification and evaluation could overcome many of the limitations in current approaches.

The challenges of implementing combinatorial biomarker discovery are many. The appropriate biomarkers may have been already identified (see [93] for a comprehensive TB biomarker review) or may still require new discovery, for example, with high-throughput methods. These are usually expensive. Moreover, assessment of combinatorial markers would require appropriate analytical expertise and computing resources. Finally, developing multi-analyte diagnostic assays may constitute an area of research in its own right, due to the difficulties of concurrently detecting signals of different nature. Nonetheless, given the rapid expansion of biological knowledge and the dizzying pace of technological advance, the success of TB biomarker discovery should be all but assured. The translation to TB control measures rests on adequate funding and political will.

Acknowledgements The authors' laboratories are currently supported by the National Institutes of Health (AI 045761; AI 063359; AI 095924; HL 106788) and by the Foundation for New and Innovative Diagnostics. Past support by the National Institutes of Health is also gratefully acknowledged.

References

1. Wallis RS, Pai M, Menzies D, Doherty TM, Walzl G, Perkins MD, Zumla A (2010) Biomarkers and diagnostics for tuberculosis: progress, needs, and translation into practice. Lancet 375(9729):1920–1937

2. Ling DI, Flores LL, Riley LW, Pai M (2008) Commercial nucleic-acid amplification tests for diagnosis of pulmonary tuberculosis in respiratory specimens: meta-analysis and meta-regression. PLoS One 3(2):e1536
3. Mase SR, Ramsay A, Ng V, Henry M, Hopewell PC, Cunningham J, Urbanczik R, Perkins MD, Aziz MA, Pai M (2007) Yield of serial sputum specimen examinations in the diagnosis of pulmonary tuberculosis: a systematic review. Int J Tuberc Lung Dis 11(5):485–495
4. Menzies D, Pai M, Comstock G (2007) Meta-analysis: new tests for the diagnosis of latent tuberculosis infection: areas of uncertainty and recommendations for research. Ann Intern Med 146(5):340–354
5. Pai M (2010) Spectrum of latent tuberculosis—existing tests cannot resolve the underlying phenotypes. Nat Rev Microbiol 8(3):242
6. Barry CE 3rd, Boshoff HI, Dartois V, Dick T, Ehrt S, Flynn J, Schnappinger D, Wilkinson RJ, Young D (2009) The spectrum of latent tuberculosis: rethinking the biology and intervention strategies. Nat Rev Microbiol 7(12):845–855
7. Fenhalls G, Stevens L, Moses L, Bezuidenhout J, Betts JC, Helden Pv P, Lukey PT, Duncan K (2002) In situ detection of Mycobacterium tuberculosis transcripts in human lung granulomas reveals differential gene expression in necrotic lesions. Infect Immun 70(11):6330–6338
8. Pheiffer C, Betts J, Lukey P, van Helden P (2002) Protein expression in Mycobacterium tuberculosis differs with growth stage and strain type. Clin Chem Lab Med 40(9):869–875
9. Pheiffer C, Betts JC, Flynn HR, Lukey PT, van Helden P (2005) Protein expression by a Beijing strain differs from that of another clinical isolate and Mycobacterium tuberculosis H37Rv. Microbiology 151(Pt 4):1139–1150
10. Yoshikai Y (2006) Immunological protection against Mycobacterium tuberculosis infection. Crit Rev Immunol 26:515–526
11. Torrado E, Robinson RT, Cooper AM (2011) Cellular response to mycobacteria: balancing protection and pathology. Trends Immunol 32:66–72
12. Ferrero E, Biswas P, Vettoretto K, Ferrarini M, Uguccioni M, Piali L, Leone BE, Moser B, Rugarli C (2003) Macrophages exposed to Mycobacterium tuberculosis release chemokines able to recruit selected leucocyte subpopulations: focus on gd cells. Immunology 108:365–374
13. Orme I (2004) Adaptive immunity to mycobacteria. Curr Opin Microbiol 7:58–61
14. Kaufmann SH, Cole ST, Mizrahi V, Rubin E, Nathan C (2005) Mycobacterium tuberculosis and the host response. J Exp Med 201:1693–1697
15. Khader SA, Gopal R (2010) IL-17 in protective immunity to intracellular pathogens. Virulence 1:423–427
16. Majlessi L, Lo-Man R, Leclerc C (2008) Regulatory B and T cells in infections. Microbes Infect 10:1030–1035
17. Lockhart E, Green AM, Flynn JL (2006) IL-17 production is dominated by gammadelta T cells rather than CD4 T cells during Mycobacterium tuberculosis infection. J Immunol 177:4662–4669
18. Behar SM, Porcelli SA (2007) CD1-restricted T cells in host defense to infectious diseases. Curr Top Microbiol Immunol 314:215–250
19. Cohen NR, Garg S, Brenner MB (2009) Antigen presentation by CD1: lipids, T cells, and NKT cells in microbial immunity. Adv Immunol 102:1–94
20. Felio K, Nguyen H, Dascher CC, Cjoi HJ, Li S, Zimmer MI, Colmone A, Moody DB, Brenner MB, Wang CR (2009) CD1-restricted adaptive immune responses to Mycobacteria in human group 1 CD1 transgenic mice. J Exp Med 206:2497–2509
21. Chiba A, Dascher CC, Besra GS, Brenner MB (2008) Rapid NKT cell responses are self-terminating during the course of microbial infection. J Immunol 181:2292–2302
22. Cruz A, Khader SA, Torrado E, Fraga A, Pearl JE, Pedrosa J, Cooper AM, Castro AG (2006) Cutting edge: IFN-gamma regulates the induction and expansion of IL-17-producing CD4 T cells during mycobacterial infection. J Immunol 177:1416–1420
23. Korf H, Vander Beken S, Romano M, Steffensen KR, Stijlemans B, Gustafsson JA, Grooten J, Huygen K (2009) Liver X receptors contribute to the protective immune responses against Mycobacterium tuberculosis in mice. J Clin Invest 119:1626–1637

24. Cui G, Qin X, Wu L, Zhang Y, Sheng X, Yu Q, Sheng H, Xi B, Zhang JZ, Zang YQ (2011) Liver X receptor (LXR) mediates negative regulation of mouse and human Th17 differentiation. J Clin Invest 121:658–670
25. Harrington LE, Hatton RD, Mangan PR, Turner H, Murphy TL, Murphy KM, Weaver CT (2005) Interleukin 17-producing CD4+ effector T cells develop via a lineage distinct from the T helper type 1 and 2 lineages. Nat Immunol 6:1123–1132
26. Park H, Li Z, Yang XO, Chang SH, Nurieva R, Wang YH, Wang Y, Hood L, Zhu Z, Tian Q, Dong C (2005) A distinct lineage of CD4 T cells regulates tissue inflammation by producing interleukin 17. Nat Immunol 6:1133–1141
27. Okamoto Yoshida Y, Umemura M, Yahagi A, O'Brien RL, Ikuta K, Kishihara K, Hara H, Nakae S, Iwakura Y, Matsuzaki G (2010) Essential role of IL-17A in the formation of a mycobacterial infection-induced granuloma in the lung. J Immunol 184:4414–4422
28. Stetson DB, Mohrs M, Reinhardt RL, Baron JL, Wang ZE, Gapin L, Kronenberg M, Locksley RM (2003) Constitutive cytokine mRNAsmark natural killer (NK) and NK T cells poised for rapid effector function. J Exp Med 198:1069–1076
29. Mycko MP, Ferrero I, Wilson I, Jiang W, Bianchi T, Trumpp A, MacDonald HR (2009) Selective requirement for c-Myc at an early stage of V(alpha)14i NKT cell development. J Immunol 182:4641–4648
30. Rolf J, Berntman E, Stenstrom M, Smith EM, Mansson R, Stenstad H, Yamagata T, Agace W, Sigvardsson M, Cardell SL (2008) Molecular profiling reveals distinct functional attributes of CD1d-restricted natural killer (NK) T cell subsets. Mol Immunol 45:2607–2620
31. Andersen P (2001) TB vaccines: progress and problems. Trends Immunol 22(3):160–168
32. Sable SB, Plikaytis BB, Shinnick TM (2007) Tuberculosis subunit vaccine development: impact of physicochemical properties of mycobacterial test antigens. Vaccine 25:1553–1566
33. Shi L, Jung YJ, Tyagi S, Gennaro ML, North RJ (2003) Expression of Th1-mediated immunity in mouse lungs induces a Mycobacterium tuberculosis transcription pattern characteristic of nonreplicating persistence. Proc Natl Acad Sci USA 100:886–891
34. Shi L, North R, Gennaro ML (2004) Effect of growth state on transcription levels of genes encoding major secreted antigens of Mycobacterium tuberculosis in the mouse lung. Infect Immun 72:2420–2424
35. Pathan AA, Wilkinson KA, Wilkinson RJ, Latif M, McShane H, Pasvol G, Hill AV, Lalvani A (2000) High frequencies of circulating IFN-γ-secreting CD8 cytotoxic T cells specific for a novel MHC class I-restricted Mycobacterium tuberculosis epitope in M. tuberculosis-infected subjects without disease. Eur J Immunol 30:2713–2721
36. Dong Y, Demaria S, Sun X, Santori FR, Jesdale BM, De Groot AS, Rom WN, Bushkin Y (2004) HLA-A2-Restricted $CD8^{+}$-cytotoxic-T-cell responses to novel epitopes in Mycobacterium tuberculosis superoxide dismutase, alanine dehydrogenase, and glutamine synthetase. Infect Immun 72:2412–2415
37. Lin MY, Ottenhoff TH (2008) Not to wake a sleeping giant: new insights into host-pathogen interactions identify new targets for vaccination against Mycobacterium tuberculosis infection. J Biol Chem 389:497–511
38. Wiker HG, Harboe M (1992) The antigen 85 complex: a major secretion product of Mycobacterium tuberculosis. Microbiol Rev 56(4):648–661
39. Sørensen AL, Nagai S, Houen G, Andersen P, Andersen ÅB (1995) Purification and characterization of a low-molecular-mass T-cell antigen secreted by *Mycobacterium tuberculosis*. Infect Immun 63:1710–1717
40. Dietrich J, Weldingh K, Andersen P (2006) Prospects for a novel vaccine against tuberculosis. Vet Microbiol 112:163–169
41. Schuck SD, Mueller H, Kunitz F, Neher A, Hoffmann H, Franken KLCM, Repsilber D, Ottenhoff THM, Kaufmann SHE, Jacobsen M (2009) Identification of T-cell antigens specific for latent Mycobctarium tuberculosis infection. PLoS One 4:e5590
42. Millington KA, Fortune SM, Low J, Garces A, Hingley-Wilson SM, Wickermasinghe M, Kon OM, Lalvani A (2011) Rv3615c is a highly immunodominant RD1 (Region of Difference

1)-dependent secreted antigen specific for Mycobacterium tuberculosis infection. Proc Natl Acad Sci USA 108:5730–5735

43. Govender L, Abel B, Hughes EJ, Scriba TJ, Kagina BMN, de Kock M, Walzl G, Black G, Rosenkrands I, Hussey GD, Mahomed H, Andersen P, Hanekom WA (2011) Higher human CD4 T cell response to novel Mycobacterium tuberculosis latency associated antigens Rv2660 and Rv2659 in latent infection compared with tuberculosis disease. Vaccine 29:51–57
44. Coomandeur S, van Meijgaarden KE, Lin MY, Franken KL, Friggen AH, Drijfhout JW, Oftung F, Korsvold GE, Geluk A, Ottenhoff TH (2011) Identification of human T-cell responses to Mycobacterium tuberculosis resuscitation-promoting factors in long-term latently infected individuals. Clin Vaccine Immunol 18:676–683
45. Wayne LG, Hayes LG (1996) An in vitro model for sequential study of shiftdown of Mycobacterium tuberculosis through two stages of nonreplicating persistence. Infect Immun 64:2062–2069
46. Pantaleo G, Harari A (2006) Functional signatures in antiviral T-cell immunity for monitoring virus-associated diseases. Nat Rev Immunol 6:417–423
47. Millington KA, Innes JA, Hackforth S, Hinks TS, Deeks JJ, Dosanjh DP, Guyot-Revol V, Gunatheesan R, Klenerman P, Lalvani A (2007) Dynamic relationship between IFN-gamma and IL-2 profile of Mycobacterium tuberculosis-specific T cells and antigen load. J Immunol 178:5217–5226
48. Sutherland JS, Young JM, Peterson KL, Sanneh B, Whittle HC, Rowland-Jones SL, Adegbol RA, Jaye A, Ota MO (2010) Polyfunctional CD4(+) and CD8(+) T cell responses to tuberculosis antigens in HIV-1-infected patients before and after anti-retroviral treatment. J Immunol 184:6537–6544
49. Caccamo N, Guggino G, Meraviglia S, Gelsomino G, Di Carlo P, Titone L, Bocchino M, Galati D, Matarese A, Nouta J, Klein MR, Salerno A, Sanduzzi A, Dieli F, Ottenhoff TH (2009) Analysis of Mycobacterium tuberculosis-specific CD8 T-cells in patients with active tuberculosis and in individuals with latent infection. PLoS One 4(5):e5528
50. Caccamo N, Guggino G, Joosten SA, Gelsomino G, Di Carlo P, Titone L, Galati D, Bocchino M, Matarese A, Salerno A, Sanduzzi A, Franken WP, Ottenhoff TH, Dieli F (2010) Multifunctional CD4(+) T cells correlate with active Mycobacterium tuberculosis infection. Eur J Immunol 40(8):2211–2220
51. Casey R, Blumenkrantz D, Millington KA, Montamat-Sicotte D, Kon OM, Wickremasinghe M, Bremang S, Magtoto M, Sridhar S, Connel D, Lalvani A (2010) Enumeration of functional T-cell subsets by fluoresecence-immunospot defines signatures of pathogen burden in tuberculosis. PLoS One 5:215619
52. Sester U, Fousse M, Dirks J, Mack U, Prasse A, Singh M, Lalvani A, Sester M (2011) Whole-blood flow-cytometric analysis of antigen-specific CD4 T-cell cytokine profiles distinguishes active tuberculosis from non-active states. PLoS One 6:e17813
53. Coomandeur S, Lin MY, van Meijgaarden KE, Friggen AH, Franken KL, Drijfhout JW, Korsvold GE, Oftung F, Geluk A, Ottenhoff TH (2011) Double- and monofunctional CD4(+) and CD8(+) T-cell responses to Mycpbacterium tuberculosis DosR antigens and peptides in long-term latently infected individuals. Eur J Immunol 41:2925–2936
54. Sallusto F, Lenig D, Forster R, Lipp M, Lanzavecchia A (1999) Two subsets of memory T lymphocytes with distinct homing potentials and effector functions. Nature 401(6754):708–712
55. Forbes EK, Sander C, Ronan EO, McShane H, Hill AV, Beverley PC, Tchilian EZ (2008) Multifunctional, high-level cytokine-producing Th1 cells in the lung, but not spleen, correlate with protection against Mycobacterium tuberculosis aerosol challenge in mice. J Immunol 181:4955–4964
56. Seder RA, Darrah PA, Roederer M (2008) T-cell quality in memory and protection: implications for vaccine design. Nat Rev Immunol 8(4):247–258
57. Belkaid Y (2007) Regulatory T cells and infection: a dangerous necessity. Nat Rev Immunol 7(11):875–888

58. Fontenot JD, Rudensky AY (2005) A well adapted regulatory contrivance: regulatory T cell development and the forkhead family transcription factor Foxp3. Nat Immunol 6(4):331–337
59. Roy S, Barnes PF, Garg A, Wu S, Cosman D, Vankayalapati R (2008) NK cells lyse T regulatory cells that expand in response to an intracellular pathogen. J Immunol 180(3):1729–1736
60. Vrabelova Z, Hrotekova Z, Hladikova Z, Bohmova K, Stechova K, Michalek J (2008) CD 127- and FoxP3+ expression on CD25+CD4+ T regulatory cells upon specific diabetogeneic stimulation in high-risk relatives of type 1 diabetes mellitus patients. Scand J Immunol 67(4):404–410
61. Chen X, Zhou B, Li M, Deng Q, Wu X, Le X, Wu C, Larmonier N, Zhang W, Zhang H, Wang H, Katsanis E (2007) CD4(+)CD25(+)FoxP3(+) regulatory T cells suppress Mycobacterium tuberculosis immunity in patients with active disease. Clin Immunol 123(1):50–59
62. Shafiani S, Tucker-Heard G, Kariyone A, Takatsu K, Urdahl KB (2010) Pathogen-specific regulatory T cells delay the arrival of effector T cells in the lung during early tuberculosis. J Exp Med 207(7):1409–1420
63. Scott-Browne JP, Shafiani S, Tucker-Heard G, Ishida-Tsubota K, Fontenot JD, Rudensky AY, Bevan MJ, Urdahl KB (2007) Expansion and function of Foxp3-expressing T regulatory cells during tuberculosis. J Exp Med 204(9):2159–2169
64. Guyot-Revol V, Innes JA, Hackforth S, Hinks T, Lalvani A (2006) Regulatory T cells are expanded in blood and disease sites in patients with tuberculosis. Am J Respir Crit Care Med 173(7):803–810
65. Marin ND, Paris SC, Velez VM, Rojas CA, Rojas M, Garcia LF (2010) Regulatory T cell frequency and modulation of IFN-gamma and IL-17 in active and latent tuberculosis. Tuberculosis (Edinb) 90(4):252–261
66. Garg A, Barnes PF, Roy S, Quiroga MF, Wu S, Garcia VE, Krutzik SR, Weis SE, Vankayalapati R (2008) Mannose-capped lipoarabinomannan- and prostaglandin E2-dependent expansion of regulatory T cells in human Mycobacterium tuberculosis infection. Eur J Immunol 38(2):459–469
67. Ozeki Y, Sugawara I, Udagawa T, Aoki T, Osada-Oka M, Tateishi Y, Hisaeda H, Nishiuchi Y, Harada N, Kobayashi K, Matsumoto S (2010) Transient role of CD4+CD25+ regulatory T cells in mycobacterial infection in mice. Int Immunol 22(3):179–189.
68. Corthay A (2009) How do regulatory T cells work? Scand J Immunol 70(4):326–336
69. Onishi Y, Fehervari Z, Yamaguchi T, Sakaguchi S (2008) Foxp3+ natural regulatory T cells preferentially form aggregates on dendritic cells in vitro and actively inhibit their maturation. Proc Natl Acad Sci USA 105(29):10113–10118
70. Divangahi M, Desjardins D, Nunes-Alves C, Remold HG, Behar SM (2010) Eicosanoid pathways regulate adaptive immunity to Mycobacterium tuberculosis. Nat Immunol 11(8):751–758
71. Mills KH, McGuirk P (2004) Antigen-specific regulatory T cells—their induction and role in infection. Semin Immunol 16(2):107–117
72. Chiacchio T, Casetti R, Butera O, Vanini V, Carrara S, Girardi E, Di Mitri D, Battistini L, Martini F, Borsellino G, Goletti D (2009) Characterization of regulatory T cells identified as CD4(+)CD25(high)CD39(+) in patients with active tuberculosis. Clin Exp Immunol 156(3):463–470
73. Hougardy JM, Verscheure V, Locht C, Mascart F (2007) In vitro expansion of CD4+CD25h ighFOXP3+CD127low/- regulatory T cells from peripheral blood lymphocytes of healthy Mycobacterium tuberculosis-infected humans. Microbes Infect 9(11):1325–1332
74. He XY, Xiao L, Chen HB, Hao J, Li J, Wang YJ, He K, Gao Y, Shi BY (2010) T regulatory cells and Th1/Th2 cytokines in peripheral blood from tuberculosis patients. Eur J Clin Microbiol Infect Dis 29(6):643–650
75. Cresswell P, Ackerman AL, Giodini A, Peaper DR, Wearsch PA (2005) Mechanisms of MHC class I-restricted antigen processing and cross-presentation. Immunol Rev 207:145–157
76. Rammensee HG (2006) Peptides made to order. Immunity 25(5):693–695

77. Purcell AW, McCluskey J, Rossjohn J (2007) More than one reason to rethink the use of peptides in vaccine design. Nat Rev Drug Discov 6(5):404–414
78. Admon A, Barnea E, Ziv T (2003) Tumor antigens and proteomics from the point of view of the major histocompatibility complex peptides. Mol Cell Proteomics 2(6):388–398
79. Bassani-Sternberg M, Barnea E, Beer I, Avivi I, Katz T, Admon A (2010) Soluble plasma HLA peptidome as a potential source for cancer biomarkers. Proc Natl Acad Sci USA 107(44):18769–18776
80. Bangia N, Ferrone S (2006) Antigen presentation machinery (APM) modulation and soluble HLA molecules in the tumor microenvironment: do they provide tumor cells with escape mechanisms from recognition by cytotoxic T lymphocytes? Immunol Invest 35(3–4):485–503
81. Campoli M, Ferrone S (2008) Tumor escape mechanisms: potential role of soluble HLA antigens and NK cells activating ligands. Tissue Antigens 72(4):321–334
82. Migliaresi S, Bresciani A, Ambrosone L, Spera M, Barbarulo D, Lombari V, Pirozzi G, Borgia G, Lombardi ML, Tirri G, Manzo C (2000) Increased serum concentrations of soluble HLA-class I antigens in hepatitis C virus related mixed cryoglobulinaemia. Ann Rheum Dis 59(1):20–25
83. Demaria S, Bushkin Y (2000) Soluble HLA proteins with bound peptides are released from the cell surface by the membrane metalloproteinase. Hum Immunol 61(12):1332–1338
84. Demaria S, Schwab R, Gottesman SR, Bushkin Y (1994) Soluble beta 2-microglobulin-free class I heavy chains are released from the surface of activated and leukemia cells by a metalloprotease. J Biol Chem 269(9):6689–6694
85. Krangel MS (1986) Secretion of HLA-A and -B antigens via an alternative RNA splicing pathway. J Exp Med 163(5):1173–1190
86. Bushkin Y, Watanabe H, Demaria S (2004) Extracellular processing of MHC class I antigens. In: Wilkes DS, Burlingham WJ (eds) Immunobiology of organ transplantation. Kluwer Academic/Plenum, New York, pp 53–65
87. Haynes LD, Bushkin Y, Love RB, Burlingham WJ (2002) Interferon-gamma drives the metalloproteinase-dependent cleavage of HLA class I soluble forms from primary human bronchial epithelial cells. Hum Immunol 63(10):893–901
88. Oelke M, Schneck JP (2010) Overview of a HLA-Ig based "Lego-like system" for T cell monitoring, modulation and expansion. Immunol Res 47(1–3):248–256
89. Chiu YL, Schneck JP, Oelke M (2011) HLA-Ig based artificial antigen presenting cells for efficient ex vivo expansion of human CTL. J Vis Exp. 2011 Apr 11;(50). pii: 2801.
90. Raj A, van den Bogaard P, Rifkin SA, van Oudenaarden A, Tyagi S (2008) Imaging individual mRNA molecules using multiple singly labeled probes. Nat Methods 5(10):877–879
91. Ruhwald M, Ravn P (2009) Biomarkers of latent TB infection. Expert Rev Respir Med 3(4):387–401
92. Wallis RS, Doherty TM, Onyebujoh P, Vahedi M, Laang H, Olesen O, Parida S, Zumla A (2009) Biomarkers for tuberculosis disease activity, cure, and relapse. Lancet Infect Dis 9(3):162–172
93. Walzl G, Ronacher K, Hanekom W, Scriba TJ, Zumla A (2011) Immunological biomarkers of tuberculosis. Nat Rev Immunol 11(5):343–354
94. Kunnath-Velayudhan S, Gennaro ML (2011) Immunodiagnosis of tuberculosis: a dynamic view of biomarker discovery. Clin Microbiol Rev 24(4):792–805
95. Flores-Batista VC, Boechat N, Lago PM, Lazzarini LC, Pessanha LR, Almeida AS, Mafort TT, Kritski AL, Ho JL, Lapa-e-Silva JR (2007) Low expression of antigen-presenting and costimulatory molecules by lung cells from tuberculosis patients. Braz J Med Biol Res 40(12):1671–1679
96. Almeida AS, Lago PM, Boechat N, Huard RC, Lazzarini LC, Santos AR, Nociari M, Zhu H, Perez-Sweeney BM, Bang H, Ni Q, Huang J, Gibson AL, Flores VC, Pecanha LR, Kritski AL, Lapa e Silva JR, Ho JL (2009) Tuberculosis is associated with a down-modulatory lung immune response that impairs Th1-type immunity. J Immunol 183(1):718–731

97. Ribeiro-Rodrigues R, Resende Co T, Johnson JL, Ribeiro F, Palaci M, Sa RT, Maciel EL, Pereira Lima FE, Dettoni V, Toossi Z, Boom WH, Dietze R, Ellner JJ, Hirsch CS (2002) Sputum cytokine levels in patients with pulmonary tuberculosis as early markers of mycobacterial clearance. Clin Diagn Lab Immunol 9(4):818–823
98. Ivashkiv LB (2000) Jak-STAT signaling pathways in cells of the immune system. Rev Immunogenet 2(2):220–230
99. Leonard WJ (2001) Role of Jak kinases and STATs in cytokine signal transduction. Int J Hematol 73(3):271–277
100. Ishihara K, Hirano T (2002) Molecular basis of the cell specificity of cytokine action. Biochim Biophys Acta 1592(3):281–296
101. Kuo HP, Yu CT (1993) Alveolar macrophage subpopulations in patients with active pulmonary tuberculosis. Chest 104(6):1773–1778
102. Shellito J, Kaltreider HB (1984) Heterogeneity of immunologic function among subfractions of normal rat alveolar macrophages. Am Rev Respir Dis 129(5):747–753
103. Ferro TJ, Kern JA, Elias JA, Kamoun M, Daniele RP, Rossman MD (1987) Alveolar macrophages, blood monocytes, and density-fractionated alveolar macrophages differ in their ability to promote lymphocyte proliferation to mitogen and antigen. Am Rev Respir Dis 135(3):682–687
104. Brannen AL, Chandler DB (1988) Alveolar macrophage subpopulations' responsiveness to chemotactic stimuli. Am J Pathol 132(1):161–166
105. Chandler DB, Fuller WC, Jackson RM, Fulmer JD (1986) Fractionation of rat alveolar macrophages by isopycnic centrifugation: morphological, cytochemical, biochemical, and functional properties. J Leukoc Biol 39(4):371–383
106. Shellito J, Kaltreider HB (1985) Heterogeneity of immunologic function among subfractions of normal rat alveolar macrophages. II. Activation as a determinant of functional activity. Am Rev Respir Dis 131(5):678–683
107. Chan J, Tanaka K, Carroll D, Flynn J, Bloom BR (1995) Effects of nitric oxide synthase inhibitors on murine infection with Mycobacterium tuberculosis. Infect Immun 63(2):736–740
108. MacMicking JD, North RJ, LaCourse R, Mudgett JS, Shah SK, Nathan CF (1997) Identification of nitric oxide synthase as a protective locus against tuberculosis. Proc Natl Acad Sci USA 94(10):5243–5248
109. Nicholson S, Bonecini-Almeida Mda G, Lapa e Silva JR, Nathan C, Xie QW, Mumford R, Weidner JR, Calaycay J, Geng J, Boechat N, Linhares C, Rom W, Ho JL (1996) Inducible nitric oxide synthase in pulmonary alveolar macrophages from patients with tuberculosis. J Exp Med 183(5):2293–2302
110. Wang CH, Liu CY, Lin HC, Yu CT, Chung KF, Kuo HP (1998) Increased exhaled nitric oxide in active pulmonary tuberculosis due to inducible NO synthase upregulation in alveolar macrophages. Eur Respir J 11(4):809–815
111. Wang CH, Kuo HP (2001) Nitric oxide modulates interleukin-1beta and tumour necrosis factor-alpha synthesis, and disease regression by alveolar macrophages in pulmonary tuberculosis. Respirology 6(1):79–84
112. Law K, Weiden M, Harkin T, Tchou-Wong K, Chi C, Rom WN (1996) Increased release of interleukin-1 beta, interleukin-6, and tumor necrosis factor-alpha by bronchoalveolar cells lavaged from involved sites in pulmonary tuberculosis. Am J Respir Crit Care Med 153(2):799–804
113. Tsao TC, Hong J, Huang C, Yang P, Liao SK, Chang KS (1999) Increased TNF-alpha, IL-1 beta and IL-6 levels in the bronchoalveolar lavage fluid with the upregulation of their mRNA in macrophages lavaged from patients with active pulmonary tuberculosis. Tuberc Lung Dis 79(5):279–285
114. Bonecini-Almeida MG, Ho JL, Boechat N, Huard RC, Chitale S, Doo H, Geng J, Rego L, Lazzarini LC, Kritski AL, Johnson WD Jr, McCaffrey TA, Silva JR (2004) Down-modulation of lung immune responses by interleukin-10 and transforming growth factor beta (TGF-beta) and analysis of TGF-beta receptors I and II in active tuberculosis. Infect Immun 72(5):2628–2634

115. Condos R, Rom WN, Liu YM, Schluger NW (1998) Local immune responses correlate with presentation and outcome in tuberculosis. Am J Respir Crit Care Med 157(3 Pt 1):729–735
116. Tsao TC, Huang CC, Chiou WK, Yang PY, Hsieh MJ, Tsao KC (2002) Levels of interferon-gamma and interleukin-2 receptor-alpha for bronchoalveolar lavage fluid and serum were correlated with clinical grade and treatment of pulmonary tuberculosis. Tuberc Lung Dis 6(8):720–727
117. Zissel G, Baumer I, Schlaak M, Muller-Quernheim J (2000) In vitro release of interleukin-15 by broncho-alveolar lavage cells and peripheral blood mononuclear cells from patients with different lung diseases. Eur Cytokine Netw 11(1):105–112
118. Hoshino Y, Tse DB, Rochford G, Prabhakar S, Hoshino S, Chitkara N, Kuwabara K, Ching E, Raju B, Gold JA, Borkowsky W, Rom WN, Pine R, Weiden M (2004) Mycobacterium tuberculosis-induced CXCR4 and chemokine expression leads to preferential X4 HIV-1 replication in human macrophages. J Immunol 172(10):6251–6258
119. Sadek MI, Sada E, Toossi Z, Schwander SK, Rich EA (1998) Chemokines induced by infection of mononuclear phagocytes with mycobacteria and present in lung alveoli during active pulmonary tuberculosis. Am J Respir Cell Mol Biol 19(3):513–521
120. Kurashima K, Mukaida N, Fujimura M, Yasui M, Nakazumi Y, Matsuda T, Matsushima K (1997) Elevated chemokine levels in bronchoalveolar lavage fluid of tuberculosis patients. Am J Respir Crit Care Med 155(4):1474–1477
121. Schwander SK, Sada E, Torres M, Escobedo D, Sierra JG, Alt S, Rich EA (1996) T lymphocytic and immature macrophage alveolitis in active pulmonary tuberculosis. J Infect Dis 173(5):1267–1272
122. Schmitt E, Meuret G, Stix L (1977) Monocyte recruitment in tuberculosis and sarcoidosis. Br J Haematol 35(1):11–17
123. LapaeSilva JR, Linhares C, Boechat N, Rego L, Almeida MG, Kriski AL, Ho JL (1996) Phenotypes of lung mononuclear phagocytes in HIV seronegative tuberculosis patients: evidence for new recruitment and cell activation. Mem Inst Oswaldo Cruz 91(3):389–394
124. Sanchez MD, Garcia Y, Montes C, Paris SC, Rojas M, Barrera LF, Arias MA, Garcia LF (2006) Functional and phenotypic changes in monocytes from patients with tuberculosis are reversed with treatment. Microbes Infect 8(9–10):2492–2500
125. Businaro R, Mannella E, Cochi S, Fioravanti D, Bettazzoni S, Granato T, Ippoliti F (1987) Inverse relationship between spontaneous interleukin-1 production and mitogen-driven proliferation in patients with pulmonary tuberculosis. Allerg Immunol (Paris) 19(5):189–196
126. Jaswal S, Dhand R, Sethi AK, Kohli KK, Ganguly NK (1993) Intracellular levels and extracellular release of lysosomal enzymes from peripheral blood monocytes in pulmonary tuberculosis patients. APMIS 101(1):50–54
127. Gomaa NI, Ahmed AS, El-Hadidy GS, Mansour MK (2004) Flow cytometric analysis of the phagocytic cells in active pulmonary tuberculosis. Egypt J Immunol 11(2):133–140
128. Vanham G, Edmonds K, Qing L, Hom D, Toossi Z, Jones B, Daley CL, Huebner B, Kestens L, Gigase P, Ellner JJ (1996) Generalized immune activation in pulmonary tuberculosis: co-activation with HIV infection. Clin Exp Immunol 103(1):30–34
129. Maertzdorf J, Ota M, Repsilber D, Mollenkopf HJ, Weiner J, Hill PC, Kaufmann SH (2011) Functional correlations of pathogenesis-driven gene expression signatures in tuberculosis. PLoS One 6(10):e26938
130. Chensue SW, Davey MP, Remick DG, Kunkel SL (1986) Release of interleukin-1 by peripheral blood mononuclear cells in patients with tuberculosis and active inflammation. Infect Immun 52(1):341–343
131. Ogawa T, Uchida H, Kusumoto Y, Mori Y, Yamamura Y, Hamada S (1991) Increase in tumor necrosis factor alpha- and interleukin-6-secreting cells in peripheral blood mononuclear cells from subjects infected with Mycobacterium tuberculosis. Infect Immun 59(9):3021–3025
132. Toossi Z, Hamilton BD, Phillips MH, Averill LE, Ellner JJ, Salvekar A (1997) Regulation of nuclear factor-kappa B and its inhibitor I kappa B-alpha/MAD-3 in monocytes by Mycobacterium tuberculosis and during human tuberculosis. J Immunol 159(8):4109–4116

133. Maertzdorf J, Repsilber D, Parida SK, Stanley K, Roberts T, Black G, Walzl G, Kaufmann SH (2011) Human gene expression profiles of susceptibility and resistance in tuberculosis. Genes Immun 12(1):15–22
134. Verbon A, Juffermans N, Van Deventer SJ, Speelman P, Van Deutekom H, Van Der Poll T (1999) Serum concentrations of cytokines in patients with active tuberculosis (TB) and after treatment. Clin Exp Immunol 115(1):110–113
135. Vankayalapati R, Wizel B, Weis SE, Klucar P, Shams H, Samten B, Barnes PF (2003) Serum cytokine concentrations do not parallel Mycobacterium tuberculosis-induced cytokine production in patients with tuberculosis. Clin Infect Dis 36(1):24–28
136. Deveci F, Akbulut HH, Turgut T, Muz MH (2005) Changes in serum cytokine levels in active tuberculosis with treatment. Mediators Inflamm 2005(5):256–262
137. Bekker LG, Maartens G, Steyn L, Kaplan G (1998) Selective increase in plasma tumor necrosis factor-alpha and concomitant clinical deterioration after initiating therapy in patients with severe tuberculosis. J Infect Dis 178(2):580–584
138. Yamada G, Shijubo N, Shigehara K, Okamura H, Kurimoto M, Abe S (2000) Increased levels of circulating interleukin-18 in patients with advanced tuberculosis. Am J Respir Crit Care Med 161(6):1786–1789
139. Dlugovitzky D, Torres-Morales A, Rateni L, Farroni MA, Largacha C, Molteni O, Bottasso O (1997) Circulating profile of Th1 and Th2 cytokines in tuberculosis patients with different degrees of pulmonary involvement. FEMS Immunol Med Microbiol 18(3):203–207
140. Djoba Siawaya JF, Beyers N, van Helden P, Walzl G (2009) Differential cytokine secretion and early treatment response in patients with pulmonary tuberculosis. Clin Exp Immunol 156(1):69–77
141. Rey AD, Mahuad CV, Bozza VV, Bogue C, Farroni MA, Bay ML, Bottasso OA, Besedovsky HO (2007) Endocrine and cytokine responses in humans with pulmonary tuberculosis. Brain Behav Immun 21(2):171–179
142. Moreno-Altamirano MM, Aguilar-Carmona I, Sanchez-Garcia FJ (2007) Expression of GM1, a marker of lipid rafts, defines two subsets of human monocytes with differential endocytic capacity and lipopolysaccharide responsiveness. Immunology 120(4):536–543
143. Sedwick CE, Altman A (2002) Ordered just so: lipid rafts and lymphocyte function. Sci STKE 2002 122:re2
144. Fessler MB, Parks JS (2011) Intracellular lipid flux and membrane microdomains as organizing principles in inflammatory cell signaling. J Immunol 187(4):1529–1535
145. Tallone T, Turconi G, Soldati G, Pedrazzini G, Moccetti T, Vassalli G (2011) Heterogeneity of human monocytes: an optimized four-color flow cytometry protocol for analysis of monocyte subsets. J Cardiovasc Transl Res 4(2):211–219
146. Arloing S (1898) Agglutination de bacille de la tuberculose vraie. Compt Rendu Acad Sci 126:1398–1400
147. Daniel T, Ferguson LE (1970) Purification and characterization of two proteins from culture filtrates of Mycobacterium tuberculosis H37Rv strain. Infect Immun 1:164–168
148. Harboe M, Wiker HG (1992) The 38-kDa protein of *Mycobacterium tuberculosis*: a review. J Infect Dis 166:874–884
149. Davies D, Liang X, Hernandez J, Randall A, Hirst S, Mu Y, Romero K, Nguyen T, Kalantari-Dehaghi M, Crotty S, Baldi P, Villarreal L, Felgner P (2005) Profiling the humoral immune response to infection using proteome microarrays: high throughput vaccine and diagnostic antigen discovery. Proc Natl Acad Sci USA 102:547–552
150. Kunnath-Velayudhan S, Salamon H, Wang HY, Davidow AL, Molina DM, Huynh VT, Cirillo DM, Michel G, Talbot EA, Perkins MD, Felgner PL, Liang X, Gennaro ML (2010) Dynamic antibody responses to the Mycobacterium tuberculosis proteome. Proc Natl Acad Sci USA 107(33):14703–14708. doi:10.1073/pnas.1009080107
151. Sonnenberg MG, Belisle JT (1997) Definition of Mycobacterium tuberculosis culture filtrate proteins by two-dimensional polyacrylamide gel electrophoresis, N-terminal amino acid sequencing, and electrospray mass spectrometry. Infect Immun 65(11):4515–4524

152. Silva VMC, Kanaujia G, Gennaro ML, Menzies D (2003) Factors associated with humoral response to ESAT-6, 38 kDa and 14 kDa antigens in patients with a spectrum of tuberculosis. Int J Tuberc Lung Dis 7:478–484
153. Wang BL, Xu Y, Li ZM, Xu YM, Weng XH, Wang HH (2005) Antibody response to four secretory proteins from Mycobacterium tuberculosis and their complex antigen in TB patients. Int J Tuberc Lung Dis 9(12):1327–1334
154. Steingart KR, Dendukuri N, Henry M, Schiller I, Nahid P, Hopewell PC, Ramsay A, Pai M, Laal S (2009) Performance of purified antigens for serodiagnosis of pulmonary tuberculosis: a meta-analysis. Clin Vaccine Immunol 16(2):260–276
155. Bothamley GH, Beck JS, Potts RC, Grange JM, Kardjito T, Ivanyi J (1992) Specificity of antibodies and tuberculin response after occupational exposure to tuberculosis. J Infect Dis 166(1):182–186
156. Capuano SV 3rd, Croix DA, Pawar S, Zinovik A, Myers A, Lin PL, Bissel S, Fuhrman C, Klein E, Flynn JL (2003) Experimental Mycobacterium tuberculosis infection of cynomolgus macaques closely resembles the various manifestations of human M. tuberculosis infection. Infect Immun 71(10):5831–5844
157. Kunnath-Velayudhan S, Davidow AL, Wang H-Y, Molina DM, Huynh VT, Salamon H, Pine R, Michel G, Perkins MD, Liang X, Felgner PL, Flynn JL, Catanzaro A, Gennaro ML (2012) Proteome-scale antibody responses and outcome of Mycobacterium tuberculosis infection in non-human primates and in tuberculosis patients. J Infect Dis J Infect Dis. 206(5):697–705
158. Cavalcante S, Kritski AL, Ferreira MA, Souza MA, Laszlo A, Werneck-Barroso EB, Fonseca LS (1997) Association between an early humoral response to Mycobacterium tuberculosis antigens and later development of tuberculosis in human immunodeficiency virus-infected individuals. Int J Tuberc Lung Dis 1(2):170–174
159. Laal S, Samanich KM, Sonnenberg MG, Belisle JT, O'Leary J, Simberkoff MS, Zolla-Pazner S (1997) Surrogate marker of preclinical tuberculosis in Human Immunodeficiency Virus infection: antibodies to an 88-kDa secreted antigen of *Mycobacterium tuberculosis*. J Infect Dis 176:133–143
160. Gennaro ML, Affouf M, Kanaujia GV, Brusasca PN, Mangura B, Reichman L (2007) Antibody markers of incident tuberculosis among HIV-infected adults in the USA: a historical prospective study. Int J Tuberc Lung Dis 11(6):624–631
161. Colangeli R, Antinori A, Cingolani A, Ortona L, Lyashchenko K, Fadda G, Gennaro ML (1999) Humoral immune responses to multiple antigens of *Mycobacterium tuberculosis* in tuberculosis patients co-infected with human immunodeficiency virus. Int J Tuberc Lung Dis 3:1127–1131
162. Bothamley GH (1995) Serological diagnosis of tuberculosis. Eur Respir J 8(suppl 20):676s–688s
163. Gennaro ML (2000) Immunologic diagnosis of tuberculosis. Clin Infect Dis 30(suppl 3):S243–246
164. Bothamley G, Gennaro ML (2008) The antibody response to infection with *Mycobacterium tuberculosis*. In: Kaufmann SHE, Britton WJ (eds) Handbook of tuberculosis: immunology and cell biology. Wiley-VHC Verlag GmbH & Co. KGaA, Weinheim, pp 227–244
165. Bothamley GH, Beck JS, Schreuder GMT, D'Amaro J, de Vries RR, Kardjito T, Ivanyi J (1989) Association of tuberculosis and M. tuberculosis-specific antibody levels with HLA. J Infect Dis 159:549–555
166. Bothamley GH, Schreuder GM, de Vries RR, Ivanyi J (1993) Association of antibody responses to the 19-kDa antigen of Mycobacterium tuberculosis and the HLA-DQ locus. J Infect Dis 167(4):992–993
167. Gao Q, Kripke KE, Saldanha AJ, Yan W, Holmes S, Small PM (2005) Gene expression diversity among Mycobacterium tuberculosis clinical isolates. Microbiology 151(Pt 1):5–14
168. Musser JM, Amin A, Ramaswamy S (2000) Negligible genetic diversity of mycobacterium tuberculosis host immune system protein targets: evidence of limited selective pressure. Genetics 155(1):7–16
169. Comas I, Chakravartti J, Small PM, Galagan J, Niemann S, Kremer K, Ernst JD, Gagneux S (2010) Human T cell epitopes of Mycobacterium tuberculosis are evolutionarily hyperconserved. Nat Genet 42(6):498–503

170. Steingart KR, Henry M, Laal S, Hopewell PC, Ramsay A, Menzies D, Cunningham J, Weldingh K, Pai M (2007) Commercial serological antibody detection tests for the diagnosis of pulmonary tuberculosis: a systematic review. PLoS Med 4(6):e202
171. Davidow A, Kanaujia GV, Shi L, Kaviar J, Guo X, Sung N, Kaplan G, Menzies D, Gennaro ML (2005) Antibody profiles characteristic of Mycobacterium tuberculosis infection state. Infect Immun 73(10):6846–6851
172. Kanaujia GV, Lam PK, Perry S, Brusasca PN, Catanzaro A, Gennaro ML (2005) Integration of microscopy and serodiagnostic tests to screen for active tuberculosis. Int J Tuberc Lung Dis 9(10):1120–1126
173. Grzybowski S, Fishaut H, Rowe J, Brown A (1971) Tuberculosis among patients with various radiologic abnormalities, followed by the chest clinic service. Am Rev Respir Dis 104(4):605–608
174. Nolan CM, Elarth AM (1988) Tuberculosis in a cohort of Southeast Asian Refugees. A five-year surveillance study. Am Rev Respir Dis 137(4):805–809
175. McNerney R, Daley P (2011) Towards a point-of-care test for active tuberculosis: obstacles and opportunities. Nat Rev Microbiol 9(3):204–213
176. Dorhoi A, Reece ST, Kaufmann SH (2011) For better or for worse: the immune response against Mycobacterium tuberculosis balances pathology and protection. Immunol Rev 240(1):235–251
177. Diel R, Loddenkemper R, Nienhaus A (2010) Evidence-based comparison of commercial interferon-gamma release assays for detecting active TB: a metaanalysis. Chest 137(4):952–968
178. Sester M, Sotgiu G, Lange C, Giehl C, Girardi E, Migliori GB, Bossink A, Dheda K, Diel R, Dominguez J, Lipman M, Nemeth J, Ravn P, Winkler S, Huitric E, Sandgren A, Manissero D (2011) Interferon-gamma release assays for the diagnosis of active tuberculosis: a systematic review and meta-analysis. Eur Respir J 37(1):100–111
179. Diel R, Goletti D, Ferrara G, Bothamley G, Cirillo D, Kampmann B, Lange C, Losi M, Markova R, Migliori GB, Nienhaus A, Ruhwald M, Wagner D, Zellweger JP, Huitric E, Sandgren A, Manissero D (2011) Interferon-gamma release assays for the diagnosis of latent Mycobacterium tuberculosis infection: a systematic review and meta-analysis. Eur Respir J 37(1):88–99

Index

J. McFadden et al. (eds.), *Systems Biology of Tuberculosis*,
DOI 10.1007/978-1-4614-4966-9, © Springer Science+Business Media, LLC 2013